Sustainable Polymer Technologies for a Circular Economy

Sustainable Polymer Technologies for a Circular Economy

Editor

Sergio Torres-Giner

MDPI • Basel • Beijing • Wuhan • Barcelona • Belgrade • Manchester • Tokyo • Cluj • Tianjin

Editor
Sergio Torres-Giner
Research Institute of Food
Engineering for Development
Universitat Politècnica
de València
Valencia
Spain

Editorial Office
MDPI
St. Alban-Anlage 66
4052 Basel, Switzerland

This is a reprint of articles from the Special Issue published online in the open access journal *Applied Sciences* (ISSN 2076-3417) (available at: www.mdpi.com/journal/applsci/special_issues/Polymer_Circular_Economy).

For citation purposes, cite each article independently as indicated on the article page online and as indicated below:

LastName, A.A.; LastName, B.B.; LastName, C.C. Article Title. *Journal Name* **Year**, *Volume Number*, Page Range.

ISBN 978-3-0365-8271-9 (Hbk)
ISBN 978-3-0365-8270-2 (PDF)

Cover image courtesy of Sergio Torres-Giner

Contents

About the Editor

Sergio Torres-Giner

Sergio Torres-Giner received a Bachelor's degree in Chemical Engineering in 2003 from the Polytechnic University of Valencia (UPV), Spain. In 2004, he attained an MSc in Process Systems Technology at Cranfield University, England, followed by an MBA in Industrial Management in 2005 at the Catholic University of Valencia "San Vincente Martir", Spain. He completed his Ph.D. in 2010 in Food Science at the University of Valencia, Spain. He currently works as a scientist and a professor at the Research Institute of Food Engineering for Development (IIAD) at UPV in the field of macromolecular science, with an interest in the applications of food packaging technologies. He has more than 18 years of experience working at both public research agencies and industrial R&D organizations. He has published over 100 peer-reviewed scientific papers indexed in JCR, 10 books and encyclopedia chapters, and 4 patents. His research activity has strongly contributed to advancing the knowledge on biopolymers and transferring such knowledge for food-related applications and products.

Preface to "Sustainable Polymer Technologies for a Circular Economy"

The current economic system of the plastics industry still follows the linear model of "make, use, and dispose", which erroneously assumes that economic growth can be based on a high abundance of resources and unlimited waste disposal. To accelerate the transition from its traditional linear economic model to a more valuable and sustainable model, the principles of circular economy should be rapidly implemented. In this scenario, the scientific community needs to consider that the game-changing strategy should be based on three pillars, namely promoting sustainable polymer technologies that decouple plastics from fossil feedstocks, drastically reducing the leakage of plastics into natural systems, and increasing the quality and uptake of plastic recycling. Only biopolymers can really "close the loop"and enable a true circular economy for the plastic packaging industry.

This reprint is divided into ten chapters and showcases original and cutting-edge research studies devoted to accelerating circularity within the plastics industry. This reprint starts with five chapters dealing with the development of bio-based polymer formulations since, to progress towards a truly sustainable model, it is imperative to first decouple polymers from fossil feedstocks. These chapters exemplify that plastic articles can be made of both thermoplastic materials, such as polysaccharides or biopolyesters, and thermoset resins, such as bio-based epoxy resins, which are produced from natural and renewable resources. These research studies also demonstrate that not only polymers but also their additives do not require fossil fuels as feedstocks. For instance, gum rosin- and cellulose-derived fillers can serve to yield highly environmentally friendly composites with a wood-like appearance and improved properties. This reprint then presents three more chapters covering the effects of plastic waste leakage on the natural environment and the sustainable solutions attained by preparing occasionally littered articles from readily biodegradable materials. In this regard, one of the most sustainable solutions is achieved via the combination of paper and biodegradable polyesters. Finally, this reprint ends with two more chapters discussing novel potential polymer technologies to maximize looping strategies and build an effective after-use plastics economy. These chapters present studies examining the reduction of costs and energy consumption during the production of plastic articles and the use of pyrolysis as a dedicated new recycling technology developed for multilayer packaging structures.

This reprint is accessible to a broad readership and can be of interest for university professors, researchers, undergraduate and postgraduate students, and engineers working in the field of polymer materials, packaging, and sustainability.

Sergio Torres-Giner
Editor

applied sciences

MDPI

Editorial

Sustainable Polymer Technologies for a Circular Economy

Sergio Torres-Giner

Research Institute of Food Engineering for Development (IIAD), Universitat Politècnica de València (UPV), Camino de Vera s/n, 46022 Valencia, Spain; storresginer@upv.es

Citation: Torres-Giner, S. Sustainable Polymer Technologies for a Circular Economy. *Appl. Sci.* **2023**, *13*, 5864. https://doi.org/10.3390/app13105864

Received: 21 April 2023
Accepted: 29 April 2023
Published: 10 May 2023

We inhabit a defining moment in history. It is a moment in which the scientific community has united to agree on an ambitious framework with which to resolve the environmental issues associated with plastic waste. Plastics are the "workhorse" material of the modern economy, offering multiple functions that help tackle several challenges facing our society. Plastics production has increased from 15 million tons in the 1960s to 311 million tons in 2014 and is expected to triple by 2050 as plastics come to serve increasingly more applications. Packaging is the dominant sectoral use of plastics globally, representing nearly 40% of the plastic market. However, after an initial short use cycle, most of the economic value of plastic packaging material is lost. Furthermore, hundreds of millions of tons of plastics escape collection systems, ending up in the environment either as microscopic particles or surviving in a recognizable form for hundreds of years. Therefore, it is high time to implement the principles of Circular Economy in the plastics industry. The Circular Economy has recently become a widespread concept, emerging in the past few decades from several schools of thought, such as "Cradle2Cradle" and "Biomimicry. It envisages a novel economic model that is restorative and regenerative by design, wherein its key aims are to eliminate the concept of waste, rebuild natural capital, and create economic value by using—not consuming—resources effectively. However, the current economic system of the plastics industry still follows the linear model 'make, use, and dispose', which erroneously assumes that economic growth can be based on the abundance of resources and unlimited waste disposal. According to the model outlined in the New Plastics Economy and other recent sustainable initiatives, strategies of Circular Economy regarding the plastics industry entail promoting sustainable polymer technologies that decouple plastics from fossil feedstocks, drastically reduce the leakage into and the effects of plastics on natural systems, and create an effective after-use plastics economy. This Special Issue focuses on recent research studies devoted to enabling better economic and environmental advances in the plastic packaging value chain that can successfully accelerate the transition of the plastics industry from its traditional linear economy to a more valuable and sustainable model.

The polymers used to produce plastics, together with the corresponding additives, should not require petroleum or other fossil fuels as feedstocks. In this context, pectin, one of the most abundant polysaccharides in nature, can be applied to produce biodegradable, biocompatible, and water-soluble films as well as coatings or interlayers in biopolymer-based structures of interest in food packaging. In this regard, Akinalan Balik et al. [1] prepared and characterized glycerol and polyethylene oxide-2000 (PEO_{2000})-containing pectin films via electrospinning and the subsequent annealing of the resultant fiber mat at temperatures below 160 °C. These newly developed pectin films were fully characterized and thereafter applied as interlayers between two external layers of poly(3-hydroxybutyrate-*co*-3-hydroxyvalerate) (PHBV). The resultant fully bio-based and biodegradable multilayer film presented a strong barrier to water and limonene vapors, revealing their promising potential in food packaging applications. Similarly, Aldas et al. [2] developed fully bio-based materials using thermoplastic starch (TPS) that was obtained from corn plasticized with water and glycerol, which was blended with up to five different pine resin derivatives as a kind of natural additive. The authors reported that all the gum-rosin-based additives were able to improve the thermal stability and mechanical performance of TPS, particularly in the case of the gum resins with higher amounts of carbonyl groups in their chemical structures due to their interaction with the hydroxyl groups of starch and glycerol. The developed TPS/pine resin blends were finally proposed for rigid packaging and disposable applications in hot foods, such as lids for containers and cups for beverages. Some relevant physical improvements in TPS films were also obtained by Freitas et al. [3] through the incorporation of cellulose microfibers (CMFs) obtained from rice straw (RS) waste. Fractions of 1, 3, and 5 wt% CMFs were incorporated into TPS by melt mixing, whereas starch was also subjected to dry heating (DH) modification, yielding TPS modified via dry heating (TPSDH). It was observed that both DH modification and 3 and 5 wt% fiber loadings interfered with starch gelatinization, leading to non-gelatinized starch granules in the biopolymer matrix. This represents a remarkable green physical methodology for enhancing the water barrier capacity and tensile strength of starch films. Bio-based polymers also include thermoset resins, as demonstrated by Lascano et al. [4]. In this research work, green composites were developed using partially bio-based epoxy resin reinforced with lignocellulosic particles obtained from by-products or wastes of the flax industry. Different particles of flaxseed flour, with sizes ranging between 100–220 μm and 40–140 μm, were incorporated at different concentrations, that is, 10, 20, 30, and 40 wt%, into the bio-based epoxy resin during molding and thereafter characterized. The most promising results in terms of mechanical properties and water absorption were obtained for the lowest reinforcement content and with the finest particle size, yielding highly environmentally friendly composites with a wood-like appearance and potential use in furniture or automotive sectors. Another example of the development of the biorefinery concept achieved by means of the valorization of agricultural and food wastes was presented by Ortiz-Barajas et al. [5]. In this study, the authors performed the torrefaction of coffee husk flour at 250 °C and then incorporated it into polylactide (PLA) pieces produced by injection molding. A content of 20 wt% of torrefied coffee husk flour (TCHF) successfully yielded pieces with balanced mechanical properties and improved hardness, thermal degradation, and thermomechanical resistance, which can be of great interest in in the design of compostable rigid packaging (e.g., food trays and containers), beverage cups, and food contact disposables (e.g., cutlery and plates).

Certainly, the best way to prevent the leakage of plastics into natural systems and its associated negative consequences is to support the growth of robust waste infrastructure. Indeed, plastic waste from land sources is continually flowing into world oceans via rivers due to accidents or carelessness, resulting in a high level of accumulation of plastics in the environment, the so-called "white pollution". Even though temperatures in nature are relatively low, different oligomers are often detected in the sea water and sand of coastal areas due to polymer degradation. In this regard, the study of Kimukai et al. [6] analyzed the formation of styrene oligomers (SOs) from polystyrene (PS) decomposition in the thermal range from 30 to 150 °C. In this temperature range, 2,4,6-triphenyl-1-hexene (styrene trimer, ST) was the dominant product, which was an intermediate yielding 1,3-diphenylpropane (styrene dimer, SD_1), 2,4-diphenyl-1-butene (styrene dimer, SD_2), and styrene (styrene monomer, SM). Based on a simulation, the authors also predicted that over 400 million metric tons of SOs would be present in the ocean by 2050, including those capable of thermal decompose at 30 °C. In addition, compostable and biodegradable polymers, which can disintegrate in the environment, represent a sustainable alternative to non-biodegradable polymers derived from petroleum. Compostable plastic articles can be collected and thus treated in municipal/industrial facilities, thereby capturing and delivering organic residues and diverting organic waste from landfills. More importantly, biodegradable polymers can reduce the impact of leaked plastics on the environment since some of them can also disintegrate in natural environments and thus be reincorporated into the natural cycle. This is the case of polyhydroxyalkanoates (PHAs), which are also renewable since they are synthesized by bacteria and other microorganisms (mainly using sugars). However, these and other thermoplastic biodegradable polymers usually lack sufficient physical properties to be properly applied in food packaging. In this context, Freitas et al. [7] prepared microstructures of microfibrillated cellulose (MFC) via atomization to reinforce films of PHBV, a PHA copolymer. The authors also explored the capacity of two reactive compatibilizers, namely, a multi-functional epoxy-based styrene-acrylic oligomer (ESAO) and a combination of triglycidyl isocyanurate (TGIC) with dicumyl peroxide (DCP), to improve the properties of the PHBV/MFC composites by enhancing filler–matrix adhesion. It was observed that the incorporation of MFC via reactive extrusion with TGIC and DCP led to green composite films with contact transparency, enhanced thermal stability, mechanical strength, and ductility, and high barrier performance with respect to aroma vapor and oxygen, thus unveiling new opportunities in the food packaging field. The environmental impacts of plastic packaging can also be mitigated by the use of paper in combination with biodegradable polymers, which are both derived from natural resources and can be easily composted together. Hernández-García et al. [8] proved that PHBV is an excellent material for paper-coating applications by developing PHBV/paper/PHBV multilayer sheets using heat-sealing technology. The authors observed that double coatings of PHBV successfully improved mechanical resistance and ductility, protected from moisture, and also reduced the aroma and oxygen permeances of paper, inducing a minimal effect on its optical and thermal properties and provoking a slight reduction in the aerobic biodegradation and disintegration of paper.

Finally, plastic materials can be successfully sorted and recycled by means of different strategies of waste management. While waste collection systems are evolving, novel postconsumer recycling technologies are being developed, including mechanical (secondary) recycling or reprocessing and chemical (tertiary) recycling or depolymerization–repolymerization. All these advanced recycling technologies can contribute to building an effective after-use plastics economy, but other options such as pre-consumer (primary) recycling in plastic factories and thermochemical (quaternary) recycling can also favor the development of a more sustainable plastics industry. In this regard, the research carried out by Klejnowska et al. [9] focused on determining the influence of the process temperature on the composition of gases produced during the pyrolysis of pre- and postconsumer waste pharmaceutical blisters (WPBs) in order to enhance waste valorization. Pyrolysis, a thermal decomposition process that occurs in the absence of oxygen, could effectively separate plastic from the metal fraction of WPBs in order to recover aluminum and generate gaseous fuel. The authors demonstrated that high temperatures favor the production of larger amounts of process gas rather than oil and wax. In addition, the calorific value of the gas at the maximum temperature (450 °C) was higher (21.96 MJ/m^3) for the mixed postconsumer waste than that for the clean pre-consumer material (20.14 MJ/m^3). Ascertaining the key factors of reducing costs and energy consumption during production can also represent another strategy to improve the economy and waste management of plastics. In this regard, Chen et al. [10] successfully analyzed the most influencing aspects during the extrusion-molding quality control of food-grade polypropylene (PP) packaging products. The study revealed that four key technical factors are involved: (1) extrusion sheet production; (2) extrusion line design; (3) forming and mold manufacturing; and (4) mold and thermoforming line equipment design. Moreover, according to their research results, these key factors are not only applicable to classical PP extrusion sheet and thermoforming production but are also related to processes of extrusion and thermoforming techniques used for other plastic products, such as expanded polypropylene (EPP) and PLA sheets. These results are considered to be capable of providing a key technical reference with which enterprises can improve quality in order to enhance the competitiveness of products, reduce production costs, and achieve sustainable development, energy savings, and carbon reductions.

Funding: This research was funded by the Spanish Ministry of Science and Innovation (MICI), grant number PID2021-128749OB-C33.

Acknowledgments: S.T.-G. acknowledges MICI for his Ramón y Cajal contract (RYC2019-027784-I).

Conflicts of Interest: The authors declare no conflict of interest.

References

1. Akinalan Balik, B.; Argin, S.; Lagaron, J.M.; Torres-Giner, S. Preparation and Characterization of Electrospun Pectin-Based Films and Their Application in Sustainable Aroma Barrier Multilayer Packaging. *Appl. Sci.* **2019**, *9*, 5136. [CrossRef]
2. Aldas, M.; Pavon, C.; López-Martínez, J.; Arrieta, M.P. Pine Resin Derivatives as Sustainable Additives to Improve the Mechanical and Thermal Properties of Injected Moulded Thermoplastic Starch. *Appl. Sci.* **2020**, *10*, 2561. [CrossRef]
3. Freitas, P.A.V.; Arias, C.I.L.F.; Torres-Giner, S.; González-Martínez, C.; Chiralt, A. Valorization of Rice Straw into Cellulose Microfibers for the Reinforcement of Thermoplastic Corn Starch Films. *Appl. Sci.* **2021**, *11*, 8433. [CrossRef]
4. Lascano, D.; Garcia-Garcia, D.; Rojas-Lema, S.; Quiles-Carrillo, L.; Balart, R.; Boronat, T. Manufacturing and Characterization of Green Composites with Partially Biobased Epoxy Resin and Flaxseed Flour Wastes. *Appl. Sci.* **2020**, *10*, 3688. [CrossRef]
5. Ortiz-Barajas, D.L.; Arévalo-Prada, J.A.; Fenollar, O.; Rueda-Ordóñez, Y.J.; Torres-Giner, S. Torrefaction of Coffee Husk Flour for the Development of Injection-Molded Green Composite Pieces of Polylactide with High Sustainability. *Appl. Sci.* **2020**, *10*, 6468. [CrossRef]
6. Kimukai, H.; Kodera, Y.; Koizumi, K.; Okada, M.; Yamada, K.; Hiaki, T.; Saido, K. Low Temperature Decomposition of Polystyrene. *Appl. Sci.* **2020**, *10*, 5100. [CrossRef]
7. Freitas, P.A.V.; Barrrasa, H.; Vargas, F.; Rivera, D.; Vargas, M.; Torres-Giner, S. Atomization of Microfibrillated Cellulose and Its Incorporation into Poly(3-hydroxybutyrate-*co*-3-hydroxyvalerate) by Reactive Extrusion. *Appl. Sci.* **2022**, *12*, 2111. [CrossRef]
8. Hernández-García, E.; Freitas, P.A.V.; Zomeño, P.; González-Martínez, C.; Torres-Giner, S. Multilayer Sheets Based on Double Coatings of Poly(3-hydroxybutyrate-*co*-3-hydroxyvalerate) on Paper Substrate for Sustainable Food Packaging Applications. *Appl. Sci.* **2023**, *13*, 179. [CrossRef]
9. Klejnowska, K.; Pikoń, K.; Ścierski, W.; Skutil, K.; Bogacka, M. Influence of Temperature on the Composition and Calorific Value of Gases Produced during the Pyrolysis of Waste Pharmaceutical Blisters. *Appl. Sci.* **2020**, *10*, 737. [CrossRef]
10. Chen, D.-C.; Chen, D.-F.; Huang, S.-M.; Shyr, W.-J. The Investigation of Key Factors in Polypropylene Extrusion Molding Production Quality. *Appl. Sci.* **2022**, *12*, 5122. [CrossRef]

applied sciences

MDPI

Article

Preparation and Characterization of Electrospun Pectin-Based Films and Their Application in Sustainable Aroma Barrier Multilayer Packaging

Busra Akinalan Balik [1], Sanem Argin [1], Jose M. Lagaron [2,*] and Sergio Torres-Giner [2,*]

1 Department of Food Engineering, Yeditepe University, Atasehir, 34755 Istanbul, Turkey; busra.akinalan@yeditepe.edu.tr (B.A.B.); sanem.argin@yeditepe.edu.tr (S.A.)

2 Novel Materials and Nanotechnology Group, Institute of Agrochemistry and Food Technology (IATA), Spanish National Research Council (CSIC), Calle Catedrático Agustín Escardino Benlloch 7, 46980 Paterna, Spain

* Correspondence: lagaron@iata.csic.es (J.M.L.); storresginer@iata.csic.es (S.T.-G.); Tel.: +34-963-900-022 (J.M.L. & S.T.-G.)

Received: 25 October 2019; Accepted: 22 November 2019; Published: 27 November 2019

Featured Application: The present study aims to develop novel pectin-based films by electrospinning. The here-prepared films were applied as aroma barrier interlayers between two biopolymer films to develop fully bio-based and biodegradable food packaging articles according to the principles of the Circular Economy.

Abstract: Pectin was first dissolved in distilled water and blended with low contents of polyethylene oxide 2000 (PEO_{2000}) as the carrier polymer to produce electrospun fibers. The electrospinning of the water solution of pectin at 9.5 wt% containing 0.5 wt% PEO_{2000} was selected as it successfully resulted in continuous and non-defected ultrathin fibers with the highest pectin content. However, annealing of the resultant pectin-based fibers, tested at different conditions, developed films with low mechanical integrity, high porosity, and also dark color due to their poor thermal stability. Then, to improve the film-forming process of the electrospun mats, two plasticizers, namely glycerol and polyethylene glycol 900 (PEG_{900}), were added to the selected pectin solution in the 2–3 wt% range. The optimal annealing conditions were found at 150 °C with a pressure of 12 kN load for 1 min when applied to the electrospun pectin mats containing 5 wt% PEO_{2000} and 30 wt% glycerol and washed previously with dichloromethane. This process led to completely homogenous films with low porosity and high transparency due to a phenomenon of fibers coalescence. Finally, the selected electrospun pectin-based film was applied as an interlayer between two external layers of poly(3-hydroxybutyrate-*co*-3-hydroxyvalerate) (PHBV) by the electrospinning coating technology and the whole structure was annealed to produce a fully bio-based and biodegradable multilayer film with enhanced barrier performance to water vapor and limonene.

Keywords: pectin; electrospinning; annealing; barrier interlayers; food packaging

1. Introduction

Pectin is one of the most abundant polysaccharides in nature, which is found in the middle lamella of cell wall, primary cell walls, and plasma membrane of plants [1,2]. It is commercially produced by the industrial waste of apple pomace, citrus peel, and sugar beet pulp [3]. Pectin is mainly composed of linear chains of α-1,4 linked D-galacturonic acid units but also different types of side chains may exist in its chemical structure such as those containing rhamnose, xylose, galactose, and arabinose [2,4]. Depending on the side chains, pectin domains are named differently such as homogalacturonan (HGA), rhamnogalacturonan-I (RG-I), rhamnogalacturonan-II (RG-II), and xylogalacturonan (XGA) [5,6]. HGA units are habitually referred as 'smooth' regions of pectin and it comprises galacturonic acid groups whereas other units are called 'hairy' regions [4,7].

Pectin can be used as gelling, thickening, and stabilizing agent in the food and pharmaceutical industry. Moreover, the polyelectrolyte nature, biodegradability, biocompatibility, and water solubility of pectin also open up new uses including coatings or edible films for food packaging applications [8,9]. Nevertheless, the intrinsically high hydrophilicity and low mechanical strength of pectin-based films compared to conventional ones such as those made of high-density polyethylene (HDPE) and polypropylene (PP) currently limit the application of this carbohydrate [10–12]. In this context, to enhance the mechanical integrity and also reduce brittleness, plasticizers are habitually added to the pectin formulations to form the films [13]. The addition of different kinds of plasticizers, for instance glycerol [14–16], sorbitol [13,17,18], polyethylene glycol (PEG) [13,19] or xylitol [20], to polysaccharide film-forming solutions during the casting process can improve both film formation and flexibility by enhancing the motion of polymer chains and reducing their intra- and intermolecular forces [21,22]. Additionally, water dissolution of pectin can be decreased by in situ cross-linking with divalent metal ions including Ca^{2+}, Zn^{2+} or Mg^{2+} [23,24]. Furthermore, the mechanical, thermal, and barrier properties of the pectin-based films can also be enhanced by blending with other biopolymers such as chitosan [25,26], cellulose, and its derivatives [27,28] or the addition of inorganics such as nanoclays [11].

In the polymer literature, neat pectin films are technically not feasible to be produced by conventional melt technologies and they have been so far obtained by the solvent casting method using large contents of plasticizers [24,25,29,30]. In this regard, electrospinning is a novel technique that provides manufacturing of ultrathin fibers with diameters extending from several nanometers to a few micrometers [31]. Electrospun nanofibers may offer many functional advantages such as superior mechanical properties, large surface-to-mass ratio, tailored fiber morphology, and the capability of encapsulation and subsequent release of active and bioactive principles [32–36]. However, aqueous solutions of neat pectin cannot be electrospun due to the limited viscoelasticity of pectin and its insufficient chain entanglements [37–39]. As a result, electrospun pectin nanofibers have been only obtained by blending with different synthetic polymers such as polyethylene oxide (PEO) [39–45], polyvinyl alcohol (PVOH) [46–48], and pullulan [38]. In addition, ternary blends of alginate/pectin/PEO [49] and chitosan/pectin/PVOH [50] have been recently successfully electrospun for biomedical purposes.

The few studies reporting the development of electrospun pectin-based nanofibers have been mainly focused on the areas of antibacterial surfaces [48], tissue engineering [40–42,50], drug delivery [44], and encapsulation [38,49], whereas their utilization for food packaging applications remains unexplored due to the inherent discontinuity and porous structure of the nanofibers mats. Interestingly, electrospun mats can be subjected to a thermal post-treatment above the glass transition (T_g) and below the melting temperature (T_m) of the polymer, also termed annealing, in order to remove or minimize their porosity and produce continuous and homogenous films [51–55]. Until now, this technology has been successfully applied to different polyester-type biopolymers with different potential applications in the food packaging field. For instance, electrospun poly(3-hydroxybutyrate) (PHB) films showed better optical properties, similar barrier performance, and higher elongation at break and toughness in comparison with equivalent films obtained by compression molding [51]. Electrospun films of PHB, PVOH, and also polylactide (PLA) were also developed by electrospinning and originally applied as coating materials on a paper-based packaging material to develop multilayers with improved barrier properties against water and limonene vapors [52]. In another study carried out by Cherpinski et al. [53], a similar strategy was followed to coat cellulose nanopapers by PHB and poly(3-hydroxybutyrate-*co*-3-hydroxyvalerate) (PHBV) electrospun layers. Similarly, electrospun ultrathin fibers of bio-waste derived PHBV were subjected to annealing by Melendez-Rodriguez et al. [54] to successfully produce continuous biopolymer films with similar barrier performance than petroleum-based polyethylene terephthalate (PET) films. Other recent studies have been focused on the incorporation of antimicrobial or antioxidant ingredients in the electrospun fibers, which can be thereafter integrated as active layers in packaging structures. For example, Figueroa-Lopez et al. [56] prepared electrospun active films of PHBV with antimicrobial and antioxidant properties by the incorporation into the fibers of different essential oils (EOs) and natural extracts (NEs). Also, Quiles-Carrillo et al. [57] recently developed multilayer bioactive films with controlled release capacity of the natural antioxidant gallic acid (GA) by the incorporation of electrospun PLA interlayers into cast-extruded PLA films. Lastly, Radusin et al. [58] recently prepared antimicrobial PLA films containing *Allium ursinum* L. extract by electrospinning.

The aim of this study was to obtain, for the first time, electrospun pectin-based films as potential candidates for food packaging applications. First, various water-based pectin solutions containing different amounts of PEO and/or in combination with the addition of two different types of plasticizers were processed by electrospinning to determine the best system to produce a film. Thereafter, the selected electrospun mats were characterized and the most promising fibers were selected and subjected to annealing to produce pectin-based films. The morphology, chemical, and thermal properties of the films were reported. The optimal electrospun film was, finally, applied as an interlayer in a multilayer structure based on PHBV and the barrier properties of the resultant multilayers were analyzed and compared to an equivalent multilayer of a cast-film pectin interlayer.

2. Materials and Methods

2.1. Materials

Low methyl esterified amidated pectin was kindly received from AROMSA Inc. (Gebze, Turkey). The product (GENU pectin, LM-104 AS-FS, degree of esterification 27%, degree of amidation 20%) was produced and delivered in powder form by CP Kelco (Copenhagen, Denmark). PEO with molecular weight (M_w) of 2000 kDa, that is, PEO_{2000}, was obtained in powder form as SENTRYTM POLYOXTM WSR N80-LEO NF grade by The Dow Chemical Company (Midland, MI, USA). PEG with M_w of 900 kDa, that is, PEG_{900}, was provided by Honeywell Fluka Chemicals Company (Bucharest, Romania). Bacterial aliphatic copolyester PHBV was ENMAT™ Y1000P, produced by Tianan Biologic Materials (Ningbo, China) and distributed by NaturePlast (Ifs, France). The product was delivered as off-white pellets packaged in plastics bags. The biopolymer resin presents a true density of 1.23 g/cm^3 and the pellets a bulk density of 0.74 g/cm^3, as determined by ISO 1183 and ISO 60, respectively. Sorbitan monolaurate was obtained from Sigma-Aldrich S.A. (Madrid, Spain) as Span® 20. According to the manufacturer, its fatty acid composition was lauric acid (C12:0) ≥ 44%; balance primarily myristic (C14:0), palmitic (C16:0), and linolenic (C18:3) acids. Calcium chloride, dichloromethane, 2,2,2-trifluoroethanol (TFE), ≥99% purity, and glycerol, ≥99.5% purity, were all also purchased from Sigma-Aldrich S.A. (Madrid, Spain).

2.2. Preparation of Solutions

The total concentration of solids in distilled water to prepare the fiber-forming solution for electrospinning was set at 10 wt%. For this, pectin was first dissolved in water at 70 °C for 3 h and the solution was gently stirred overnight at room temperature. Then, PEO_{2000} with or without a plasticizer, that is, glycerol or PEG_{900}, was added to the pectin solution and it was further stirred for 24 h. In all cases, Span® 20 was added as a surfactant to the pectin solutions at 2 wt% with respect to the total solid weight content of the solution. Table 1 summarizes the compositions of the water-based solutions prepared for electrospinning. For the electrospinning of the PHBV layers, the copolyester resin was dissolved at 10% (w/v) in TFE at room conditions during 24 h.

Table 1. Different solutions prepared according to the weight content (wt%) of pectin, polyethylene oxide 2000 (PEO_{2000}), glycerol, and polyethylene glycol 900 (PEG_{900}) in distilled water.

Solution	Pectin (wt%)	PEO_{2000} (wt%)	Glycerol (wt%)	PEG_{900} (wt%)	Water (wt%)
S1	9.9	0.1	-	-	90
S2	9.75	0.25	-	-	90
S3	9.5	0.5	-	-	90
S4	9.0	1.0	-	-	90
S5	7.5	0.5	2.0	-	90
S6	7.0	0.5	2.5	-	90
S7	6.5	0.5	3.0	-	90
S8	7.5	0.5	-	2.0	90
S9	7.0	0.5	-	2.5	90
S10	6.5	0.5	-	3.0	90

2.3. Characterization of Solutions

Prior to electrospinning, all the prepared pectin solutions were characterized in terms of their viscosity, surface tension, and conductivity. Solution viscosity was determined by a VISCO BASIC Plus L rotational viscosity meter equipped with a low-viscosity adapter (LCP) from Fungilab S.A. (San Feliu de Llobregat, Spain). Conductivity was measured in a conductivity meter XS Con6 from Lab-box (Barcelona, Spain). Surface tension was determined in an EasyDyne tensiometer K20 model Krüss GmbH (Hamburg, Germany) following the Wilhelmy plate method. All measurements were carried out at room temperature in triplicate.

2.4. Electrospinning

An electrospinning machine Fluidnatek® LE500 (Bioinicia S.L., Valencia, Spain) placed in a closed chamber and connected to an environmental control unit was used in the lab mode. Fibers were collected in vertical mode on a rectangular plate collector covered with aluminum foil. The most suitable conditions for the electrospinning of pectin were determined during the process. The flow-rate of the pectin-base solutions was set at the highest value possible in order to attain the maximum yield. Flow-rate was optimal at 3 mL/h since higher values led to some droplets on the collector. The tip-to-collector distance was also adjusted by decreasing it until the fibers were formed, being optimal at 25 cm. Finally, the applied voltage was smoothly increased up to the point a stable jet was obtained. The most optimal values of voltage ranged from 16 to 20 kV. For the electrospinning of PHBV, the voltage was set at 10 kV, the tip-to-collector distance was 15 cm, and the flow-rate was 6 mL/h. These values were selected based on our previous study [54]. All the experiments were conducted at 25 °C and 30% relative humidity (RH).

2.5. Washing and Drying

The resultant electrospun pectin-based fibers were washed by soaking the mats into dichloromethane for 60 s. Dichloromethane was chosen since pectin is not soluble in this solvent [39], but it could facilitate fiber coalescence by reducing the porosity of the electrospun mats. The washed mats were then placed in a Vaciotem-TV (P. Selecta, Barcelona, Spain) vacuum drying oven connected to a Vacuubrand vacuum pump at 27 °C and 100 mmHg pressure for 18 h in order to remove the organic solvent.

2.6. Films Preparation

The washed and dried fibers were then subjected to annealing in a 4122-model press from Carver, Inc. (Wabash, IN, USA). A set of experiments were conducted to select the optimal temperature, time, and load to produce homogenous and transparent films. To this end, the electrospun mats were post-treated in the temperature range of 50–240 °C and the pressure range of 6–30 kN for times ranging from 5 s to 120 s. A pectin film was also prepared by casting as a control material. To this end, 2 g of pectin powder was dissolved in 100 mL of distilled water and then 0.92 g of glycerol was added. After 24 h of mixing, 10 mL of solution was poured into polystyrene (PS) petri dishes and left at room conditions, that is, 25 °C and 40%, for 3 days.

The multilayer films were prepared by placing either the solvent-casted or electrospun pectin-based films as an interlayer between two electrospun layers of PHBV. This was accomplished by electrospinning PHBV fibers on one side of the previously prepared pectin films. The resultant coated films were turned down and coated on the other side. The two side coated films were then placed in the press and annealed at 160 °C for 10 s, without pressure, based on our previous research [55]. These annealing conditions were selected since the film-forming process is controlled by the external layers, which are habitually thicker. Control films made of two electrospun layers of PHBV without pectin were also prepared in the same conditions.

2.7. Characterization of the Electrospun Fibers and Films

2.7.1. Thickness and Conditioning

Prior to testing, the thickness of the electrospun mats and films was measured using a digital micrometer series S00014 (Mitutoyo Corporation, Kawasaki, Japan) with ±0.001 mm accuracy. Measurements were performed at five random positions and values were averaged. The samples were stored in a desiccator at 25 °C and 0% RH for 24 h before characterization.

2.7.2. Morphology

The morphologies of the electrospun fibers and the top views and cross-sections of the pectin-based films were investigated by a scanning electron microscope (SEM, Hitachi S-4800, Tokyo, Japan). The samples were cryo-fractures using nitrogen liquid. Prior to analysis, all the samples were coated with a gold/palladium alloy for 2 min by a Polaron sputter coater (Quarum Technologies, Kent, UK). A 5 kV voltage was applied during SEM analysis. Fiber diameters and layer thicknesses were determined by the software ImageJ, Java v.1.52a from the measurement of, at least, 50 fibers.

2.7.3. Thermal Analysis

The thermal properties of PEO_{2000} and pectin powders as well as the electrospun pectin-based fibers obtained from S3 and from S5 to S10 were determined by differential scanning calorimetry (DSC) and Thermogravimetric analysis (TGA). Thermal transitions were determined in a DSC-7 analyzer from PerkinElmer, Inc. (Waltham, MA, USA), equipped with a cooling accessory Intracooler 2 from PerkinElmer, Inc. Approximately 3 mg of sample was placed into the aluminum pan, while an empty pan was used as reference. Calibration was previously conducted using an indium sample. The samples were first heated from −70 °C to 160 °C, then cooled back to −70 °C, and then heated again to 300 °C. The heating and cooling rates were set at 10 °C/min. The experiments were conducted under nitrogen atmosphere and all DSC tests were performed in triplicate.

To ascertain their thermal stability, TGA was performed under nitrogen atmosphere in a Thermobalance TG-STDA Mettler Toledo model TGA/STDA851e/LF/1600 analyzer (Greifensee, Switzerland). TGA curves were obtained after conditioning the samples in the sensor for 5 min at 30 °C. The samples were heated from 25 °C to 700 °C at a heating rate of 10 °C/min. All TGA tests were also carried out in triplicate.

2.7.4. Fourier Transform Infrared Spectroscopy

Fourier transform infrared spectroscopy (FTIR) spectra of the raw materials, that is, the pectin powder, glycerol, and PEO_{2000}, and also the electrospun fibers and film obtained from S7 solution were obtained from an average of 20 scans by a Bruker Tensor 37 (Rheinstetten, Germany) spectrometer connected with a Golden Gate of Specac, Ltd. (Orpington, UK) attenuated total reflection (ATR) accessory. ATR-FTIR was performed in order to analyze the existence of any chemical interactions between the components. The scans were collected in the wavelength values from 4000 cm^{-1} to 600 cm^{-1} at a resolution of 4 cm^{-1}.

2.7.5. Color Measurements

The color of the pectin-based interlayers and the resultant multilayers in PHBV was carried out in a benchtop spectrophotometer Konica Minolta CM-5, from Hunter Associates Laboratory, Inc. (Reston, VA, USA). The Commission Internationale de l'Eclairage (CIE) standard illuminant D65 was used to assess the CIE Lab color space coordinates L*a*b* using an observer angle of 10°. L* represents the luminance (black to white), a* indicates the change between green and red, and b* represents the change from blue to yellow. The colorimeter was calibrated with a white standard tile and a mirror device for the black (no light reflection). The L*a*b* coordinate values were obtained on five different samples and the color difference (ΔE*) was calculated following Equation (1).

$$\Delta E^* = \left[(\Delta L^*)^2 + (\Delta a^*)^2 + (\Delta b^*)^2\right]^{0.5} \quad (1)$$

where ΔE*, Δa*, and Δb* corresponded to the differences between the color parameters of the multilayer samples and the values of the PHBV/PHBV multilayer. Color change was evaluated as follows: Unnoticeable ($\Delta E^*_{ab} < 1$), only an experienced observer can notice the difference ($\Delta E^*_{ab} \geq 1$ and <2), an unexperienced observer notices the difference ($\Delta E^*_{ab} \geq 2$ and < 3.5), clear noticeable difference ($\Delta E^*_{ab} \geq 3.5$ and < 5), and the observer notices different colors ($\Delta E^*_{ab} \geq 5$) [59].

2.7.6. Permeability Tests

Water Vapor Permeance

The water vapor permeance of the multilayer films was measured according to the ASTM 2011 gravimetric method. In order to conduct this test, 5 mL of distilled water was put inside a Payne permeability cup (Inner diameter = 3.5 cm) (Elcometer Sprl, Belgium). The films were located in the cups so that on one side they were exposed to 100% RH, avoiding direct contact with water. Then, the cups were locked with silicon rings and kept in a conditioned desiccator (25 °C and 0% RH). The control samples were cups with aluminum films to estimate solvent loss through the sealing. The cups were weighed regularly for every 24 h using an analytical balance having an accuracy of ±0.0001 g, until the values reached a plateau. Water vapor permeation rate corresponded to the slope value of the steady state line of time versus weight loss per unit area and the weight loss was calculated as the total loss minus the loss through the sealing. Water permeance was obtained by correcting the water vapor permeation rate for the permeant partial pressure. Tests were conducted in triplicate.

Limonene Vapor Permeance

Limonene vapor permeance of the multilayer films was determined as similar as described above for water vapor. For this 5 mL of D-limonene was placed inside the Payne permeability cups and the cups containing the films were placed at controlled room conditions of 25 °C and 40% RH. The limonene vapor permeation rates were estimated from the steady-state permeation slopes and the weight loss was calculated as the total cell loss minus the loss through the sealing. Limonene permeance was obtained by correcting the limonene vapor permeation rate for the permeant partial pressure. Tests were conducted in triplicate.

2.8. Statistical Analysis

All data were analyzed statistically by SPSS Statistics 17.0 (IBM, Chicago, IL, USA). Tukey's HSD test was used to determine the significant differences among samples ($p < 0.05$). Different superscripts show statistically different results. Unlike groups of letters were used to express each group of properties.

3. Results and Discussion

3.1. Preparation of Electrospun Pectin-Based Fibers

The morphology of the resultant electrospun fibers is shown in Figure 1. A solution of pure pectin was initially also tested but, instead of continuous jets, it formed large droplets when subjected to high voltages during electrospinning due to the limited viscoelasticity and insufficient chain entanglements of the carbohydrate [37–39]. A similar phenomenon was faced by Liu et al. [38] during the electrospinning of pectin. In order to increase the viscoelasticity of pectin, different quantities of PEO_{2000} from 0.1 to 1.0 wt% were added to the pectin solutions for electrospinning, the here so-called S1 to S4. The addition of PEO can help reducing the repulsive forces among negatively charged pectin chains, enhancing chain entanglement and fiber formation [45]. The primary intention was to keep the PEO_{2000} content at a minimum value in order to produce electrospun fibers with the highest content of pectin. In Figure 1a it can be observed that the electrospinning of the pectin solution having 0.1 wt% PEO_{2000}, that is, S1, resulted in fibers with a discontinuous and beaded morphology. The fibers produced from S2, shown in Figure 1b, which is based on 0.25 wt% PEO_{2000}, were more continuous and non-beaded. However, the resultant electrospun mat showed poor integrity as these easily fractured during detachment from the collector. In the case of S3, made of a pectin solution with 0.5 wt% PEO_{2000}, neat fibers free of beads and with a uniform diameter were produced as it can be observed in Figure 1c. Furthermore, branched and thick pectin fibers were obtained for the electrospinning of S4, which contained 1 wt% of PEO_{2000} and are shown in Figure 1d. These results confirm that the use of a carrier polymer in an appropriate amount is a key parameter in order to achieve continuous and non-defected fibers during electrospinning. Based on these results, S3 was selected due to it contained the optimal amount of PEO_{2000} that yielded the fibers with the highest uniformity and a relatively low diameter, whereas S1, S2, and S4 were discarded from the study. Although there was not any particular requirement for the diameter, it was necessary to obtain continuous and free-bead fibers in order to produce films. Thereafter, different types and amount of plasticizers were also tested to improve the film-forming capacity of the electrospun pectin mats. Therefore, glycerol or PEG_{900} were added to the pectin solution based on the S3 composition in which the PEO_{2000} content was kept constant at 0.5 wt%. One can observe in Figure 1e–j that similar morphologies, based on smooth and continuous fibers but slightly thinner in the case of PEG_{900}-containing fibers, were generated when the plasticizers were added. The average diameters of the electrospun pectin-based fibers are summarized in Table 2.

Table 2. Properties of the pectin-based solutions and mean diameters of their resultant electrospun fibers.

Solution	Viscosity (cP)	Surface Tension (mN/m)	Conductivity (μS·cm)	Fiber Diameter (nm)
S1	1892 ± 78 ^{a}	28.8 ± 0.2 ^{a}	7.25 ± 0.12 ^{a}	-
S2	3066 ± 48 ^{b}	28.4 ± 0.3 ^{a}	6.37 ± 0.15 ^{a,b}	156 ± 33 ^{a}
S3	5950 ± 219 ^{c,e}	30.3 ± 0.2 ^{b,d}	6.10 ± 0.10 ^{c}	186 ± 32 ^{b}
S4	12,155 ± 1660 ^{d}	32.9 ± 0.2 ^{c}	5.99 ± 0.25 ^{b,c}	299 ± 41 ^{c}
S5	5511 ± 299 ^{c}	31.6 ± 0.2 ^{b,e,f}	5.64 ± 0.37 ^{c,d}	272 ± 43 ^{d}
S6	5751 ± 401 ^{c}	30.9 ± 0.3 ^{d,e}	5.30 ± 0.13 ^{c,d}	304 ± 56 ^{c,d}
S7	6440 ± 327 ^{c,e}	32.1 ± 0.2 ^{g,f}	4.98 ± 0.09 ^{c,d}	329 ± 42 ^{c,d}
S8	6047 ± 219 ^{e}	31.2 ± 0.3 ^{e}	5.11 ± 0.20 ^{c,e}	189 ± 40 ^{b}
S9	6106 ± 124 ^{e}	31.1 ± 0.3 ^{e}	4.82 ± 0.21 ^{e}	197 ± 44 ^{b}
S10	6317 ± 364 ^{e}	32.5 ± 0.2 ^{g,c}	4.72 ± 0.22 ^{e}	223 ± 33 ^{b}

$^{a–g}$ Different letters in the same column indicate a significant difference among the samples ($p < 0.05$).

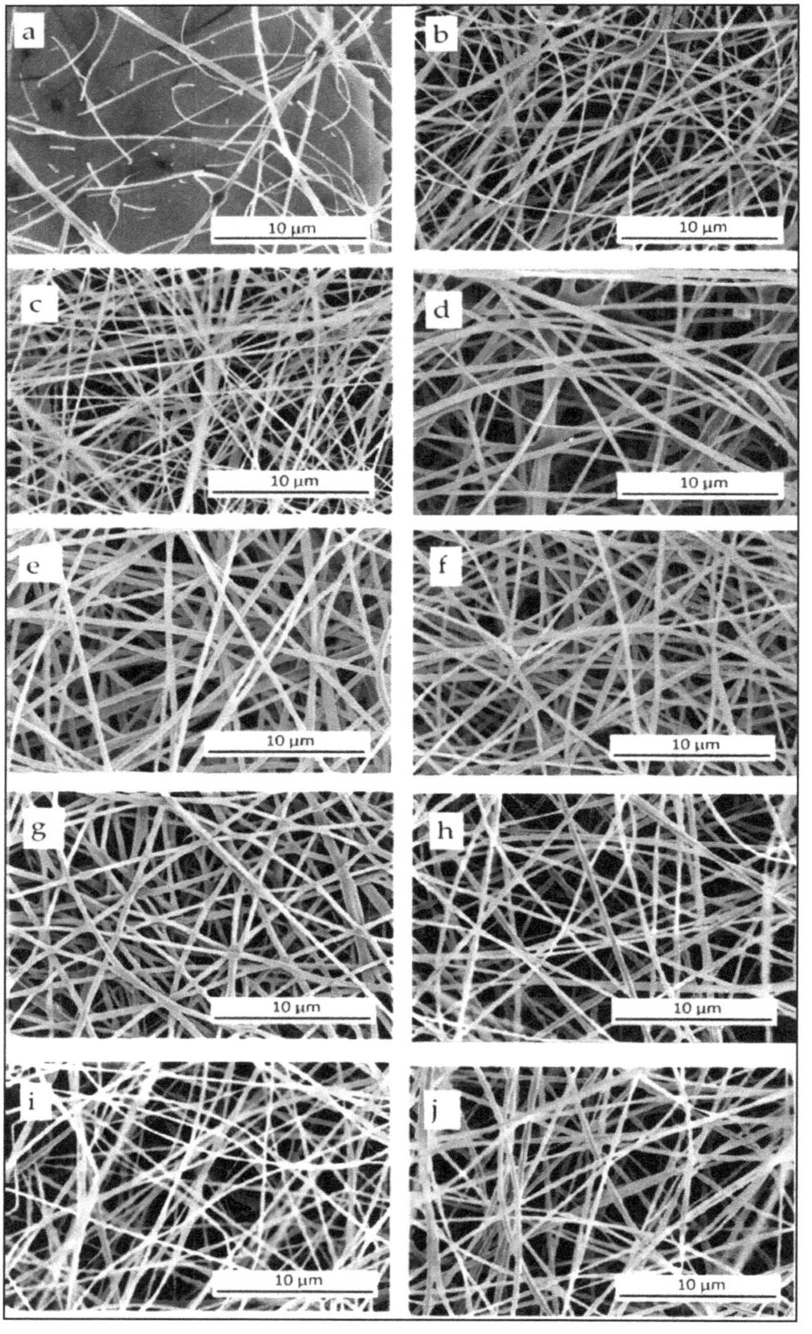

Figure 1. Scanning electron microscopy (SEM) images of the electrospun mats obtained from the pectin-based solutions of: (**a**) S1; (**b**) S2; (**c**) S3; (**d**) S4; (**e**) S5; (**f**) S6; (**g**) S7; (**h**) S8; (**i**). S9; (**j**) S10.

The solution properties were determined to better understand the morphologies of the ultrathin pectin-based fibers in the attained electrospun mats. The values of viscosity, surface tension, and conductivity of the fiber-forming solutions are also given in Table 2. For the non-plasticized solutions, that is, S1–S4, viscosity of the pectin-based solutions increased significantly when the amount of PEO_{2000} was increased. One can observe that the addition of plasticizers did not create any significant difference in the solution viscosity. Even when the content of glycerol or PEG_{900} was increased to 3 wt%, the solution viscosity did not change considerably, showing a value around 6000–6400 cP. The surface tension slightly increased from approximately 29 mN/m to values in the range of 31–33 mN/m when the PEO_{2000} content was increased above 0.5 wt%. It was not observed a clear correlation between the plasticizer content and the surface tension of pectin-based solutions. It can also be observed that, due to the inherent polyelectrolyte nature of pectin [60], its aqueous solutions were highly conductive, showing values in the 6–8 μS.cm range. However, the conductivity values decreased up to values close to 4 μS.cm when the PEO_{2000} content was increased and, particularly, when the plasticizers were added since the amount of pectin in the solution was reduced. Therefore, the present results suggest that the fiber formation was attained due to a combined effect of viscosity increase and conductivity decrease. The optimal values were particularly attained in the range of ~5500–6500 cP and ~5–6 μS.cm of viscosity and conductivity, respectively. Therefore, moderate-to-high viscosities in combination with relatively low conductivities tended to produce the fibers with the most optimal morphology, while the effect of surface tension was negligible. As previously indicated in other works describing the role of solution properties in the electrospinnability of biopolymers [32], it was difficult to elucidate the effect of a single property without considering the impact of the other ones.

In relation to the plasticizers, one can observe that the diameters of the fibers varied from 156 ± 33 nm to 329 ± 42 nm, which could be related to differences in the solution properties described above. Significantly thicker fibers were attained with the increase of the PEO_{2000} content (from S2 to S4) for the non-plasticized samples since the viscosity of these solutions was increased with the PEO_{2000} content [49]. Also, the incorporation of glycerol to the PEO_{2000} containing fibers resulted in significantly thicker fibers. This increase in the fiber diameter can be related to the increase in the solution viscosity. As shown in Table 2, S3 showed a viscosity of 5950 cP and it yielded fibers with average diameter of 186 ± 32 nm, whereas fibers with an average diameter of 329 ± 42 nm was attained in S7 with a viscosity of 6440 cP. This observation can be ascribed to the plasticizing effect of glycerol, which favored the molecular entanglements of the pectin chains by reducing their intermolecular interaction. The latter effect is based on the fact that intermolecular H-bonds of pectin can be substituted by glycerol-pectin H-bonds and covalent esters, thus the attraction between pectin molecules is reduced [29]. A similar phenomenon was also reported by Cui et al. [40] where glycerol-containing solutions resulted in pectin-based fibers with higher diameters than those obtained from solutions using dimethylformamide (DMF) or dimethyl sulfoxide (DMSO) as co-solvents. Also, if one compares the solutions containing 2.5 wt% of plasticizer, which were labelled as S5 and S8, the glycerol-containing ones yielded fibers with an average diameter of 272 ± 43 nm, while the solutions with PEG_{900} produced fibers in the range of 189 ± 40 nm. This can be explained by the fact that PEG_{900} has a higher M_w than glycerol and, thus, less substitution of H-bonds could have been occurred and therefore less chain entanglements were formed. As mentioned earlier, electrospinning of the 1 wt% PEO containing solution, that is, S4, resulted in pectin-based fibers with a discontinuous and beaded morphology, which is a consequence of the formation of a solution with low viscosity and high conductivity.

3.2. Thermal Properties of Electrospun Pectin-Based Fibers

The DSC curves corresponding to the cooling and second heating steps of the as-received PEO_{2000} and pectin powders and the electrospun pectin-based fibers obtained from S3 and S5–S10 are gathered in Figure 2. During the cooling process, shown in Figure 2a, one can observe that PEO_{2000} crystallized from the melt showing a crystallization temperature (T_c) of approximately 40 °C [61]. Alternatively, the pectin powder showed no crystallization during cooling in the whole tested thermal range. The crystallization peak attributed to PEO_{2000} was not observed in the electrospun PEO_{2000}-containing pectin fibers due to its relative low content. However, a small exothermic peak was observed at approximately 23 °C for the electrospun pectin fibers obtained from S10, which can be attributed to the crystallization of the PEG_{900} confined in the carbohydrate [61,62]. In Figure 2b one can observe the melting temperature (T_m) of PEO_{2000} at nearly 68 °C and also two low-intense melting peaks corresponding to the melting of the PEG_{900} fraction in the pectin fibers obtained from S10 at approximately 24 °C and 40 °C [62]. Pectin did not exhibit any thermal transition of first order, that is, crystallization or melting, thus confirming the carbohydrate is fully amorphous [13,63]. Moreover, no glass transition was observed up to nearly 230 °C, temperature from which the carbohydrate started thermal degradation. Previous studies on the thermal properties of pectin reported that thermal degradation of this carbohydrate is an exothermic process [63,64]. It is worthy to note that the degradation temperature (T_{deg}) of pectin shifted from 232 °C, for the neat powder, to 236 °C, for the electrospun fibers obtained from S10, which suggests that the thermal stability of the carbohydrate was slightly enhanced by the addition of PEG_{900}. This improvement can be related to the newly formed intermolecular PEG-pectin H-bonds and covalent esters, which delayed the thermal degradation of the pectin macromolecule [29].

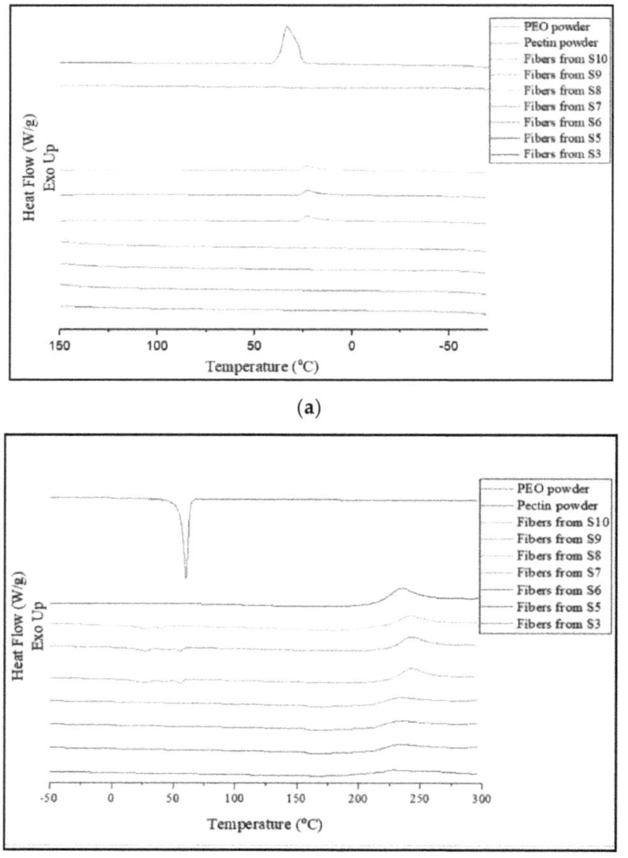

(**b**)

Figure 2. Differential scanning calorimetry (DSC) curves of polyethylene oxide 2000 (PEO_{2000}) powder, pectin powder, and electrospun pectin-based fibers during: (**a**) cooling and (**b**) second heating.

The TGA curves of the neat pectin and PEO_{2000} powders as well as the electrospun pectin-based fibers obtained from S3 and S5–S10 are shown in Figure 3. The most relevant thermal parameters obtained from the TGA curves are given in Table 3. One can notice that pectin exhibited three significant main weight losses, occurring at approximately 100 °C, 217 °C, and 240 °C. The first mass loss, which took place in the 75–110 °C range, can be ascribed to the removal of bound water from the carbohydrate due to its highly hydrophilic nature. In this regard, Gloyna et al. [65] indicated that water evaporation of pectin observed between 50 °C and 150 °C, with a maximum at 100 °C, is identical to the dehydration process of other polysaccharides. Kastner et al. [66] and Nisar et al. [67] also observed a second weight loss between 195–350 °C. According to their studies, pectin shows a two-step degradation process in this range that might be related to the cracking of bonds or functional groups, structural depolymerization, and chain breaking of the polysaccharide. The carbohydrate then lost 60% of its total weight up to 350 °C whereas, in the 350–697 °C range, the mass loss was 15% of its original weight and the maximum rate of weight loss was observed at 235 °C. During thermal degradation of pectin different depolymerization reactions occur, including demethoxylation, depolymerization by backbone hydrolysis and hydrolytic cleavage of neutral sugar side chains [66–68]. The most instable bonds, formed by the neutral sugars, thermally degrade first. The polygalacturonic acid units of high M_W hydrolyze later whereas the glycoside bonds formed by uronic acids also degrade at higher temperatures. Depending on the pectin properties, that is, pH, source, degree and pattern of methyl esterification, acid hydrolysis or β-elimination reactions take place during the thermal degradation [69]. In particular, the here-used low methyl esterified amidated pectin results in acid hydrolysis reactions with the temperature increase [70,71]. During hydrolysis, longer chains break up to the shorter as the cleavage of α-(1,4)-glycosidic bond connecting two uronic acids by the addition of a water molecule [72]. In our study, above 250 °C, a secondary degradation of pectin has been reported to occur, including release of functional side groups and chains break, while gasification of char residues arises at temperatures around 600 °C [73].

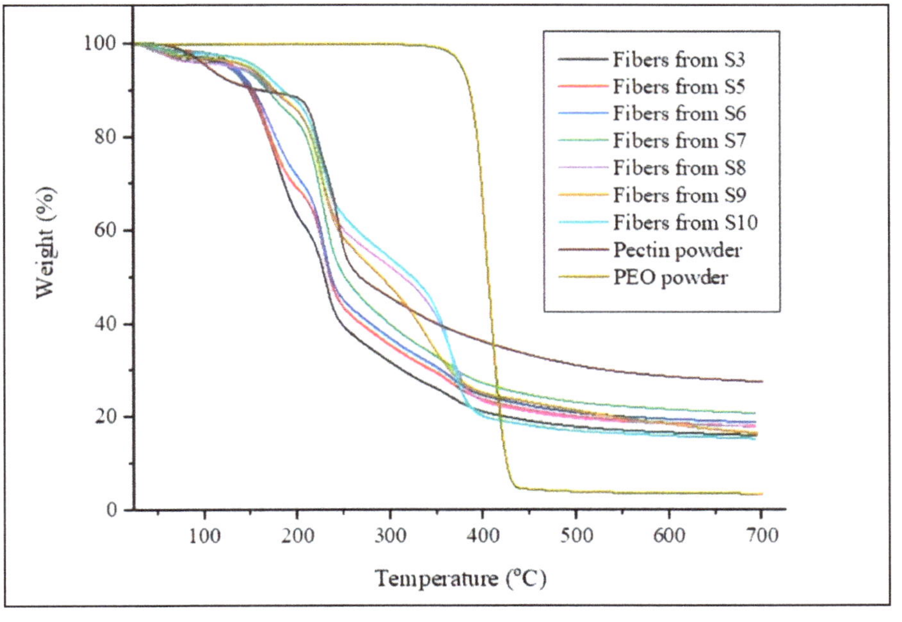

Figure 3. Thermogravimetric analysis (TGA) curves of the neat pectin powder, PEO_{2000} powder, and electrospun pectin-based fibers.

Table 3. Main thermal parameters of the neat pectin powder, PEO_{2000} powder, and the electrospun pectin-based fibers in terms of: onset temperature of degradation (T_{onset}), degradation temperature (T_{deg}), and residual mass at 700 °C.

Sample	T_{onset} (°C)	T_{deg1} (°C)	T_{deg2} (°C)	T_{deg3} (°C)	Residual Mass (%)
Pectin powder	98.0 ± 5.2 ^{a}	217.4 ± 3.2 ^{a}	240.2 ± 2.1 ^{a}	-	27.4 ± 3.2 ^{a}
PEO_{2000} powder	375.3 ± 2.4 ^{b}	-	-	400.0 ± 2.1 ^{a}	23.7 ± 1.2 ^{b}
Fibers from S3	175.4 ± 2.5 ^{c}	230.5 ± 3.2 ^{b}	-	367.5 ± 3.2 ^{b}	19.2 ± 1.1 ^{c}
Fibers from S5	169.3 ± 3.7 ^{c,d}	230.3 ± 3.1 ^{b}	297.3 ± 1.1 ^{b}	367.3 ± 2.1 ^{b}	20.2 ± 1.9 ^{c}
Fibers from S6	169.5 ± 2.1 ^{c,d}	229.8 ± 2.0 ^{b}	297.5 ± 1.2 ^{b}	367.1 ± 2.0 ^{b}	21.2 ± 2.0 ^{b,c}
Fibers from S7	170.3 ± 4.1 ^{c,d}	227.5 ± 2.1 ^{b,c}	297.4 ± 1.1 ^{b}	367.3 ± 3.2 ^{b}	23.4 ± 1.5 ^{b}
Fibers from S8	165.5 ± 5.2 ^{d}	226.4 ± 2.2 ^{c}	-	366.4 ± 2.6 ^{b}	21.2 ± 1.3 ^{b,c}
Fibers from S9	168.2 ± 4.4 ^{d}	226.2 ± 2.1 ^{c}	-	367.4 ± 3.1 ^{b}	20.0 ± 1.6 ^{c}
Fibers from S10	173.6 ± 5.0 ^{c,d}	227.2 ± 3.0 ^{b,c}	-	369.5 ± 2.0 ^{b}	17.3 ± 1.2 ^{d}

$^{a-d}$ Different letters in the same column indicate a significant difference among the samples ($p < 0.05$).

Alternatively, it can be observed that PEO_{2000} was highly thermally stable, showing the values of onset degradation temperature (T_{onset}) and maximum degradation temperature (T_{deg}) of approximately 375 °C and 400 °C, respectively. One can also observe that the incorporation of PEO_{2000} successfully delayed the thermal degradation of pectin up to nearly 175 °C. The reason of this behavior can be explained by the occurrence of more chain entanglements and the newly formed intermolecular PEO-pectin H-bonds and covalent esters, leading to the formation of a more stable and less volatile material. The pectin-based fiber obtained from the electrospinning of S3, S5, and S6 showed a similar thermal degradation profile than the neat pectin powder but with even lower values of T_{deg}. This result suggests that, even though PEO_{2000} delays the onset of degradation of pectin, it also catalyzes its thermal degradation. Interestingly, the presence of both plasticizers contributed to increasing the thermal degradation of the PEO_{2000}-containing pectin fibers. The pectin-based fibers obtained from the electrospinning of S5–S10, which included glycerol or PEG_{900}, showed a similar thermal degradation profile in comparison with the PEO_{2000}-containing pectin fibers but with slightly lower values of T_{onset}. Additionally, the new weight loss obtained nearly 290 °C can be attributed to glycerol degradation. One can then consider that the plasticizers played a role in the pectin chains motion that favored chain scission during thermal degradation. Similar results were found for chia mucilage-glycerol films in the study performed by Dick et al. [74], showing that the glycerol addition lowered the heat resistance of the carbohydrate. Moreover, lower residual masses were observed for the plasticizer-containing fibers compared with the non-plasticized fibers. This effect was particularly intense for the electrospun pectin fibers obtained from S10, that is, the solution containing 3 wt% PEG_{900}, which showed a residual mass of 17 wt%.

3.3. Film-Forming Process of Electrospun Pectin-Based Fibers

The electrospun mats of pectin fibers were subjected to annealing in order to eliminate or minimize the porosity and then produce homogenous and continuous films. Based on the DSC results shown above, pure pectin is fully amorphous. Therefore, a heat treatment in a wide range of temperatures, that is, 50–220 °C, was needed to be applied for different processing times, that is, from 5s to 120 s, to find out the best conditions. Figure 4 shows the visual aspect of the different electrospun films obtained after annealing and Figure 5 includes the SEM micrographs of their surface fracture. However, one can observe that, at any of these conditions, the annealing failed to provide homogenous films using the fibers obtained from the non-plasticized pectin solution, that is, from S3. Moreover, the color of films became darker when subjected to temperatures above 160 °C due to the low onset temperature of degradation of pectin fibers, which is shown above in Table 3. As seen in Figure 4a,b, by increasing the temperature or time, not only the pectin materials developed a dark color but also the fiber mats lost their integrity. Therefore, based on these results, one can consider that the non-plasticized pectin-based fibers were too rigid to gain flexibility by the only application of heat. Therefore, pressure was also applied during annealing to promote fibers rearrangement and the removal of porosity. In Figure 4c it can be observed that some transparent regions along the pectin-based materials were successfully attained when the mats were subjected to temperatures above 165 °C with an applied pressure of 12–24 kN for 1 min. Nevertheless, the resultant materials also exhibited an intense brownish color, partially lost their integrity and also became too brittle to be applied as packaging materials.

Figure 4. Effect of the post-treatment conditions on the pectin-based materials obtained from S3. (**A**) Different temperatures were applied for 1 min without pressure. (**B**) Different times were applied at 190 °C without pressure. (**C**) Different pressures were applied at 165 °C for 60 s.

The above-reported morphological change was further confirmed in Figure 5 by comparison of the SEM images of the cross-sections of the electrospun pectin-based mats from S3 prior to annealing, shown in Figure 5a, and post-processed at 220 °C for 20 s, shown in Figure 5b. This phenomenon can be explained by the oxidation and degradation of pectin compounds at high temperatures. Nevertheless, interestingly, it can be observed that a compact packing rearrangement of the electrospun fibers was observed by a process of fibers coalescence. This phenomenon successfully resulted in the formation of a continuous film. However, some voids were also formed that could result from the evaporation of gases during thermal decomposition.

Figure 5. Scanning electron microscopy (SEM) images of the fracture surface of the electrospun pectin-based fibers from S3: (**a**) Without any post-treatment and (**b**) Processed at 220 °C for 20 s with a pressure of 24 kN.

Based on the results shown above, the effect on the film-forming process of the two plasticizers, that is, glycerol and PEG_{900}, was analyzed. The SEM images of the pectin-based films obtained from the fibers produced with the solutions containing the plasticizers are gathered in Figure 6. The electrospinning of the PEO_{2000}-containing pectin solutions with 20–30 wt% glycerol, that is, S5, S6, and S7, resulted in electrospun mats that, after annealing at 150 °C, produced softer and more flexible films. A similar improvement was attained for the electrospun PEO_{2000}-containing pectin mats with 20–30 wt% PEG_{900}, labelled as S8, S9, and S10, after annealing at 155 °C, though the films were less homogenous. As mentioned earlier, the role of plasticizers in pectin is based on increasing the chain mobility and free volume by creating H-bond interactions with the biopolymer chains [21]. The use of plasticizers thus successfully opened up the post-processability of the electrospun mats by annealing. However, temperatures around 160 °C were still needed, which are certainly close to the T_{deg} values previously measured for pectin and are also responsible for darkening the film samples. Also, for all the formulations, the annealed mats still comprised some porosity, as one can observe in the SEM images gathered in the left column of Figure 6.

To increase the homogeneity of the resultant pectin materials, the electrospun pectin-based fibers were washed with dichloromethane prior to annealing. As seen in the SEM images shown in the middle column of previous Figure 6, the washed fibers partially coalesced, which could potentially enable to reduce the energy requirement for annealing. As mentioned earlier, dichloromethane is a solvent that does not dissolve pectin, but it dissolves glycerol whereas PEG_{900} and PEO_{2000} are slightly soluble. When the pectin fiber mats containing glycerol were immersed in dichloromethane, they did not lose their integrity but, due to the removal of glycerol, the electrospun fibers partially coalesced. Thereafter, annealing of the washed fibers was successful when applied at 150 °C and 140 °C for the fibers obtained from the solutions containing glycerol, that is, S5, S6, and S7, and PEG_{900}, that is, S8, S9, and S10, respectively. As it can be observed in the right SEM images of Figure 6, the most homogenous film structures were obtained for the pectin fibers mats with 25 wt% and 30 wt% glycerol and 5 wt% PEO_{2000} were annealed. These results were confirmed by the observation of the optical images of the film samples gathered in Figure 7. Due to the low porosity and completely homogenous film structure, the pectin-based film obtained from S7, that is the pectin formulation with 30 wt% glycerol and 5 wt% PEO_{2000}, was selected as the most appropriate candidate for multilayer films.

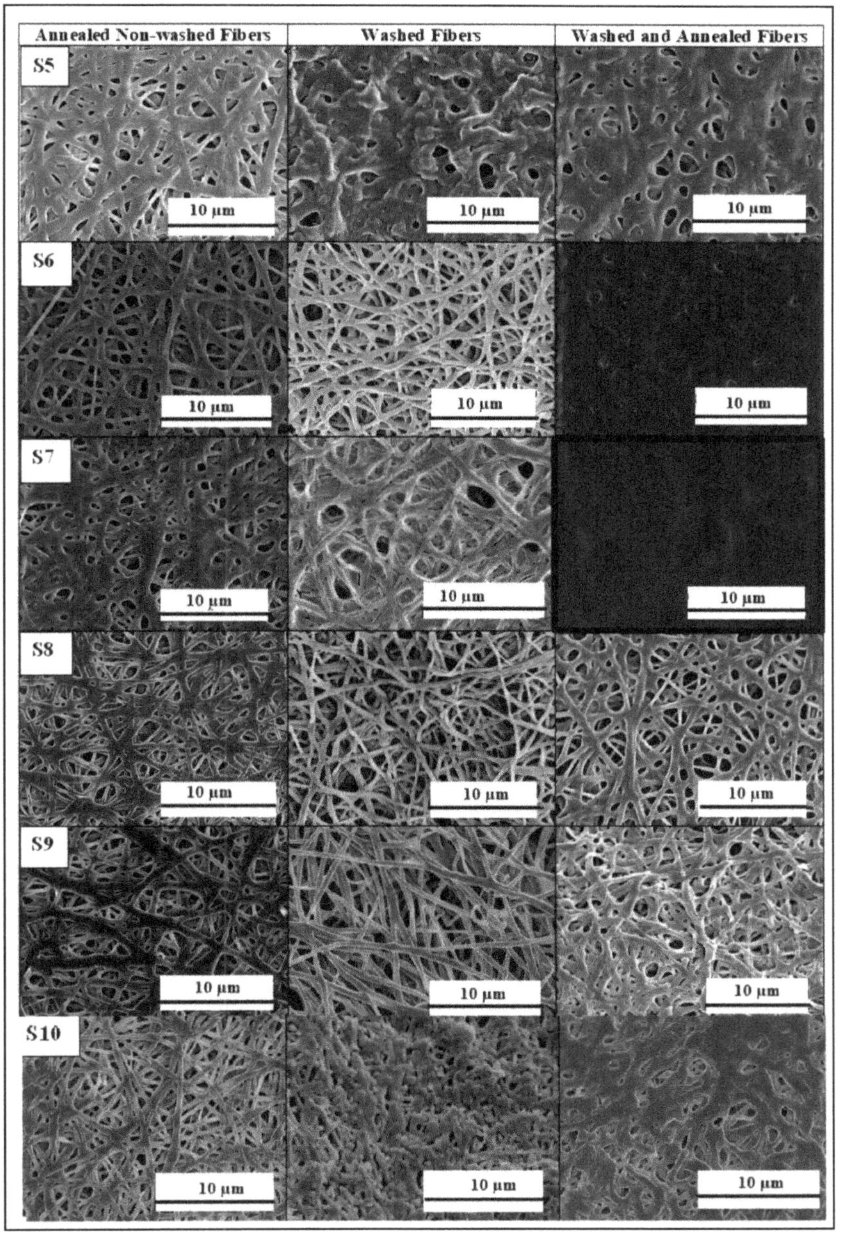

Figure 6. Scanning electron microscopy (SEM) images of the electrospun mats after annealing at 150 °C of the pectin-based fibers obtained from S5, S6, and S7 and at 155 °C of the pectin-based fibers obtained from S8, S9, and S10, all processed with a pressure of 24 kN load for 1 min (**left**). Same pectin-based fibers after washing with dichloromethane (**middle**). Washed pectin-based fibers after annealing at 140 °C for the fibers obtained from S5, S6, and S7 and at 150 °C for the fibers obtained from S8, S9, and S10, all processed with a pressure of 12 kN load for 1 minute (**right**). Scale markers of 10 µm in all cases.

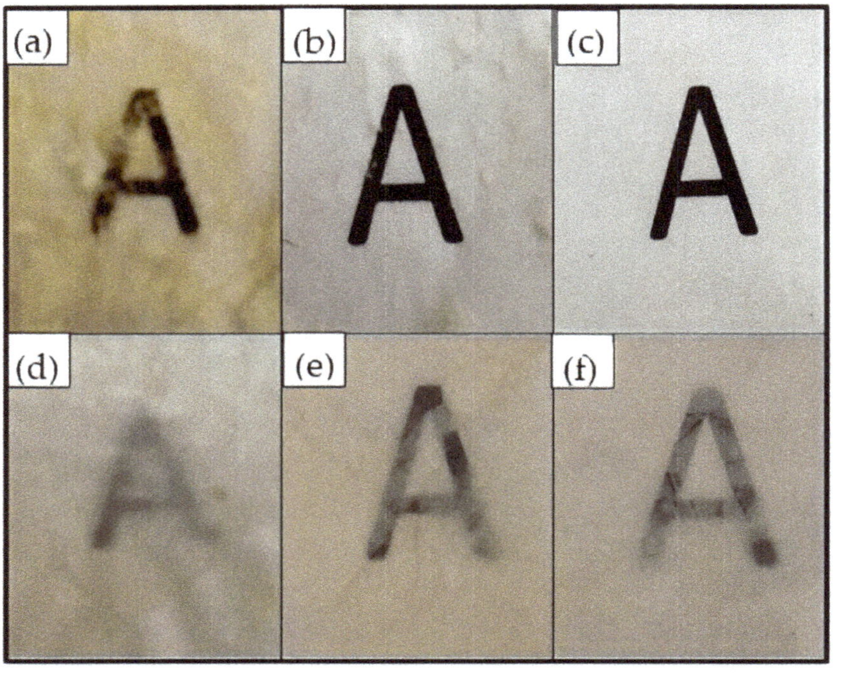

Figure 7. Optical images of the electrospun pectin films obtained from the washed fibers of: (**a**) S5; (**b**) S6; (**c**) S7; (**d**) S8; (**e**) S9; (**f**) S10.

3.4. Chemical Characterization of Pectin-Based Electrospun Film

ATR-FTIR was carried out on the selected film, obtained from the pectin-based fibers of S7, to ascertain the effect of PEO_{2000} and glycerol on pectin and the thermal post-treatment. The FTIR spectra of the pure components, that is, the pectin powder, glycerol, and PEO_{2000}, were also collected. Figure 8 gathers the FTIR spectra of these components and of the pectin-based fibers and film. Table 4 summarizes the band attribution of each peak observed in the FTIR spectra of the pectin materials. One of the characteristic peaks of pectin was seen at 1741 cm^{-1}, which refers to the C–H stretching of carboxyl groups (COOH). Also, the two characteristic bands at 1672 cm^{-1} (amide I) and 1595 cm^{-1} (amide II) relate to the presence of amide groups [75,76] of the amidated pectin. The peaks centered at 1132 cm^{-1} and 1070 cm^{-1} may be ascribed to the contribution of the C–C and C–O bonds in secondary alcohol groups of –CH–OH. For glycerol, five characteristic bands at the wavenumbers in the 1150–850 cm^{-1} range arise from the vibrations of C–C and C–O linkages [77]. The FTIR spectrum of PEO_{2000} exhibited CH_2 scissoring at 1465 cm^{-1}, CH_2 wagging at 1357 cm^{-1} and 1340 cm^{-1}, CH_2 bending at 1280 cm^{-1} and 1240 cm^{-1}, C–O–C stretching at 1093 cm^{-1} and 1145 cm^{-1}, and CH_2 rocking at 962 cm^{-1} and 847 cm^{-1} [78]. The shift in the wavenumber values related to the—CH bending from 1423 cm^{-1}, for the pectin powder, to 1411 cm^{-1}, for the pectin fibers and film, could be related to a decrease in the interaction between pectin molecules that can be ascribed to the above-described plasticizing effect of glycerol and also to the interaction with PEO_{2000}. No changes were observed in the FTIR spectrum of the pectin-based material from fibers to film after the annealing, which proves the absence of chemical reactions and degradation during the applied thermal post-treatment.

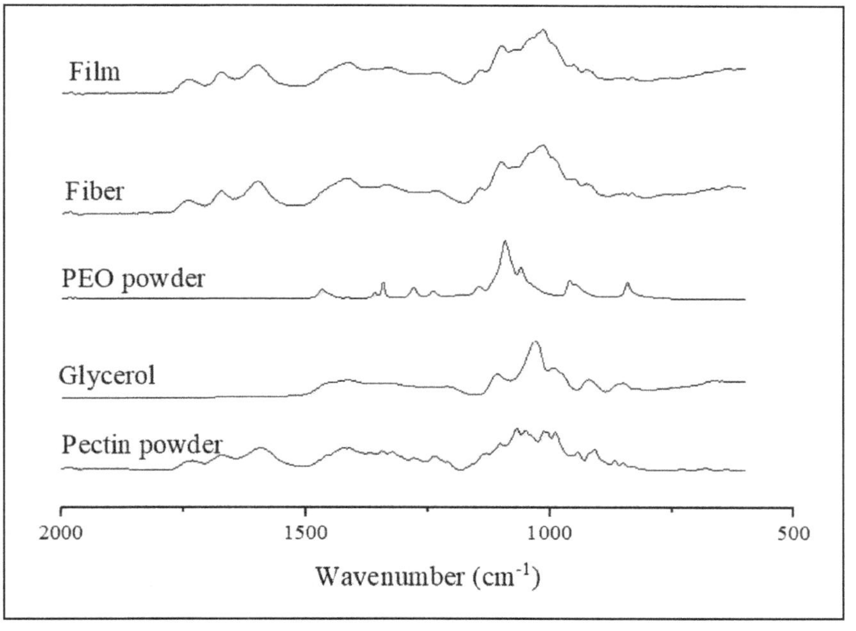

Figure 8. Fourier transform infrared spectroscopy (FTIR) spectra, from bottom to top, of pectin powder, glycerol, and polyethylene oxide 2000 (PEO_{2000}), and electrospun fibers and film obtained from S7.

Table 4. Band attribution of the main peaks observed in the Fourier transform infrared spectroscopy (FTIR) spectra of the pectin-based materials.

Wavenumber (cm^{-1})	Chemical Group
1741	C–H stretching
1672	Amide I
1595	Amide II
1132	C–C
1070	C–O

3.5. Application of Pectin-Based Electrospun Film in Multilayers

The electrospun pectin film was integrated as an inner layer into two electrospun PHBV films, which were prepared as described previously by Cherpinski et al. [51]. The objective of this multilayer was to apply the here-prepared pectin film, which was prepared by the annealing of fibers from S7 solution, as a barrier interlayer that was protected from humidity by two external electrospun layers of PHBV, a bio-based and biodegradable hydrophobic polyester. The whole multilayer structure was formed by electrospinning and subsequent annealing in order to achieve a high adhesion between the layers. A cast film of pectin was also applied in the same conditions for comparison purposes. Figure 9 shows the visual aspect of the electrospun multilayer films to evaluate their contact transparency. The effect of the inner pectin-based interlayer on the color coordinates L*a*b* and ΔE* of the electrospun PHBV films are shown in Table 5. One can observe that all the here-prepared multilayer films presented a good contact transparency. However, a slightly yellowish was developed when the pectin layer was incorporated, particularly for the material obtained by electrospinning and annealing. In particular, the b* values of the PHBV/PHBV multilayer increased from −0.36 to 4.95 and 0.83 for the PHBV/solvent-casted pectin/PHBV and PHBV/electrospun pectin/PHBV multilayer films, respectively. Furthermore, in the case of the PHBV/electrospun pectin/PHBV multilayer, the slight increase in the a* value suggest that the films became brown. This effect can be related to the original color of the electrospun pectin layers, which was shown in previous Figure 7 and their color values are gathered in Table 5. Therefore, both pectin interlayers generated multilayer films in which an observer can notice different colors ($\Delta E^*_{ab} \geq 5$). In particular, the ΔE* values for the PHBV/solvent-casted and PHBV/electrospun pectin/PHBV multilayer films were 9.60 and 11.48, respectively. Additionally, it is worthy to mention that the neat PHBV/PHBV films showed lower transparency than the PHBV monolayers films that were reported in our previous research study [56]. This can be related to possible air entrapment between both PHBV layers during annealing.

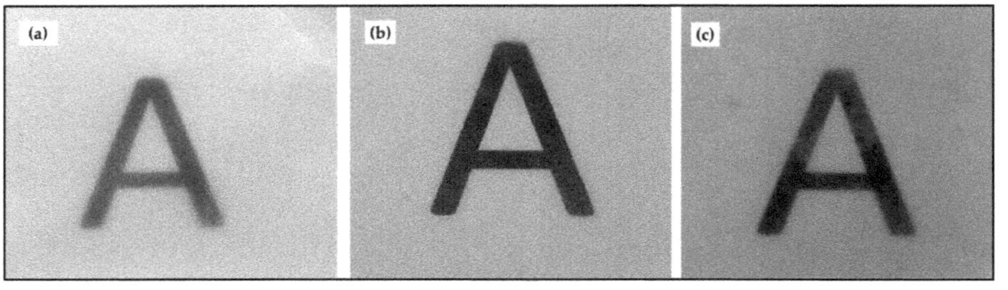

Figure 9. Visual aspect of the films based on pectin and poly(3-hydroxybutyrate-*co*-3-hydroxyvalerate) (PHBV): (**a**) PHBV/PHBV multilayer; (**b**) PHBV/solvent-casted pectin/PHBV; (**c**) PHBV/electrospun pectin/PHBV.

Table 5. Color coordinates L*, a*, b* and color difference (ΔE*) of the films based on pectin and poly(3-hydroxybutyrate-*co*-3-hydroxyvalerate) (PHBV).

Films	L*	a*	b*	ΔE*
Solvent-casted pectin	33.21 ± 0.28 [b]	0.47 ± 0.05 [b]	1.28 ± 0.12 [b]	-
Electrospun pectin	34.98 ± 0.64 [a]	−0.19 ± 0.03 [a]	1.53 ± 0.16 [a]	-
PHBV/PHBV	58.26 ± 1.20 [c]	−0.70 ± 0.02 [c]	−0.36 ± 0.05 [c]	-
PHBV/solvent-casted pectin/PHBV	46.82 ± 0.54 [e]	−0.73 ± 0.04 [c]	0.83 ± 0.09 [e]	9.60 ± 0.25 [b]
PHBV/electrospun pectin/PHBV	50.58 ± 0.48 [d]	−0.27 ± 0.08 [d]	4.95 ± 1.17 [d]	11.48 ± 0.44 [a]

a*: red/green coordinates (+a red, −a green), **b***: yellow/blue coordinates (+b yellow, −b blue), **L***: Luminosity (+L luminous, −L dark). [a–e] Different letters in the same column indicate a significant difference ($p < 0.05$).

The cross-sectional SEM images of the obtained multilayer structures are displayed in Figure 10. One can observe in Figure 10a that the bilayer control of PHBV/PHBV formed a continuous structure in which both layers could not be discerned. In relation to the PHBV/pectin/PHBV multilayers, it can be seen that the structure obtained from the electrospun-based pectin film, shown in Figure 10b, presented higher adhesion between layers. However, the multilayer structure based on the cast film of pectin, seen in Figure 10c, easily delaminated during the preparation and observation by SEM. Therefore, multilayer assemblies with a high interlayer adhesion were successfully attained by combining electrospinning and annealing treatments. This effect has been recently attributed to the high aspect ratio of the electrospun fibers, which coalesce during annealing and thus highly adhere to the material substrate [53].

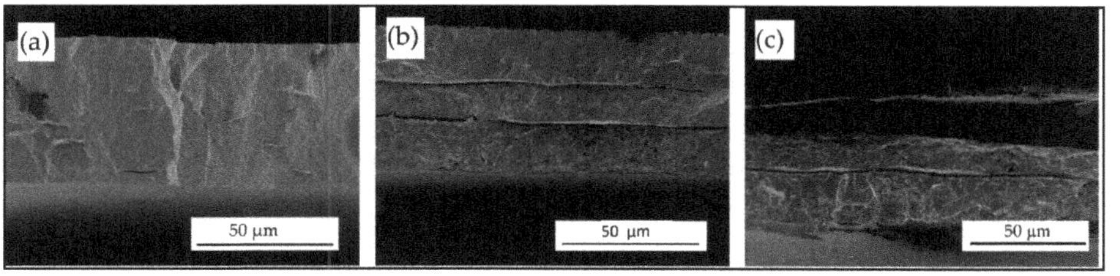

Figure 10. Scanning electron microscopy (SEM) images of cross-sections of the multilayer films based on pectin and poly(3-hydroxybutyrate-*co*-3-hydroxyvalerate) (PHBV): (**a**) PHBV/PHBV; (**b**) PHBV/electrospun pectin/PHBV; (**c**) PHBV/solvent-casted pectin/PHBV. Scale markers of 50 μm.

Finally, the permeance of the multilayer films of PHBV/PHBV, PHBV/electrospun pectin/PHBV, and PHBV/solvent-casted pectin/PHBV multilayer films to water and limonene vapors was measured. Permeance is the permeability expression with elimination of the thickness factor used to ascertain the barrier of multilayer structures. On account of comparing the permeance values of multilayer films, thicknesses of the inner and outer layers remained in the same range. Table 6 shows the water and limonene permeance values and the thickness of each layer and the whole structure. In terms of water vapor, the PHBV/PHBV and PHBV/solvent-casted pectin/PHBV multilayer films exhibited somewhat higher permeance values, that is, $5.00 \pm 0.83 \times 10^{-10}$ kg·m^{-2}·Pa^{-1}·s^{-1} and $3.94 \pm 0.56 \times 10^{-10}$ kg·m^{-2}·Pa^{-1}·s^{-1}, respectively, than the PHBV/electrospun pectin/PHBV multilayers, that is, $1.75 \pm 0.14 \times 10^{-10}$ kg·m^{-2}·Pa^{-1}·s^{-1}. As a result, the incorporation of the cast but more in particular of the electrospun pectin interlayer into PHBV provided enhanced barrier properties against water vapor, even though pectin has intrinsically a hydrophilic character. The reason why the cast film did not have comparable barrier performance to the electrospun film may be ascribed to the lower adhesion at the interphase of the cast material and also to the fact that the casting usually results in less dense film materials due to the slow forming process.

Table 6. Water vapor and limonene permeance of the multilayer films based on pectin and poly(3-hydroxybutyrate-*co*-3-hydroxyvalerate) (PHBV).

Multilayer Structure	Water Vapor Permeance × 10^{10} (kg·m^{-2}·Pa^{-1}·s^{-1})	Limonene Permeance × 10^{10} (kg·m^{-2}·Pa^{-1}·s^{-1})	Thickness (μm)		
			PHBV Layers	Pectin Layer	Total
PHBV/PHBV	5.00 ± 0.83 [a]	3.81 ± 0.47 [a]	72 ± 9 [a]	-	72 ± 9 [a]
PHBV/electrospun pectin/PHBV	1.75 ± 0.14 [b]	0.22 ± 0.11 [b]	73 ± 7 [a]	25 ± 5 [a]	98 ± 8 [b]
PHBV/solvent-casted pectin/PHBV	3.94 ± 0.56 [c]	0.22 ± 0.08 [b]	70 ± 5 [a]	24 ± 5 [a]	95 ± 5 [b]

[a–f] Different letters in the same column indicate a significant difference ($p < 0.05$).

Limonene permeance analysis was also carried out since this organic vapor is often used as an standard for aroma barrier and is also used as an indication for gas barrier properties. From the results, one can observe that the multilayers containing the pectin-based interlayers significantly reduced the limonene permeance, having both the electrospun and solution-casted pectin films the same performance. Limonene permeability is strongly governed by solubility in PHA. Indeed, solubility of limonene in PHBV is relatively high and 100-μm PHBV films can uptake up to 12.7 wt% of limonene [79]. Therefore, it can be considered that the presence of the high barrier to organic vapors, pectin interlayer blocked the passage of the aroma molecules due to its inherent low solubility to limonene. The developed multilayers containing the pectin-based interlayers showed lower permeance than monolayers of polylactide (PLA) electrospun film, that is, $2.62 \pm 1.54 \times 10^{-10}$ kg·m^{-2}·Pa^{-1}·s^{-1}, and PET electrospun film, that is, $0.64 \pm 0.11 \times 10^{-10}$ kg·m^{-2}·Pa^{-1}·s^{-1} [80].

4. Conclusions

Pectin-based electrospun films were successfully produced by electrospinning followed by annealing process. Among the different formulations tested, the water solution of 6.5 wt% pectin, 3.0 wt% glycerol, and 0.5 wt% PEO_{2000} resulted in the production of an electrospun mat composed of defect-free ultrathin fibers that, after annealing, produced the most homogenous and transparent film. Washing fibers with dichloromethane had also a positive effect on the pectin fibers coalescence. The annealing conditions were found optimal at 140 °C and 12 kN for 1 min. The resultant electrospun pectin-based films were finally incorporated as interlayers between two external layers of electrospun PHBV to produce multilayer structures with high barrier properties. The electrospun pectin interlayer successfully decreased the water and limonene vapors barrier values of PHBV and it also showed higher barrier performance when compared with an equivalent multilayer based on a solution-casted pectin interlayer. In this context, the produced electrospun pectin-based films can be considered as sustainable materials to be used for packaging applications of aromatic products.

Author Contributions: Conceptualization was carried out by S.A., J.M.L., and S.T.-G.; methodology by B.A.B.; validation and formal analysis, B.A.B. and S.T.-G.; investigation and data curation, B.A.B.; writing—original draft preparation, B.A.B.; writing—review and editing, S.T.-G.; supervision and project administration, S.A., S.T.-G., and J.M.L.

Funding: This study was supported by the Turkish Scientific and Technological Research Council (TUBITAK) 2214-A International Research Fellowship Programme for PhD Students and by the Spanish Ministry of Science, Innovation, and Universities (MICIU) project numbers AGL2015-63855-C2-1-R.

Acknowledgments: S.T.-G. is a recipient of a Juan de la Cierva—Incorporación contract (IJCI-2016-29675) from MICIU.

Conflicts of Interest: The authors have declared no conflict of interest.

References

1. Kertesz, Z.I. *The Pectic Substances*, 1st ed.; Interscience Publishers, Inc.: New York, NY, USA, 1951.
2. Thakur, B.R.; Singh, R.K.; Handa, A.K.; Rao, M.A. Chemistry and uses of pectin-A review. *Crit. Rev. Food Sci. Nutr.* **1997**, *37*, 47–73. [CrossRef] [PubMed]
3. Sriamornsak, P. Chemistry of pectin and its pharmaceutical uses: A review. *Silpakorn Univ. Int. J.* **2003**, *3*, 206–228.
4. Ridley, B.L.; O'Neill, M.A.; Mohnen, D. Pectins: Structure, biosynthesis, and oligogalacturonide-related signaling. *Phytochemistry* **2001**, *57*, 929–967. [CrossRef]
5. Mukhiddinov, Z.K.; Khalikov, D.K.; Abdusamiev, F.T.; Avloev, C.C. Isolation and structural characterization of a pectin homo and ramnogalacturonan. *Talanta* **2000**, *53*, 171–176. [CrossRef]
6. Pérez, S.; Rodríguez-Carvajal, M.A.; Doco, T. A complex plant cell wall polysaccharide: Rhamnogalacturonan II. A structure in quest of a function. *Biochimie* **2003**, *85*, 109–121. [CrossRef]
7. Mohnen, D. Pectin structure and biosynthesis. *Curr. Opin. Plant Biol.* **2008**, *11*, 266–277. [CrossRef]
8. Naqash, F.; Masoodi, F.A.; Rather, S.A.; Wani, S.M.; Gani, A. Emerging concepts in the nutraceutical and functional properties of pectin-A Review. *Carbohydr. Polym.* **2017**, *168*, 227–239. [CrossRef]
9. Noreen, A.; Akram, J.; Rasul, I.; Mansha, A.; Yaqoob, N.; Iqbal, R.; Tabasum, S.; Zuber, M.; Zia, K.M. Pectins functionalized biomaterials; a new viable approach for biomedical applications: A review. *Int. J. Biol. Macromol.* **2017**, *101*, 254–272. [CrossRef]
10. Bernhardt, D.C.; Pérez, C.D.; Fissore, E.N.; De'Nobili, M.D.; Rojas, A.M. Pectin-based composite film: Effect of corn husk fiber concentration on their properties.: Effect of corn husk fiber concentration on their properties. *Carbohydr. Polym.* **2017**, *164*, 13–22. [CrossRef]
11. Yu, W.X.; Wang, Z.W.; Hu, C.Y.; Wang, L. Properties of low methoxyl pectin-carboxymethyl cellulose based on montmorillonite nanocomposite films. *Int. J. Food Sci. Technol.* **2014**, *49*, 2592–2601. [CrossRef]
12. Da Silva, M.A.; Bierhalz, A.C.K.; Kieckbusch, T.G. Alginate and pectin composite films crosslinked with Ca^{2+} ions: Effect of the plasticizer concentration. *Carbohydr. Polym.* **2009**, *77*, 736–742. [CrossRef]
13. Jantrawut, P.; Chaiwarit, T.; Jantanasakulwong, K.; Brachais, C.H.; Chambin, O. Effect of plasticizer type on tensile property and in vitro indomethacin release of thin films based on low-methoxyl pectin. *Polymers* **2017**, *9*, 289. [CrossRef] [PubMed]

14. Chaiwarit, T.; Ruksiriwanich, W.; Jantanasakulwong, K.; Jantrawut, P. Use of orange oil loaded pectin films as antibacterial material for food packaging. *Polymers* **2018**, *10*, 1144. [CrossRef] [PubMed]
15. Ma, X.; Chang, P.R.; Yu, J.; Stumborg, M. Properties of biodegradable citric acid-modified granular starch/thermoplastic pea starch composites. *Carbohydr. Polym.* **2009**, *75*, 1–8. [CrossRef]
16. Argin, S.; Gulerim, M.; Sahin, F. Development of antimicrobial gelatin films with boron derivatives. *Turk. J. Biol.* **2019**, *43*, 47–57. [CrossRef] [PubMed]
17. Kowalczyk, D.; Baraniak, B. Effects of plasticizers, pH and heating of film-forming solution on the properties of pea protein isolate films. *J. Food Eng.* **2011**, *105*, 295–305. [CrossRef]
18. Kowalczyk, D.; Gustaw, W.; Świeca, M.; Baraniak, B. A study on the mechanical properties of pea protein isolate films. *J. Food Process. Preserv.* **2014**, *38*, 1726–1736. [CrossRef]
19. Suyatma, N.E.; Tighzert, L.; Copinet, A.; Coma, V. Effects of hydrophilic plasticizers on mechanical, thermal, and surface properties of chitosan films. *J. Agric. Food Chem.* **2005**, *53*, 3950–3957. [CrossRef]
20. Talja, R.A.; Helén, H.; Roos, Y.H.; Jouppila, K. Effect of various polyols and polyol contents on physical and mechanical properties of potato starch-based films. *Carbohydr. Polym.* **2007**, *67*, 288–295. [CrossRef]
21. Vieira, M.G.A.; Da Silva, M.A.; Dos Santos, L.O.; Beppu, M.M. Natural-based plasticizers and biopolymer films: A review. *Eur. Polym. J.* **2011**, *47*, 254–263. [CrossRef]
22. Sanyang, M.L.; Sapuan, S.M.; Jawaid, M.; Ishak, M.R.; Sahari, J. Effect of plasticizer type and concentration on physical properties of biodegradable films based on sugar palm (*Arenga pinnata*) starch for food packaging. *J. Food Sci. Technol.* **2016**, *53*, 326–336. [CrossRef] [PubMed]
23. Pavlath, A.E.; Voisin, A.; Robertson, G.H. Pectin-based biodegradable water insoluble films. *Am. Chem. Soc. Polym. Prepr. Div. Polym. Chem.* **1998**, *39*, 692. [CrossRef]
24. Penhasi, A.; Meidan, V.M. Preparation and characterization of in situ ionic cross-linked pectin films: Unique biodegradable polymers. *Carbohydr. Polym.* **2014**, *102*, 254–260. [CrossRef] [PubMed]
25. Lorevice, M.V.; Otoni, C.G.; de Moura, M.R.; Mattoso, L.H.C. Chitosan nanoparticles on the improvement of thermal, barrier, and mechanical properties of high-and low-methyl pectin films. *Food Hydrocoll.* **2016**, *52*, 732–740. [CrossRef]
26. Martelli, M.R.; Barros, T.T.; de Moura, M.R.; Mattoso, L.H.C.; Assis, O.B.G. Effect of chitosan nanoparticles and pectin content on mechanical properties and water vapor permeability of banana puree films. *J. Food Sci.* **2013**, *78*, 98–104. [CrossRef]
27. Chaichi, M.; Hashemi, M.; Badii, F.; Mohammadi, A. Preparation and characterization of a novel bionanocomposite edible film based on pectin and crystalline nanocellulose. *Carbohydr. Polym.* **2017**, *157*, 167–175. [CrossRef]
28. Šešlija, S.; Nešić, A.; Škorić, M.L.; Krušić, M.K.; Santagata, G.; Malinconico, M. Pectin/carboxymethylcellulose films as a potential food packaging material. *Macromol. Symp.* **2018**, *378*, 160–163. [CrossRef]
29. Pasini Cabello, S.D.; Takara, E.A.; Marchese, J.; Ochoa, N.A. Influence of plasticizers in pectin films: Microstructural changes. *Mater. Chem. Phys.* **2015**, *162*, 491–497. [CrossRef]
30. Espitia, P.J.P.; Du, W.X.; de Jesus Avena-Bustillos, R.; de Fatima Ferrira Soares, N.; McHugh, T.H. Edible films from pectin: Physical-mechanical and antimicrobial properties-A review. *Food Hydrocoll.* **2014**, *35*, 287–296. [CrossRef]
31. Bhardwaj, N.; Kundu, S.C. Electrospinning: A fascinating fiber fabrication technique. *Biotechnol. Adv.* **2010**, *28*, 325–347. [CrossRef]
32. Torres-Giner, S.; Wilkanowicz, S.; Melendez-Rodriguez, B.; Lagaron, J.M. Nanoencapsulation of *Aloe vera* in synthetic and naturally occurring polymers by electrohydrodynamic processing of interest in food technology and bioactive packaging. *J. Agric. Food Chem.* **2017**, *65*, 4439–4448. [CrossRef] [PubMed]
33. Torres-Giner, S.; Pérez-Masiá, R.; Lagaron, J.M. A review on electrospun polymer nanostructures as advanced bioactive platforms. *Polym. Eng. Sci.* **2016**, *56*, 500–527. [CrossRef]
34. Frenot, A.; Chronakis, I.S. Polymer nanofibers assembled by electrospinning. *Curr. Opin. Colloid Interface Sci.* **2003**, *8*, 64–75. [CrossRef]
35. Kim, J.S.; Reneker, D.H. Mechanical properties of composites using ultrafine electrospun fibers. *Polym. Compos.* **1999**, *20*, 124–131. [CrossRef]
36. Torres-Giner, S.; Lagaron, J.M. Zein-based ultrathin fibers containing ceramic nanofillers obtained by electrospinning. I. Morphology and thermal properties. *J. Appl. Polym. Sci.* **2010**, *118*, 778–789.

37. Furlan, R.; Rosado, J.A.M.; Rodriguez, G.G.; Fachini, E.R.; da Silva, A.N.R.; da Silva, M.L.P. Formation and characterization of oriented micro- and nanofibers containing poly(ethylene oxide) and pectin. *J. Electrochem. Soc.* **2012**, *159*, 66–71. [CrossRef]
38. Liu, S.C.; Li, R.; Tomasula, P.M.; Sousa, A.M.M.; Liu, L. Electrospun food-grade ultrafine fibers from pectin and pullulan blends. *Food Nutr. Sci.* **2016**, *7*, 636–646. [CrossRef]
39. Cui, S.; Yao, B.; Sun, X.; Hu, J.; Zhou, Y.; Liu, Y. Reducing the content of carrier polymer in pectin nanofibers by electrospinning at low loading followed with selective washing. *Mater. Sci. Eng. C* **2016**, *59*, 885–893. [CrossRef]
40. Cui, S.; Sun, X.; Yao, B.; Peng, X.X.; Zhang, X.T.; Zhou, Y.F.; Hu, J.-L.; Liu, Y.C. Size-tunable low molecular weight pectin-based electrospun nanofibers blended with low content of poly(ethylene oxide). *J. Nanosci. Nanotechnol.* **2017**, *17*, 681–689. [CrossRef]
41. McCune, D.; Guo, X.; Shi, T.; Stealey, S.; Antrobus, R.; Kaltchev, M.; Chen, J.; Kumpaty, S.; Hua, X.; Ren, W.; et al. Electrospinning pectin-based nanofibers: A parametric and cross-linker study. *Appl. Nanosci.* **2018**, *8*, 33–40. [CrossRef]
42. Chen, S.; Cui, S.; Zhang, H.; Pei, X.; Hu, J.; Zhou, Y.; Liu, Y. Cross-linked pectin nanofibers with enhanced cell adhesion. *Biomacromolecules* **2018**, *19*, 490–498. [CrossRef] [PubMed]
43. Li, K.; Cui, S.; Hu, J.; Zhou, Y.; Liu, Y. Crosslinked pectin nanofibers with well-dispersed Ag nanoparticles: Preparation and characterization. *Carbohydr. Polym.* **2018**, *199*, 68–74. [CrossRef] [PubMed]
44. Rockwell, P.L.; Kiechel, M.A.; Atchison, J.S.; Toth, L.J.; Schauer, C.L. Various-sourced pectin and polyethylene oxide electrospun fibers. *Carbohydr. Polym.* **2014**, *107*, 110–118. [CrossRef] [PubMed]
45. Akinalan Balik, B.; Argin, S. Role of rheology on the formation of Nanofibers from pectin and polyethylene oxide blends. *J. Appl. Polym. Sci.* **2020**, *137*. [CrossRef]
46. Patra, M.; Salerno, N.; Cernik, M. Electrospun polyvinyl alcohol/pectin composite nanofibers. In *Electrospun Nanofibers*, 1st ed.; Afshar, M., Ed.; Woodhead Publishing: Cambridge, UK, 2017; pp. 599–608.
47. Roy, P.; Dutta, R.K. Electrospun PVA-Pectin-Magnetite nanofiber as a novel drug carrier matrix. *Int. J. Appl. Eng. Res.* **2014**, *9*, 629–635.
48. Ye, X.; Zhan, Y.; Li, T.; Shi, X.; Deng, H.; Du, Y. Pectin based composite nanofabrics incorporated with layered silicate and their cytotoxicity. *Int. J. Biol. Macromol.* **2016**, *93*, 123–130. [CrossRef]
49. Alborzi, S.; Lim, L.T.; Kakuda, Y. Electrospinning of sodium alginate-pectin ultrafine fibers. *J. Food Sci.* **2010**, *75*, 100–107. [CrossRef]
50. Lin, H.Y.; Chen, H.H.; Chang, S.H.; Ni, T.S. Pectin-chitosan-PVA nanofibrous scaffold made by electrospinning and its potential use as a skin tissue scaffold. *J. Biomater. Sci. Polym.* **2013**, *24*, 470–484. [CrossRef]
51. Cherpinski, A.; Torres-Giner, S.; Cabedo, L.; Lagaron, J.M. Post-processing optimization of electrospun submicron poly(3-hydroxybutyrate) fibers to obtain continuous films of interest in food packaging applications. *Food Addit. Contam.-Part A Chem. Anal. Control. Expo. Risk Assess.* **2017**, *34*, 1817–1830. [CrossRef]
52. Cherpinski, A.; Torres-Giner, S.; Cabedo, L.; Méndez, J.A.; Lagaron, J.M. Multilayer structures based on annealed electrospun biopolymer coatings of interest in water and aroma barrier fiber-based food packaging applications. *J. Appl. Polym. Sci.* **2018**, *135*, 1–11. [CrossRef]
53. Cherpinski, A.; Torres-Giner, S.; Vartiainen, J.; Peresin, M.S.; Lahtinen, P.; Lagaron, J.M. Improving the water resistance of nanocellulose-based films with polyhydroxyalkanoates processed by the electrospinning coating technique. *Cellulose* **2018**, *25*, 1291–1307. [CrossRef]
54. Melendez-Rodriguez, B.; Castro-Mayorga, J.L.; Reis, M.A.; Sammon, C.; Cabedo, L.; Torres-Giner, S.; Lagaron, J.M. Preparation and characterization of electrospun food biopackaging films of poly (3-hydroxybutyrate-*co*-3-hydroxyvalerate) derived from fruit pulp biowaste. *Front. Sustain. Food Syst.* **2018**, *2*, 1–16. [CrossRef]
55. Melendez-Rodriguez, B.; Figueroa-Lopez, K.J.; Bernardos, A.; Martínez-Máñez, R.; Cabedo, L.; Torres-Giner, S.; Lagaron, J.M. Electrospun antimicrobial films of poly (3-hydroxybutyrate-*co*-3-hydroxyvalerate) containing eugenol essential oil encapsulated in mesoporous silica nanoparticles. *Nanomaterials* **2019**, *9*, 227. [CrossRef] [PubMed]

56. Figueroa-Lopez, K.J.; Vicente, A.A.; Reis, M.A.; Torres-Giner, S.; Lagaron, J.M. Antimicrobial and antioxidant performance of various essential oils and natural extracts and their incorporation into biowaste derived poly (3-hydroxybutyrate-*co*-3-hydroxyvalerate) layers made from electrospun ultrathin fibers. *Nanomaterials* **2019**, *9*, 144. [CrossRef] [PubMed]
57. Quiles-Carrillo, L.; Montanes, N.; Lagaron, J.M.; Balart, R.; Torres-Giner, S. Bioactive multilayer polylactide films with controlled release capacity of gallic acid accomplished by incorporating electrospun nanostructured coatings and interlayers. *Appl. Sci.* **2019**, *9*, 533. [CrossRef]
58. Radusin, T.; Torres-Giner, S.; Stupar, A.; Ristic, I.; Miletic, A.; Novakovic, A.; Lagaron, J.M. Preparation,characterization and antimicrobial properties of electrospun polylactide films containing *Allium ursinum L.* extract. *Food Packag. Shelf Life* **2019**, *21*. [CrossRef]
59. Agüero, A.; Morcillo, M.C.; Quiles-Carrillo, L.; Balart, R.; Boronat, T.; Lascano, D.; Torres-Giner, S.; Fenollar, O. Study of the Influence of the Reprocessing Cycles on the Final Properties of Polylactide Pieces Obtained by Injection Molding. *Polymers* **2019**, *11*, 1908. [CrossRef]
60. Ralet, M.C.; Crépeau, M.J.; Buchholt, H.C.; Thibault, J.F. Polyelectrolyte behaviour and calcium binding properties of sugar beet pectins differing in their degrees of methylation and acetylation. *Biochem. Eng. J.* **2003**, *16*, 191–201. [CrossRef]
61. Bizarria, M.T.M.; D'Ávila, M.A.; Mei, L.H.I. Non-woven nanofiber chitosan/PEO membranes obtained by electrospinning. *Braz. J. Chem. Eng.* **2014**, *31*, 57–68. [CrossRef]
62. Gerasimov, A.V.; Ziganshin, M.A.; Gorbatchuk, L.S.; Usmanova, V.V. Low molecular weight polyethylene glycols as matrix to obtain solid dispersions of sulfanilamide. *Int. J. Pharm. Pharm. Sci.* **2014**, *6*, 372–377.
63. Einhorn-Stoll, U.; Kunzek, H.; Dongowski, G. Thermal analysis of chemically and mechanically modified pectins. *Food Hydrocoll.* **2007**, *21*, 1101–1112. [CrossRef]
64. Godeck, R.; Kunzek, H.; Kabbert, R. Thermal analysis of plant cell wall materials depending on the chemical structure and pre-treatment prior to drying. *Eur. Food Res. Technol.* **2001**, *213*, 395–404. [CrossRef]
65. Gloyna, D.; Kunzek, H. Thermal solid phase degradation of plant cell wall polysaccharides. *Pol. J. Food Nutr. Sci.* **1998**, *7*, 55–60.
66. Kastner, H.; Einhorn-Stoll, U.; Senge, B. Structure formation in sugar containing pectin gels-Influence of Ca^{2+} on the gelation of low-methoxylated pectin at acidic pH. *Food Hydrocoll.* **2012**, *27*, 42–49. [CrossRef]
67. Nisar, T.; Wang, Z.C.; Yang, X.; Tian, Y.; Iqbal, M.; Guo, Y. Characterization of citrus pectin films integrated with clove bud essential oil: Physical, thermal, barrier, antioxidant and antibacterial properties. *Int. J. Biol. Macromol.* **2018**, *106*, 670–680. [CrossRef]
68. Van Buren, J.P. The chemistry of texture in fruits and vegetables. *J. Texture Stud.* **1979**, *10*, 1–23. [CrossRef]
69. Sila, D.N.; Smout, C.; Elliot, F.; Loey, A.V.; Hendrickx, M. Non-enzymatic depolymerization of carrot pectin: Toward a better understanding of carrot texture during thermal processing. *J. Food Sci.* **2006**, *71*, 1–9. [CrossRef]
70. Diaz, J.V.; Anthon, G.E.; Barrett, D.M. Nonenzymatic degradation of citrus pectin and pectate during prolonged heating: Effects of pH, temperature, and degree of methyl esterification. *J. Agric. Food Chem.* **2007**, *55*, 5131–5136. [CrossRef]
71. Krall, S.M.; McFeeters, R.F. Pectin hydrolysis: Effect of temperature, degree of methylation, pH, and calcium on hydrolysis rates. *J. Agric. Food Chem.* **1998**, *46*, 1311–1315. [CrossRef]
72. Rodrigo, D.; Cortés, C.; Clynen, E.; Schoofs, L.; Van Loey, A.; Hendrickx, M. Thermal and high-pressure stability of purified polygalacturonase and pectinmethylesterase from four different tomato processing varieties. *Food Res. Int.* **2006**, *39*, 440–448. [CrossRef]
73. Aburto, J.; Moran, M.; Galano, A.; Torres-García, E. Non-isothermal pyrolysis of pectin: A thermochemical and kinetic approach. *J. Anal. Appl. Pyrolysis* **2015**, *112*, 94–104. [CrossRef]
74. Dick, M.; Costa, T.M.H.; Gomaa, A.; Subirade, M.; de Oliveira Rios, A.; Flôres, S.H. Edible film production from chia seed mucilage: Effect of glycerol concentration on its physicochemical and mechanical properties. *Carbohydr. Polym.* **2015**, *130*, 198–205. [CrossRef]
75. Mishra, R.K.; Anis, A.; Mondal, S.; Dutt, M.; Banthia, A.K. Reparation and Characterization of Amidated Pectin Based Polymer Electrolyte Membranes. *Chin. J. Polym. Sci.* **2009**, *27*, 639–646. [CrossRef]
76. Villanova, J.C.O.; Ayres, E.; Oréfice, R.L. Design, characterization and preliminary *in vitro* evaluation of a mucoadhesive polymer based on modified pectin and acrylic monomers with potential use as a pharmaceutical excipient. *Carbohydr. Polym.* **2015**, *121*, 372–381. [CrossRef]

77. Basiak, E.; Lenart, A.; Debeaufort, F. How glycerol and water contents affect the structural and functional properties of starch-based edible films. *Polymers* **2018**, *10*, 412. [CrossRef]
78. Gondaliya, N.; Kanchan, D.K.; Sharma, P.; Joge, P. Structural and Conductivity Studies of Poly (Ethylene Oxide)-Silver Triflate Polymer Electrolyte System. *Mater. Sci. Appl.* **2011**, *2*, 1639–1643. [CrossRef]
79. Sanchez-Garcia, M.D.; Gimenez, E.; Lagaron, J.M. Morphology and barrier properties of solvent cast composites of thermoplastic biopolymers and purified cellulose fibers. *Carbohydr. Polym.* **2008**, *71*, 235–244. [CrossRef]
80. Fabra, M.J.; Lopez-Rubio, A.; Lagaron, J.M. Nanostructured interlayers of zein to improve the barrier properties of high barrier polyhydroxyalkanoates and other polyesters. *J. Food Eng.* **2014**, *127*, 1–9. [CrossRef]

applied sciences

MDPI

Article

Pine Resin Derivatives as Sustainable Additives to Improve the Mechanical and Thermal Properties of Injected Moulded Thermoplastic Starch

Miguel Aldas [1,2,*], Cristina Pavon [2], Juan López-Martínez [2] and Marina Patricia Arrieta [3,*]

[1] Departamento de Ciencia de Alimentos y Biotecnología, Facultad de Ingeniería Química y Agroindustria, Escuela Politécnica Nacional (EPN), Quito 170517, Ecuador

[2] Instituto de Tecnología de Materiales, Universitat Politècnica de València (UPV), 03801 Alcoy-Alicante, Spain; crisppavonv@gmail.com (C.P.); jlopezm@mcm.upv.es (J.L.-M.)

[3] Facultad de Óptica y Optometría, Universidad Complutense de Madrid (UCM), Arcos de Jalón 118, 28037 Madrid, Spain

* Correspondence: miguel.aldas@epn.edu.ec (M.A.); marrie06@ucm.es (M.P.A.); Tel.: +593-999-736-444 (M.A.); +34-913-946-885 (M.P.A.)

Received: 24 March 2020; Accepted: 3 April 2020; Published: 8 April 2020

Abstract: Fully bio-based materials based on thermoplastic starch (TPS) were developed starting from corn starch plasticized with glycerol. The obtained TPS was further blended with five pine resin derivatives: gum rosin (GR), disproportionated gum rosin (dehydroabietic acid, RD), maleic anhydride modified gum rosin (CM), pentaerythritol ester of gum rosin (LF), and glycerol ester of gum rosin (UG). The TPS–resin blend formulations were processed by melt extrusion and further by injection moulding to simulate the industrial conditions. The obtained materials were characterized in terms of mechanical, thermal and structural properties. The results showed that all gum rosin-based additives were able to improve the thermal stability of TPS, increasing the degradation onset temperature. The carbonyl groups of gum rosin derivatives were able to interact with the hydroxyl groups of starch and glycerol by means of hydrogen bond interactions producing a significant increase of the glass transition temperature with a consequent stiffening effect, which in turn improve the overall mechanical performance of the TPS-resin injected moulded blends. The developed TPS–resin blends are of interest for rigid packaging applications.

Keywords: bioplastic; corn starch; glycerol; thermoplastic starch; gum rosin; injection-moulding

1. Introduction

Polymers have become essential materials in our lives mainly due to their unique properties featuring lightness and durability and, as a consequence, their consumption has increased during the last decades [1]. However, the mass production of plastics and the limitation of non-renewable sources have led to problems with their final disposal and in end-of-life options [1,2]. This has promoted the search for alternatives to the use of fossil-based polymers with materials that present better environmental performance, particularly for short term applications (i.e., food packaging, disposable cutlery, agricultural applications) [2,3]. In recent years, biopolymers have risen as an alternative to traditional plastics, mainly for short term applications. Many research and industrial efforts have been focused on the development of sustainable polymers, mostly for single use disposal applications such as bottles, cold drink cups, thermoformed trays and container lids, blister packages, overwrap as well as flexible films, which are currently commercialized [2–4]. However, many commercially available biopolymers possess a low glass transition temperature, above which the polymeric matrix loses its rigidity leading to plastic deformation, which make them unsuitable for hot beverages and/or hot food applications.

Biopolymers are materials whose synthesis involves micro-organisms and that have a reduced lifespan after final disposition [5,6]. Biopolymers can be produced naturally or obtained from biomass. They can be produced from: (i) wood derivatives such as cellulose [7,8], lignin [9] or terpenes [10], (ii) lipids as fatty acids [11] and (iii) polysaccharides as sugar [12], chitin, chitosan [12] or starch [3,13]. Polysaccharides account for 75% of the world's annual biomass production, with 170 trillion metric tons. Starch is the second most abundant and available polysaccharide in nature [14]. It is composed of two D-glucose polysaccharides: amylose, a linear polymer and amylopectin, a branched polymer [3]. Starch can be obtained from various sources including cassava, wheat, rice, potato, pea, and corn. Corn is the largest source of starch in the world [3,14]. Starch is widely used in the production of biopolymers given its abundance and low cost [3,15]. Starch is used as a main polymeric matrix as well as minor component or filler in polymer blends [16]. Moreover, some bioplastics in the market are produced from starch, such as poly(lactic acid) (PLA) [2,4]. Starch in its native state does not exhibit a thermoplastic behaviour. Hence, to use starch in the plastic industry it is necessary to break down the structure of granular starch [17]. The molecular order of the starch granules is destroyed by converting its molecular structure to thermoplastic through the gelatinization process, changing its crystalline structure to an amorphous structure [14]. To produce plasticization in the starch structure these features are required: (1) temperatures greater than 70 °C, (2) presence of plasticizers (water or polyols) and, (3) shear stress [17,18]. These conditions can be achieved in the extrusion process, which will allow an amorphous paste, known as thermoplastic starch (TPS) to be obtained, where most of the inter-macromolecular hydrogen bonds of starch are destroyed [3]. TPS can melt and flow allowing it to be processed by conventional moulding and extrusion processing techniques used for most thermoplastic polymers [18]. Glycerol is the most widely used plasticizer for the production of TPS [3]. From an environmental point of view, glycerol is a by-product of biodiesel production and the revalorization of glycerol increases its added value from a low-grade by-product to a useful biopolymer plasticizer [19]. TPS has gained industrial attention, compared with other biopolymers, due to its high availability and low cost, being one of the cheapest biopolymers in the market [5,15,16]. In fact, the next generation of packaging materials should fulfil the requirements to support the transition to a circular economy, which comprise a ban on single-use traditional plastics by the end of 2020 [20,21]. Moreover, from an environmental point of view, TPS is not only bio-based but it is also biodegradable, allowing the loop of circular economy to be closed. However, TPS industrial application is limited by its fragility, low water resistance, and the re-crystallization and retrogradation that its structure is subject to, which leads to undesirable changes in the thermomechanical properties of the material [5,17]. In this context, several strategies have been focused on improving biopolymer performance for extending its industrial applications in the packaging field, such as blending, copolymerization and/or the addition of either micro- or nano-fillers for the development of composites and nanocomposites [2,22]. Among all strategies, physical blending seems to be one of the most convenient routes to create new bio-based polymeric formulations, with the required combination of properties, by simply mixing two or more bio-based polymeric matrices in the melt state with no chemical reactions taking place [8,23]. Thus, melt-blending strategies have many advantages since they offer the opportunity to improve the final performance of biopolymers through relatively simple and readily available processing methods that are cost-effective for the industrial sector with respect to the development of new starting monomers and/or polymerization techniques [8,23]. Therefore, to improve TPS performance and extend its industrial applications, the blending approach has been widely studied. In fact, TPS have been blended with either fossil-based polymers (i.e., polypropylene (PP) [24], polystyrene (PS) [25], polyethylene (PE) [26] and ethylene-vinyl acetate (EVA) [27]) as well as with other biopolymers (i.e., PLA [28], and polyhydroxybutyrate (PHB) [29]) to improve its performance and increase its range of applications.

Gum rosin (GR) is a natural alternative to fossil-based polymers obtained from heating and evaporation of pine resin [30–32]. It possesses many advantages for the plastic industry since it is abundant in the nature, has a relatively low cost, and is easily converted into high performance macromolecular materials [30,33]. Untreated rosin is rigid and brittle and has a thermoplastic behaviour. It is insoluble in water and soluble in acids and most organic solvents (i.e., glacial acetic acid, chloroform, ether and alcohol) [30]. GR is composed of 10–20% of neutral compounds and 80–90% of resin acids [32]. Resin acids structure is composed by a tricyclic skeleton with two double bonds and a carboxylic functional group. This structure provides GR the possibility to be chemically modified in order to obtain a wide range of derivatives, including polymerizable structures for both linear and crosslinked materials [10].

Chemical derivatives of GR can be obtained from oxidation, hydrogenation, dehydrogenation, isomerization, Diels-Alder couplings, esterification, saponification reactions with formaldehyde and phenol [30,32,34–36]. These GR derivatives can be used for the synthesis of homopolymers and copolymers with traditional monomers in order to modulate the properties of the resulting material [35]. In our previous studies, GR and two GR esters derivatives have been used as additives in a Mater-Bi® type biopolymer, a commercial blend based on TPS developed by Novamont. It has been determined that neat GR provided a plasticising, compatibilising and solubilising effect in the polymeric matrix which facilitates the processability of the material by decreasing the processing temperature. Meanwhile, the GR esters used in previous works improve the miscibility of the TPS-based matrix components and improve its mechanical performance [5,37]. The present work aims to study the effect of GR and some GR derivatives on the TPS matrix obtained from native corn starch. In a first step, native corn starch was mixed with glycerol and water and further melt-extruded at a maximum temperature of 130 °C at high pressure and high shear to obtain TPS. The obtained TPS was then blended either with GR or GR derivatives by melt-extrusion and further injection moulding process. The obtained formulations were mechanically, thermally and structurally characterised to get information of the possible application of these sustainable materials at an industrial level.

2. Materials and Methods

2.1. Materials

Food grade corn starch, containing 27% amylose was supplied by Cargill (Barcelona, Spain). Distilled water and glycerol (Panreac 99% of purity—Barcelona, Spain) were used as plasticisers. As additives, five pine resins derivatives were used with the following molecular structure (Figure 1):

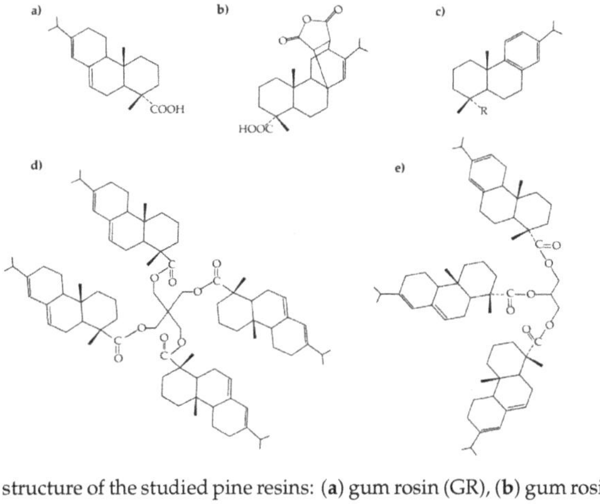

Figure 1. Molecular structure of the studied pine resins: (**a**) gum rosin (GR), (**b**) gum rosin modified with maleic anhydride (CM), (**c**) disproportionated gum rosin (dehydroabietic acid) (RD), (**d**) pentaerythritol ester of gum rosin (LF), and (**e**) glycerol ester of gum rosin (UG).

Gum rosin (GR, softening point of 76 °C and acid number 167) was kindly supplied by Sigma-Aldrich (Mostoles, Spain). Gum rosin modified with maleic anhydride under the trade name Colmodif R-330 (CM, softening point of 123 °C and acid number 252), a disproportionated gum rosin (dehydroabietic acid) under the trade name of Residis 455 (RD, softening point of 74.6 °C and acid number 157), a pentaerythritol ester of gum rosin under the trade name of Lurefor 125 (LF, softening point of 125 °C and acid number 11.9) were kindly supplied by LureSA (Segovia, Spain). Glycerol ester of gum rosin under the trade name of Unik Gum G88 (UG, food grade, softening point of 87 °C and acid number 7) was kindly supplied by United Resins (Figueira da Foz, Portugal).

2.2. Methods

2.2.1. TPS-Resin Blends Preparation

Figure 2 shows a schematic representation of the preparation and processing of the developed materials and involves: a manual mixing, a first extrusion process to obtain TPS, a second extrusion process to obtain TPS-resin blends and a final injection moulding step to obtain the injected moulded parts.

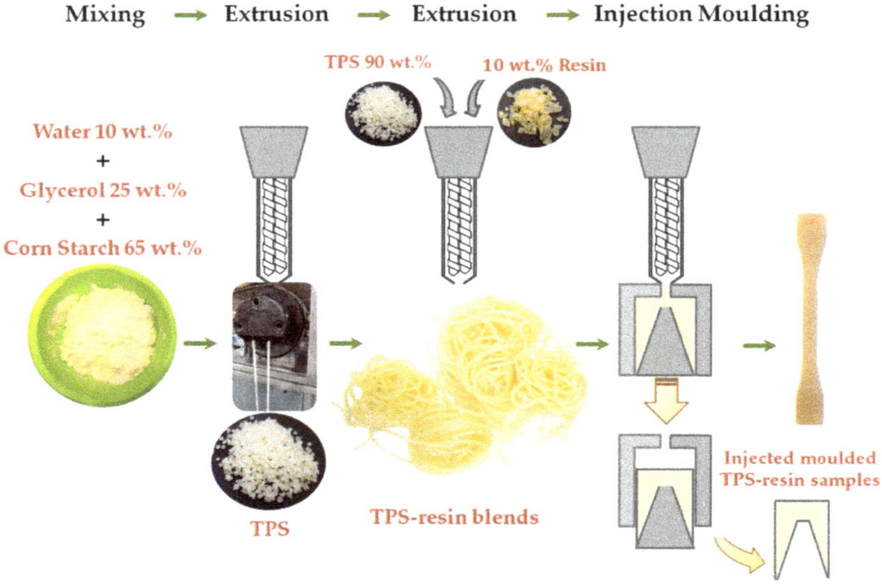

Figure 2. Schematic representation of the thermoplastic starch (TPS) preparation and the processing of TPS–resin blend formulations.

In brief, TPS was prepared manually premixing in a hermetically sealed polyethylene bag, 65% of native corn starch with 25 wt. % of glycerol and 10 wt. % of distilled water, 24 h before the extrusion process to ensure homogeneity of the material and to allow the correct diffusion of the plasticisers in the starch matrix [38]. Thus, the corn starch, glycerol and water mixture was processed in a co-rotating twin-screw extruder, L/D ratio of 25 from Dupra S.L (Castalla, Spain) at 20 rpm with a temperature profile of 130, 110, 100, 90, 80 °C (from die to hopper) to obtain TPS. A second extrusion step was performed to prepare the TPS-resin based formulations, adding each pine resin derivatives in 10 wt. % to the previously prepared TPS. Neat TPS was also melt extrusion processed a second time to have the same processing conditions. Thus, six formulations were prepared and named as summarised in Table 1. After the melt extrusion process, the TPS and TPS–resin materials were pelletised and conditioned at 25 °C and 50 ± 5% of relative humidity (RH) [5]. Then, the sample test specimens were obtained by injection moulding process in an injection moulding machine Sprinter-11, Erinca S.L. (Barcelona, Spain), with a temperature profile of 130, 110, 100 °C, from die to hopper. Prior to characterization, injected moulded samples were conditioned at 25 °C and 50 ± 5% RH [5].

Table 1. TPS–resin blend studied formulations and their composition.

Formulations	Type of Resin in the Blend	Resin Commercial Name	TPS (wt. %)	Resin (wt. %)
TPS	-	-	100	0
TPS-GR	Gum rosin or unmodified colophony	Gum rosin	90	10
TPS-CM	Maleic anhydride modified gum rosin	Colmodif R-330	90	10
TPS-LF	Pentaerythritol ester of gum rosin	Lurefor 125	90	10
TPS-RD	Disproportionated gum rosin	Residis 455	90	10
TPS-UG	Glycerol ester of gum rosin	Unik Gum G88	90	10

2.2.2. Mechanical Characterization

Tensile test was performed according to standard tests methods ISO 527 [39]. It was carried out in a universal test machine Ibertest ELIB-50-W (Madrid, Spain) with a load cell of 5 kN and a crosshead speed of 100 mm/min. The samples were injected in dog-bone-shaped specimens "1BA" type (80 mm × 10 mm × 4 mm) according to ISO 527 [39]. At least five specimens for each formulation were characterized and the mean and standard deviation of the values are reported. The Young's modulus, the tensile strength and the elongation at break were determined. In addition, to calculate the toughness of the materials, the area under the typical stress-strain curve was calculated. The area was calculated using the OriginPro2015 software from OriginLab (Northampton, MA, USA).

Shore D hardness of all studied formulations was measured on a durometer Model 673-D from Instrument J.Bot, S.A. (Barcelona, Spain) using samples of 4 mm thickness, according to standard tests methods ISO 868 [40]. The mean of 20 measurements at random positions of the samples was reported as the hardness values.

The significant differences in the mechanical parameters were statistically assessed at 95% confidence level according to Tukey's test using a one-way analysis of variance (ANOVA) by means of OriginPro2015 software.

2.2.3. Thermal Characterization

Differential scanning calorimetry (DSC) was carried out in a Mettler DSC821e (Toledo, Spain) with a thermal cycles program that consisted in a first heating scan from −50 to 160 °C, followed by a cooling cycle from 160 to −50 °C and a second heating scan from −50 to 200 °C. The heating rate for all cycles was 10 °C/min and the test were performed under a nitrogen atmosphere with a flow of 30 mL/min. The glass transition temperature (T_g) was calculated on the second heating scan of the DSC curve.

Thermogravimetric analysis (TGA) was conducted in a TGA PT1000 from Linseis (Selb, Germany). The heating was carried out from 30 to 700 °C at a heating rate of 10 °C/min in a nitrogen atmosphere with a flow rate of 30 mL/min. The degradation onset temperature ($T_{5\%}$) was determined at 5% of mass loss. As well, the degradation endset temperature ($T_{95\%}$) was assessed at 95% of mass loss. The maximum degradation temperature (T_{max}) was determined at the peak of the first derivative of the TGA curve (DTG).

2.2.4. Structural Characterization

Attenuated total reflectance–Fourier transform infrared spectroscopy (FTIR-ATR) was applied to study the chemical interactions between TPS, GR and gum rosin derivatives (CM, LF, RD and UG). The analysis was performed in a Perkin Elmer Spectrum BX (FTIR system) coupled with an ATR Pike MIRacle ™ (Beaconsfield, United Kingdom). All the formulations as well as neat TPS and the raw resins were assessed within the range of 4000–700 cm^{-1} with a resolution of 16 cm^{-1} and 20 scans.

Scanning electron microscopy (SEM) of the cryo-fractured surface of each formulation was carried out on a Phenon SEM equipment of FEI (Eindhoven, The Netherlands), with a voltage of 5kV. Prior to SEM observation, the samples were coated with a gold-palladium alloy layer to make the sample surface conductive in a Sputter Mod Coater Emitech SC7620, Quorum Technologies (East Sussex, UK).

3. Results

TPS was successfully produced from corn starch plasticised with glycerol and water and then it was further used to prepare TPS-resin blends by melt extrusion using GR and GR derivatives. Both neat TPS as well as each TPS-resin compounding were further processed by injection moulding. Injection moulded samples of TPS-resin blend formulations were successfully obtained without any change in the injection moulding processing parameters with respect to neat TPS.

It is known that during the melt-extrusion and injection moulding process, bio-based polymeric matrices can suffer degradation due to the strong shear stresses that act in the viscous molten polymer [22,41]. In fact, injection moulding parameters (i.e., injection speed, barrel temperature and back pressure) significantly affect the impact strength of the polymeric matrices [42]. Thus, injection moulded polymers for rigid packaging are required for high mechanical performance to overcome the strong shear stresses during processing to successfully obtain injected moulded parts as well as to offer the desired mechanical resistance during service. Thus, tensile tests measurements up-to-rupture as well as Shore D hardness were assayed on injected TPS and TPS–resin blend formulations (Figure 3). Results show that TPS is a brittle polymer with a Young's modulus of 11.93 MPa (Figure 3a), a tensile strength of 2 MPa (Figure 3b) and a strain at break of 63.46% (Figure 3c). Young's modulus of TPS-GR, TPS-CM, TPS-RD and TPS-UG have no statistical differences among them nor with TPS (Figure 3a, $p > 0.05$). TPS-LF is the only formulation that presents a statistically higher modulus with respect to TPS ($p < 0.05$), indicating a stiffening effect due to LF addition to the TPS matrix. These results show that LF increased the mechanical properties in the formulation due to a reinforcement effect, which increases the TPS modulus by 55%. This reinforcement effect is because LF increases the cohesion between the starch phases [37]. The ester groups in LF interacted with the starch hydroxyls groups by hydrogen bonding interactions which leads to the compatibilisation of both components [43]. In a previous work we have blended Mater-Bi® type bioplastic (Mater-Bi® NF 866 which is composed of TPS, polybutylene adipate-co-terephthalate (PBAT) and poly(ε-caprolactone) (PCL) with LF and we observed that 10 wt. % was not able to significantly improve the Young's modulus of the neat polymeric matrix, while it requires higher amounts (i.e.,: 15 wt. %) to significantly increase it [5]. The higher Young's modulus obtained here shows that LF possess good miscibility with starch matrix, being the only resin able to significantly increase the Young's modulus of TPS. In fact, in a more detailed microscopic analysis it was observed that LF not only possess good miscibility with starch phase of Mater-Bi® NF 866, but also it was able to increase the miscibility between the semi-crystalline and the amorphous phases of thermoplastic starch (TPS) [37]. The TPS tensile strength (Figure 3b) was significantly reduced when all the additives were incorporated to the polymeric matrix, being TPS-LF and TPS-UG the formulations with the highest tensile strength among the studied samples. Regarding the flexibility of the materials, the results show that the elongation at break significant (Figure 3c, $p < 0.05$) increased only when unmodified GR was added to the TPS matrix. GR act as a plasticiser as it increased neat TPS elongation at break in 43%. The plasticising effect of gum rosin has been already observed for other bioplastics such as PLA [44], PCL [45] and Mater-Bi® NF 866 [5]. Meanwhile, the other gum rosin derivatives produce no significant effect on the elongation at break of neat TPS (Figure 3c, $p > 0.05$).

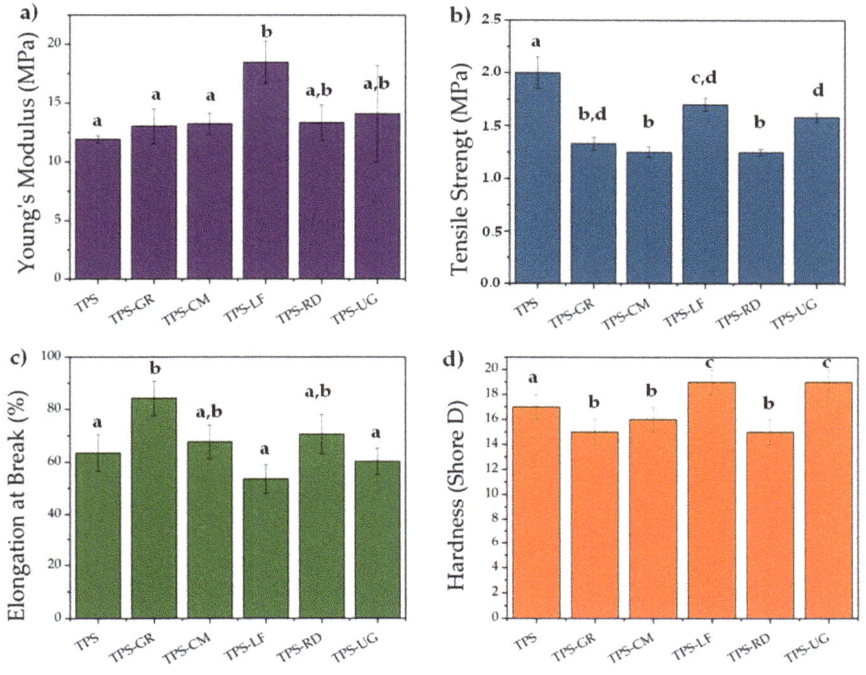

Figure 3. Tensile and hardness properties of TPS and TPS-resin formulations: (**a**) Young modulus, (**b**) Tensile Strength, (**c**) elongation at break and (**d**) Shore D. [a–d] Different letters within the same property show statistically significant differences between formulations ($p < 0.05$).

Hardness properties of TPS and the formulations with GR and rosin derivatives were also measured (Figure 3d). All the formulations presented significant differences with TPS Shore D hardness ($p < 0.05$). While the addition of an ester resin (LF or UG) increased TPS hardness values in 12% (from 17 in neat TPS to 19 for TPS-LF and/or TPS-UG), the materials that contain the other resin additives (GR, CM, and RD) reduced TPS hardness values in 12% (from 17 in neat TPS to values up to 15) (Figure 3d). Although the hardness of polymers do not necessarily provide correlate results with other fundamental properties, since it is an empirical quantity related to the inherent indentation resistance [46], it was observed that the TPS-resin blends showed similar tend than the tensile strength values (Figure 3b). Similarly, Ferri et al. (2018) blended PLA with TPS in 70:30 proportion and further added increasing amounts of maleinised linseed oil (from 2 to 8 phr) to increase the compatibility between both biopolymeric matrices, and observed a decrease of Shore D values accompanied by a related decrease on the tensile strength values [28]. Comparing these results with those previously obtained for Mater-Bi® NF 866 blended with 10 wt. % of GR and LF, it was observed that GR decreased the Shore D values of Mater-Bi® NF 866 (around 8%), but it was maintained in Mater-Bi® blended with 10% of LF [5]. These results confirm once again the good miscibility of LF with starch polymeric matrix.

The typical stress–strain curves of the materials with the toughness values of each formulation is shown in Figure 4. It was found that the addition of either GR or rosin derivatives significantly reduced the TPS toughness ($p < 0.05$). The results show that GR has the highest toughness among the studied formulation and that the toughness values present no statistical differences when LF or RD were added ($p > 0.05$). It is important to notice that TPS-LF presents statistically the same elongation at break and a higher Young's modulus than neat TPS (Figure 3a). Therefore, LF provides the highest strength between the studied materials. The same trend was exhibited by UG, since it is an ester of gum rosin. On the other hand, GR statistically maintains TPS Young's modulus and increases its elongation at break. Thus, TPS-GR would withstand greater tension that neat TPS. The other modified gum rosin additives (CM and RD) exhibit behaviour similar to GR. These results confirm a stiffening and a plasticisation effect produced by LF and GR addition, respectively.

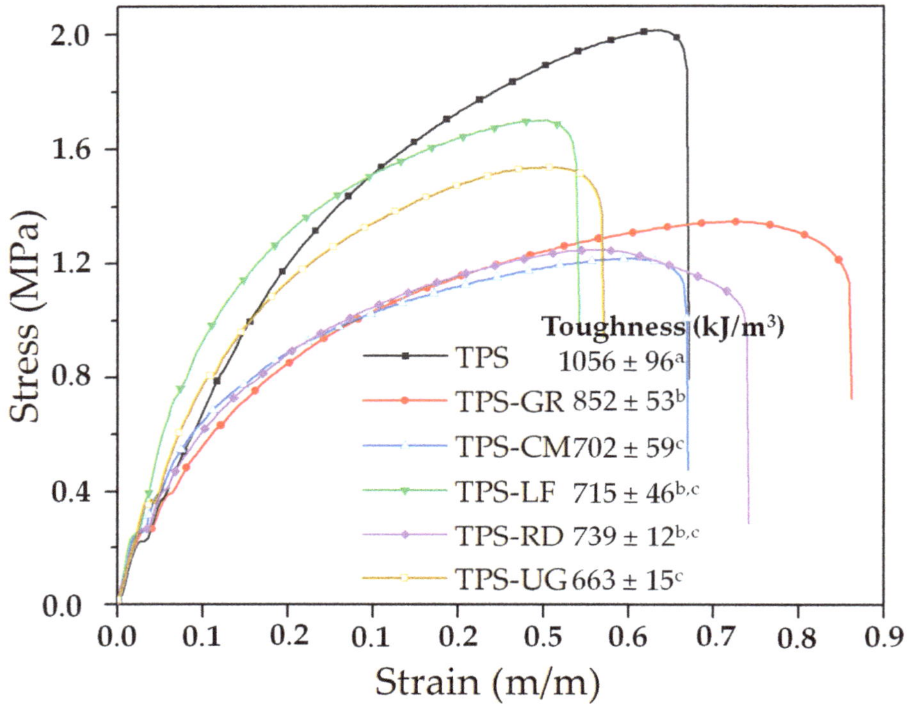

Figure 4. Stress–strain curves of TPS and TPS-resin blend formulations with the toughness values. [a–d] Different letters within the same property show statistically significant differences between formulations ($p < 0.05$).

The DSC curves of TPS and its TPS-resin formulations with 10 wt. % of GR and rosin derivatives are shown in Figure 5. T_g of TPS is dependent on glycerol and water contents and it could be found between 34 °C and 70 °C [47,48]. The studied TPS exhibits the T_g at 45.7 °C. However, the addition of gum rosin and rosin derivatives shifts the glass transition temperature to higher temperatures. For TPS-GR and TPS-RD, T_g was found at 56.5 °C and 58.1 °C, respectively. Meanwhile for TPS-LF, TPS-CM and TPS-UG it was found between 70 °C and 80 °C. These results suggest that the formulations present a stiffening effect since GR and its derivatives increase the T_g values, but it was more marked in those formulations blended with UG, CM and LF. The increase on T_g values is very interesting for several applications (i.e., a cup for hot beverages and/or lids for hot food).

GR and RD possess similar chemical structure (see Figure 1). GR is mainly made up of abietane-type acid while RD is made up of pimarane-type acids and is more stable [49]. Therefore, RD side groups are not modified [49], and its chains can move freely between TPS chains lubricating them [5]. On the other hand, CM, LF and UG have chemicals modifications in its chemical side groups (see Figure 1) that hinder its mobility, mainly because they are able to interact with TPS hydroxyl groups [43,49]. However, these chemicals modifications produced a different effect on the mobility of the chains that will further affect the final performance of TPS-resin based formulations. LF is a pentaerythritol ester while UG is a glycerol ester. Thus, TPS-LF would have a higher T_g than TPS-UG, as a pentaerythritol ester has higher cohesion, adhesion, viscosity at fusion and glass transition temperature than a glycerol ester. On the other hand, CM is a modified gum rosin with maleic anhydride that have reactive sites that differ from those in unmodified gum rosin [49], as CM has three carboxylic acid groups, which increases its acidity [49]. As a result, CM, LF and UG shift T_g to higher temperatures (70 °C or more). The resins with group modifications had a more marked stiffening effect. In fact, the tendency of increment in T_g values follows the same tendency as the Young's modulus. The shifts in TPS glass transition temperatures suggest that GR and its derivatives have good miscibility with TPS matrix [37,50]. With DSC results, it was not possible to establish differences between the T_g of the TPS-GR and TPS-RD. However, it is possible to determine that LF is the modified resin with the highest stiffening effect because it produced the highest displacement of T_g. In fact, it shifted the T_g of TPS from 45 °C to 79.3 °C, which is 9 °C more than UG and 7 °C more than CM. The obtained DSC results are consistent with TPS-LF increment of Young's modulus, TPS-LF and TPS-UG increase in hardness, as well as with the plasticisation behaviour of TPS-GR, already discussed.

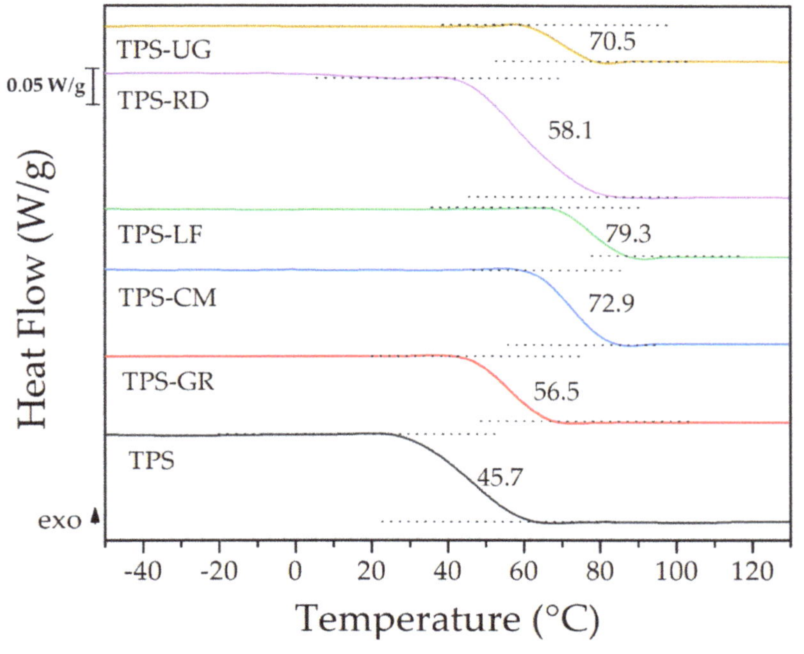

Figure 5. Differential scanning calorimetry (DSC) second heating scan curves of neat TPS and TPS-resins blend formulations.

TGA curves and their derivatives (DTG) for TPS and its TPS-resin blend formulations are shown in Figure 6. The thermal values at different degradations stages, determined by TGA analysis, are shown in Table 2. It can be noticed that the formulated samples were more thermally stable than the neat TPS in all the degradation stages. TGA curves showed a three-step degradation process (Figure 6a). Between 100 °C and 170 °C an initial weight loss of 12% occurs, which is related with the bound water content in the TPS samples [51,52]. The degradation onset temperature ($T_{5\%}$) of TPS (Table 2) increased with the addition of GR and rosin derivatives. GR rises $T_{5\%}$ by 29 °C, CM by 23 °C, LF does by 25 °C, while RD and UG produce the lowest increase in $T_{5\%}$ by 16 °C and 11 °C, respectively. Therefore, the chemical structures of the additives allow them to positively interact with the TPS matrix, providing a thermal stabilising effect [5]. Nevertheless, GR and LF create a higher protection to thermal degradation than the other additives. The increased thermal stability has been related to the positive interaction between the carbonyl groups of the pine resins and the hydroxyl groups of the starch material (i.e., hydrogen bonding interactions) [53].

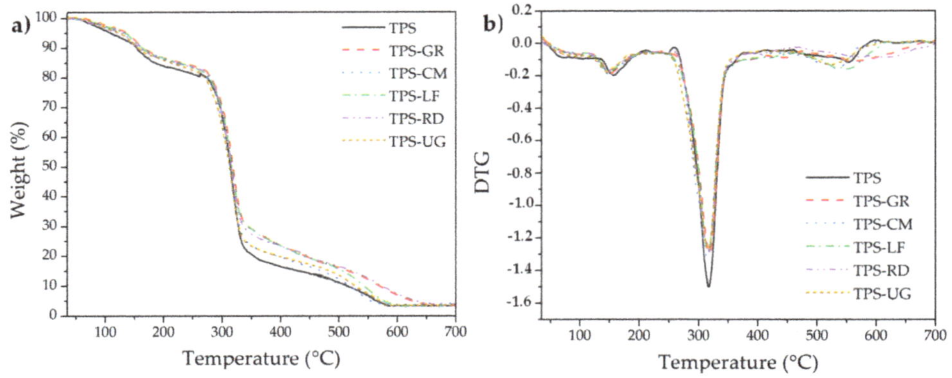

Figure 6. (**a**) TGA and (**b**) their derivatives (DTG) curves of neat TPS and TPS-resin blend formulations.

Table 2. TGA thermal parameters degradation onset temperature ($T_{5\%}$), temperature of maximum degradation rate (T_{max}) and degradation endset temperature ($T_{95\%}$) for TPS and the TPS-resin blend formulations.

Material	$T_{5\%}$ (°C)	T_{max} (°C)	$T_{95\%}$ (°C)
TPS	109.3 ± 2.1	317.3 ± 1.9	564.8 ± 2.3
TPS-GR	138.3 ± 1.9	318.3 ± 1.7	624.3 ± 2.1
TPS-CM	132.8 ± 2.0	315.8 ± 1.9	553.8 ± 1.9
TPS-LF	133.8 ± 1.9	316.3 ± 1.7	576.3 ± 2.3
TPS-RD	126.3 ± 1.8	316.8 ± 2.0	632.2 ± 1.8
TPS-UG	121.8 ± 2.0	315.8 ± 1.8	570.2 ± 2.0

The second step of degradation starts at 260 °C and it is the main degradation process. In this step the degradation is produced because of the depolymerization of starch [51], that is the cleavage of ether linkages in starch backbone and it is also related with the degradation of the starch/glycerol reach phase [52]. During this degradation step the dehydration of neighbouring hydroxyl groups in the glucose ring take place, resulting in the formation of C-C bonds or breakdown of the glucose ring with the further formation of aldehyde end groups [27].

The DTG curve (Figure 6b) shows that the maximum degradation temperature (T_{max}) did not present differences between all the materials with a mean value around 316 °C (316.6 ± 1.8 °C). Even so, the effect in the mass loss was appreciable in the formulations that contain gum rosin and derivatives. This effect can be seen in Figure 6a, at 350 °C. The TPS reports a mass loss of 20%. In the case of TPS blended with GR, LF and RD reduction in the mass loss of 29%, 29% and 27% were observed, respectively. The incorporation of CM and UG to TPS showed the mass loss of TPS in 23%. GR, LF and RD created interactions and links with TPS matrix, because of the presence of carbonyl and ester groups, which protect the polymeric matrix from thermal degradation [5]. Besides, CM resin produced less effect in the thermal protection because the spatial distribution of its molecules avoids the creation of a lot of links between resin and TPS polymeric matrix. It is important to notice that even when UG resin has carbonyl groups on its structure, the effect in the thermal stability is lower that the LF resin which possesses one more carbonyl group and higher molecular weight. At 335 °C the final degradation step takes place, where organic residues decompose and turn into ashes. It is seen that all the pine resins protected TPS matrix from thermal degradation, with GR, LF and RD providing the greatest stability in the degradation endset temperature.

The FTIR spectra comparison of neat TPS and TPS-resin blends formulations with 10 wt. % of gum rosin (GR) and rosin derivatives (CM, LF, RD, UG) is presented in Figure 7.

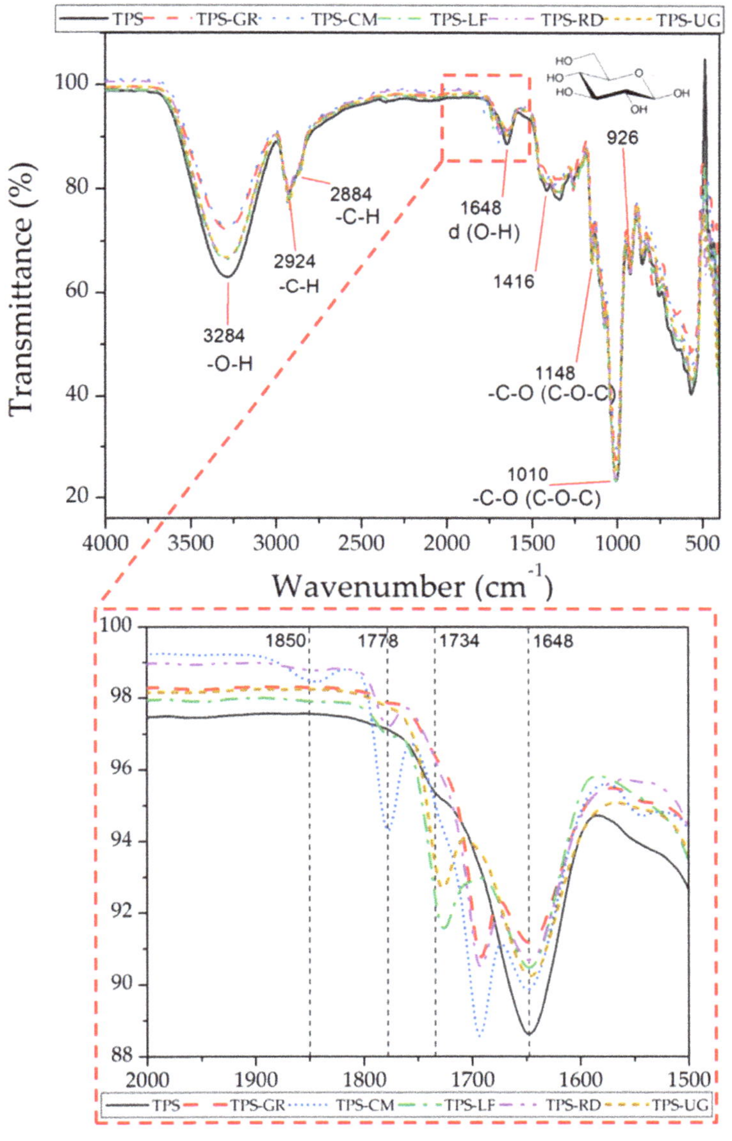

Figure 7. Fourier transform infrared spectroscopy (FTIR) spectra with representative peaks of net TPS and its comparison with TPS-resin blend formulations. A zoom in the range between 2000 cm^{-1} and 1500 cm^{-1} is also shown.

The FTIR spectrum of neat TPS presents the typical absorption bands corresponding to the functional groups of starch and glycerol. The band corresponding to C-O stretching of C-O-C bond was found at 1010 cm^{-1} and 1148 cm^{-1} and the C-O stretching of pyranose rings at 926 cm^{-1}. The bound water band was located at 1648 cm^{-1} (δ (O-H)). The O-H groups are found at 3284 cm^{-1} and C-H stretching is located at 2924 cm^{-1} [54–57]. The representative bands of the glycerol plasticizer were displayed at 2884 cm^{-1}(C-H), associated with the hydroxyl groups, as well as at 1416 cm^{-1} and at 1148 cm^{-1} (C-O stretching), associated with carbon-oxygen (C–O) absorption peaks characteristic of primary and secondary alcohols [19,54,56].

It is seen that the spectra of the formulations are similar to that of neat TPS. This is expected since the resins content is 10 wt. %. Dang and Yoksan (2015) have reported a similar behaviour when working with starch films blended with chitosan added in 0.37–1.45 wt. % [57]. Further, there are no significant shifts in the characteristic peaks of TPS, which indicates that the interaction with the resins is not strong enough to be detectable with this technique in accordance with Mendes et al. (2017) [54]. Nevertheless, two main differences can be observed in the spectra of TPS-resin blend formulations with respect to neat TPS. The first one is referred to the reduction of the -OH group band related with the bound water, found at 3284 cm^{-1} and at 1648 cm^{-1}. A reduction in this peak indicates that when TPS is blended with resins, the -OH groups of TPS have more affinity with resins than with water. Therefore, -OH groups are involved in positives interaction with GR and/or rosin derivatives such as hydrogen bonding [57]. The second difference was found in the region between 2000 cm^{-1} and 1500 cm^{-1} (see zoom in Figure 7). In TPS-CM signals of -C=O stretch of maleic anhydride appeared as low intensity peaks at 1780 cm^{-1} and 1850 cm^{-1} [58]. In TPS-LF and TPS-UG a low intensity band was found at 1727 cm^{-1}. This band is representative of the -C=O stretching of the ester group [43]. In TPS-GR and TPS-RD the peak of stretching of the carboxylic group (-C=O) is found at 1692 cm^{-1} [5]. Those displacement of carbonyl group band suggest an hydrogen bonding interactions between TPS and the respective resins [5,59].

Finally, the morphology and microstructure of the cryo-fractured surfaces of TPS as well as the effect the gum rosin presence on the TPS matrix were studied by SEM (Figure 8). No apparent phase separation is observed in the SEM image of the cryo-fractured surfaces of plasticized corn starch with glycerol and water (Figure 8a), confirming the homogeneous dispersion of glycerol and water in corn starch matrix and the successful formation of TPS. Moreover, neat TPS shows a smooth fracture surface typical of TPS with amylose content around 25–30% [52]. No apparent phase separation was observed in TPS-resin blends regardless the resin used, indicating that all resins are well incorporated to the TPS matrix (Figure 8b–f). All TPS-resin blend formulations showed a much rougher behaviour in accordance with the increased stiffening effect. Among al TPS-resin blend formulations, TPS-GR (Figure 8b) and TPS-RD (Figure 8e) show much cracked surface. Moreover, in TPS-GR (Figure 8b), some micro-voids are observed (see inset Figure 8b), which are probably responsible of the reduction of the mechanical resistance of this formulation as reveals the marked reduction of the tensile strength (Figure 3b) and Shore D hardness (Figure 3d). Similarly, TPS-RD (Figure 7e) shows some micro-domains that in some cases shows phase debonding (see inset Figure 8e), with the corresponding reduction of the tensile strength (Figure 3b) as well as Shore D hardness (Figure 3d). Meanwhile, in TPS-CM (Figure 8c), TPS-LF (Figure 8d) and TPS-UG (Figure 8f) formulations, resin particles appear homogeneously dispersed as small spherical domains (see inset Figure 8c for TPS-CM, inset Figure 8d for TPS-LF and inset Figure 8f for TPS-UG). In a previous work, Mater-Bi® NF 866 was blended with 10 wt. % of LF and SEM micrographs shows poor interfacial adhesion between the polymeric phases since small domains with empty interface were observed [5]. Thus, the results obtained here suggest that LF possess good miscibility with starchy matrix (see inset Figure 8d). In fact, in a more detailed microscopic analysis somewhat improvement in the miscibility between the amorphous and semi-crystalline phases of thermoplastic starch due to the LF gum rosin derivate addition was observed for the already commented Mater-Bi® NF 866 blended with 10 wt. % of LF [37].

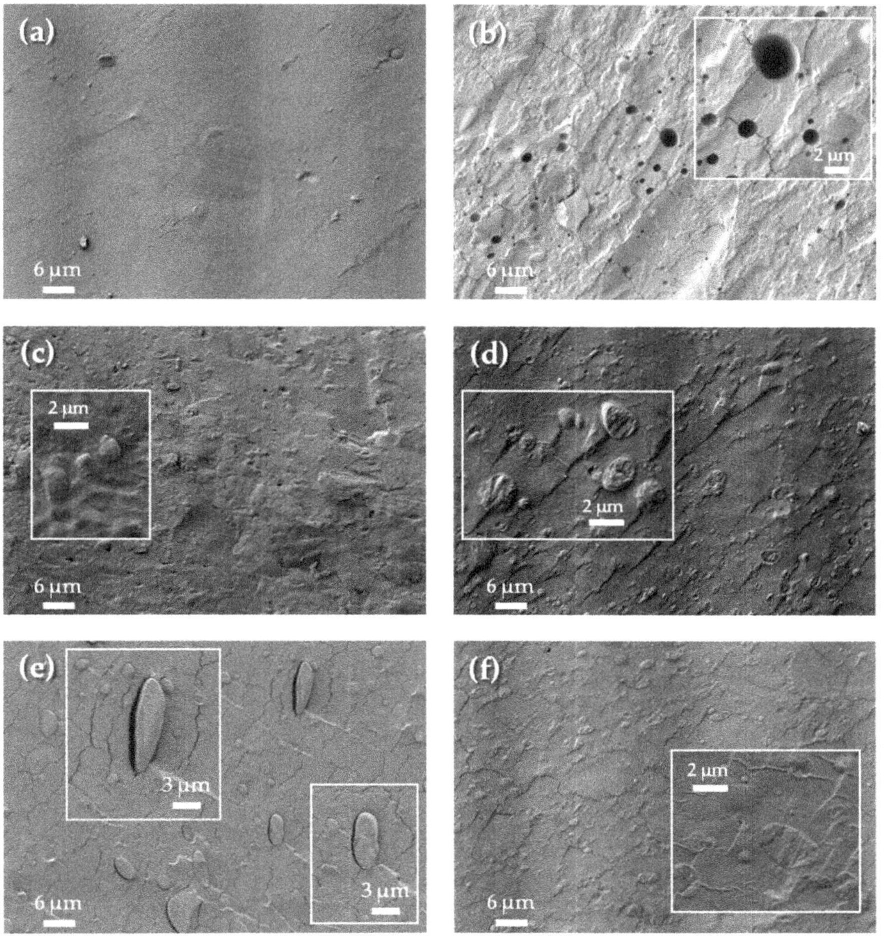

Figure 8. Scanning electron microscopy (SEM) images of the crio-fractured surface of: (**a**) TPS, (**b**) TPS-GR, (**c**) TPS-CM, (**d**) TPS-LF, (**e**) TPS-RD and (**f**) TPS-UG.

4. Conclusions

Homogeneous TPS based materials were prepared starting with corn starch and using water as well as glycerol as plasticisers. The obtained TPS was then blended with pine resin derivatives by melt extrusion and further by injection moulding process. All TPS-resin blend formulations showed higher thermal stability than TPS counterpart, improving the processing window of TPS. FTIR spectra showed that gum rosin and its derivatives established hydrogen bond interactions with the hydroxyl groups of starch and glycerol in TPS matrix. Interactions were stronger in the case of blends with higher amounts of carbonyl groups in their chemical structure such as CM, UG and particularly LF. In fact, although all resins produce a stiffening effect, CM, UG and LF generated a significant stiffening effect, shifting the T_g of TPS from 45 °C to higher temperatures greater than 70 °C. Consequently, in TPS-CM, TPS-UG and TPS-LF the Young's modulus slightly increase while were among the TPS-resin based formulations with the highest hardness values following the same trend as the T_g increment.

Therefore, it can be concluded that gum rosin and particularly modified gum rosin derivatives showed good compatibility with the developed TPS from corn starch and allowed to develop homogeneous fully bio-based materials, ensuring thermal stability for melt-extrusion and injection moulding processing as well as a significant stiffening effect. Consequently, these bio-based materials show their potential for industrial applications such as disposable cups for hot beverage and/or lids containers for hot food applications.

Author Contributions: Conceptualization, M.A. and M.P.A.; methodology, M.A.; validation, M.A. and C.P.; formal analysis, M.A. and C.P.; investigation, M.A., C.P., M.P.A. and J.L.-M.; resources, J.L.-M.; data curation, M.A., C.P. and M.P.A.; writing—original draft preparation, M.A., C.P. and M.P.A.; writing—review and editing, M.A. and M.P.A.; visualization, M.A., C.P., M.P.A. and J.L.-M.; supervision, J.L.-M.; project administration, M.P.A. and J.L.-M. funding acquisition, M.P.A. and J.L.-M. All authors have read and agreed to the published version of the manuscript.

Funding: This research was funded by the Spanish Ministry of Economy and Competitiveness (MINECO), project: PROMADEPCOL (MAT2017-84909-C2-2-R) as well as by Santander-UCM (PR87/19-22628) project. M.A. thanks Secretaria Nacional de Educación Superior, Ciencia, Tecnología e Innovación (SENESCYT-Ecuador) and Escuela Politécnica Nacional. C.P. thanks Santiago Grisolía fellowship (GRISOLIAP/2019/113) from Generalitat Valenciana and M.P.A. thanks MINECO for her postdoctoral contract: Juan de la Cierva-Incorporación (FJCI-2017-33536).

Conflicts of Interest: The authors declare no conflict of interest.

References

1. Plastics Europe—The Facts 2018. An Analysis of European Plastics Production, Demand and Waste Data. 2018. Available online: https://www.plasticseurope.org/application/files/6315/4510/9658/Plastics_the_facts_2018_AF_web.pdf (accessed on 7 April 2020).
2. Arrieta, M.P.; Samper, M.D.; Aldas, M.; López, J. On the use of PLA-PHB blends for sustainable food packaging applications. *Materials* **2017**, *10*, 1008. [CrossRef] [PubMed]
3. Averous, L. Biodegradable multiphase systems based on plasticized starch: A review. *J. Macromol. Sci. Part C Polym. Rev.* **2004**, *44*, 231–274. [CrossRef]
4. Auras, R.; Harte, B.; Selke, S. An overview of polylactides as packaging materials. *Macromol. Biosci.* **2004**, *4*, 835–864. [CrossRef] [PubMed]
5. Aldas, M.; Ferri, J.; Lopez-Martinez, J.; Samper, M.; Arrieta, M. Effect of pine resin derivatives on the structural, thermal, and mechanical properties of Mater-Bi type bioplastic. *J. Appl. Polym. Sci.* **2020**, *137*, 48236. [CrossRef]
6. Aldas, M.; Paladines, A.; Valle, V.; Pazmiño, M.; Quiroz, F. Effect of the prodegradant-additive plastics incorporated on the polyethylene recycling. *Int. J. Polym. Sci.* **2018**, *2018*. [CrossRef]
7. Simon, J.; Müller, H.; Koch, R.; Müller, V. Thermoplastic and biodegradable polymers of cellulose. *Polym. Degrad. Stab.* **1998**, *59*, 107–115. [CrossRef]
8. Arrieta, M.P.; Fortunati, E.; Burgos, N.; Peltzer, M.A.; López, J.; Peponi, L. Nanocellulose-Based Polymeric Blends for Food Packaging Applications. In *Multifunctional Polymeric Nanocomposites Based on Cellulosic Reinforcements*; Elsevier: Oxford, UK, 2016; pp. 205–252. [CrossRef]
9. Sadeghifar, H.; Cui, C.; Argyropoulos, D.S. Toward thermoplastic lignin polymers. Part 1. Selective masking of phenolic hydroxyl groups in kraft lignins via methylation and oxypropylation chemistries. *Ind. Eng. Chem. Res.* **2012**, *51*, 16713–16720. [CrossRef]
10. Wilbon, P.A.; Chu, F.; Tang, C. Progress in renewable polymers from natural terpenes, terpenoids, and rosin. *Macromol. Rapid Commun.* **2013**, *34*, 8–37. [CrossRef]
11. Liu, C.; Liu, F.; Cai, J.; Xie, W.; Long, T.E.; Turner, S.R.; Lyons, A.; Gross, R.A. Polymers from fatty acids: Poly (ω-hydroxyl tetradecanoic acid) synthesis and physico-mechanical studies. *Biomacromolecules* **2011**, *12*, 3291–3298. [CrossRef]
12. Galbis, J.A.; García-Martín, M.d.G.; de Paz, M.V.; Galbis, E. Synthetic polymers from sugar-based monomers. *Chem. Rev.* **2016**, *116*, 1600–1636. [CrossRef]
13. Lu, D.; Xiao, C.; Xu, S. Starch-based completely biodegradable polymer materials. *Express Polym. Lett.* **2009**, *3*, 366–375. [CrossRef]
14. Ciardelli, F.; Bertoldo, M.; Bronco, S.; Passaglia, E. The obtainment of bioplastics. In *Polymers from Fossil and Renewable Resources*; Springer: Cham, Switzerland, 2019; pp. 107–132.

15. Ghanbari, A.; Tabarsa, T.; Ashori, A.; Shakeri, A.; Mashkour, M. Thermoplastic starch foamed composites reinforced with cellulose nanofibers: Thermal and mechanical properties. *Carbohydr. Polym.* **2018**, *197*, 305–311. [CrossRef] [PubMed]
16. Sessini, V.; Arrieta, M.P.; Fernández-Torres, A.; Peponi, L. Humidity-activated shape memory effect on plasticized starch-based biomaterials. *Carbohydr. Polym.* **2018**, *179*, 93–99. [CrossRef]
17. Angellier, H.; Molina-Boisseau, S.; Dole, P.; Dufresne, A. Thermoplastic starch– waxy maize starch nanocrystals nanocomposites. *Biomacromolecules* **2006**, *7*, 531–539. [CrossRef]
18. Jiugao, Y.; Ning, W.; Xiaofei, M. The effects of citric acid on the properties of thermoplastic starch plasticized by glycerol. *Starch-Stärke* **2005**, *57*, 494–504. [CrossRef]
19. Arrieta, M.P.; Peltzer, M.A.; Garrigós, M.d.C.; Jiménez, A. Structure and mechanical properties of sodium and calcium caseinate edible active films with carvacrol. *J. Food Eng.* **2013**, *114*, 486–494. [CrossRef]
20. Montava-Jordà, S.; Torres-Giner, S.; Ferrandiz-Bou, S.; Quiles-Carrillo, L.; Montanes, N. Development of sustainable and cost-competitive injection-molded pieces of partially bio-based polyethylene terephthalate through the valorization of cotton textile waste. *Int. J. Mol. Sci.* **2019**, *20*, 1378. [CrossRef] [PubMed]
21. Bourguignon, D. European Parliamentary Research Service. 2018. Available online: http://www.europarl.europa.eu/RegData/etudes/ATAG/2018/625163/EPRS_ATA_ATA(2018)625163_EN.pdf (accessed on 31 March 2020).
22. Sarasini, F.; Puglia, D.; Fortunati, E.; Kenny, J.; Santulli, C. Effect of fiber surface treatments on thermo-mechanical behavior of poly (lactic acid)/phormium tenax composites. *J. Polym. Environ.* **2013**, *21*, 881–891. [CrossRef]
23. Imre, B.; Pukánszky, B. Compatibilization in bio-based and biodegradable polymer blends. *Eur. Polym. J.* **2013**, *49*, 1215–1233. [CrossRef]
24. Samper, M.D.; Bertomeu, D.; Arrieta, M.P.; Ferri, J.M.; López-Martínez, J. Interference of biodegradable plastics in the polypropylene recycling process. *Materials* **2018**, *11*, 1886. [CrossRef]
25. Dolores, S.M.; Patricia, A.M.; Santiago, F.; Juan, L. Influence of biodegradable materials in the recycled polystyrene. *J. Appl. Polym. Sci.* **2014**, *131*. [CrossRef]
26. Samper-Madrigal, M.D.; Fenollar, O.; Dominici, F.; Balart, R.; Kenny, J.M. The effect of sepiolite on the compatibilization of polyethylene–thermoplastic starch blends for environmentally friendly films. *J. Mater. Sci.* **2014**, *50*, 863–872. [CrossRef]
27. Sessini, V.; Arrieta, M.P.; Raquez, J.-M.; Dubois, P.; Kenny, J.M.; Peponi, L. Thermal and composting degradation of EVA/Thermoplastic starch blends and their nanocomposites. *Polym. Degrad. Stab.* **2019**, *159*, 184–198. [CrossRef]
28. Ferri, J.; Garcia-Garcia, D.; Carbonell-Verdu, A.; Fenollar, O.; Balart, R. Poly (lactic acid) formulations with improved toughness by physical blending with thermoplastic starch. *J. Appl. Polym. Sci.* **2018**, *135*, 45751. [CrossRef]
29. Garrido-Miranda, K.A.; Rivas, B.L.; Pérez-Rivera, M.A.; Sanfuentes, E.A.; Peña-Farfal, C. Antioxidant and antifungal effects of eugenol incorporated in bionanocomposites of poly (3-hydroxybutyrate)-thermoplastic starch. *LWT* **2018**, *98*, 260–267. [CrossRef]
30. Yadav, B.K.; Gidwani, B.; Vyas, A. Rosin: Recent advances and potential applications in novel drug delivery system. *J. Bioact. Compat. Polym.* **2016**, *31*, 111–126. [CrossRef]
31. Rodríguez-García, A.; Martín, J.A.; López, R.; Sanz, A.; Gil, L. Effect of four tapping methods on anatomical traits and resin yield in Maritime pine (Pinus pinaster Ait). *Ind. Crop. Prod.* **2016**, *86*, 143–154. [CrossRef]
32. Llevot, A.; Grau, E.; Carlotti, S.; Grelier, S.; Cramail, H. Dimerization of abietic acid for the design of renewable polymers by ADMET. *Eur. Polym. J.* **2015**, *67*, 409–417. [CrossRef]
33. Arrieta, M.P.; Samper, M.D.; Jiménez-López, M.; Aldas, M.; López, J. Combined effect of linseed oil and gum rosin as natural additives for PVC. *Ind. Crop. Prod.* **2017**, *99*, 196–204. [CrossRef]
34. Yao, K.; Tang, C. Controlled polymerization of next-generation renewable monomers and beyond. *Macromolecules* **2013**, *46*, 1689–1712. [CrossRef]
35. Gandini, A.; Lacerda, T.M. From monomers to polymers from renewable resources: Recent advances. *Prog. Polym. Sci.* **2015**, *48*, 1–39. [CrossRef]

36. Kumooka, Y. Analysis of rosin and modified rosin esters in adhesives by matrix-assisted laser desorption/ionization time-of-flight mass spectrometry (MALDI-TOF-MS). *Forensic Sci. Int.* **2008**, *176*, 111–120. [CrossRef] [PubMed]
37. Aldas, M.; Rayón, E.; López-Martínez, J.; Arrieta, M.P. A deeper microscopic study of the interaction between gum rosin derivatives and a Mater-Bi type bioplastic. *Polymers* **2020**, *12*, 226. [CrossRef]
38. Oniszczuk, T.; Janssen, L. Influence of addition of fiber on the mechanical properties of TPS moldings. In *Thermoplastic Starch: A Green Material for Various Industries*; Wiley-VCH Verlag GmbH & Co.: Winheim, Germany, 2009; pp. 197–208.
39. International Standards Organization. ISO 527-1:2012—Plastics—Determination of tensile properties—Part 1: General principles. 2012.
40. International Standards Organization. ISO 868:2003—Plastics and ebonite—Determination of indentation hardness by means of a durometer (Shore hardness). 2003.
41. Bucci, D.Z.; Tavares, L.B.B.; Sell, I. PHB packaging for the storage of food products. *Polym. Test.* **2005**, *24*, 564–571. [CrossRef]
42. Barbosa, S.E.; Kenny, J.M. Processing of short fiber reinforced polypropylene. II: Statistical study of the effects of processing conditions on the impact strength. *Polym. Eng. Sci.* **1999**, *39*, 1880–1890. [CrossRef]
43. Olivato, J.; Grossmann, M.; Bilck, A.; Yamashita, F. Effect of organic acids as additives on the performance of thermoplastic starch/polyester blown films. *Carbohydr. Polym.* **2012**, *90*, 159–164. [CrossRef]
44. Narayanan, M.; Loganathan, S.; Valapa, R.B.; Thomas, S.; Varghese, T.O. UV protective poly(lactic acid)/rosin films for sustainable packaging. *Int. J. Biol. Macromol.* **2017**, *99*, 37–45. [CrossRef]
45. Pavon, C.; Aldas, M.; López-Martínez, J.; Ferrándiz, S. New materials for 3D-printing based on polycaprolactone with gum rosin and beeswax as additives. *Polymers* **2020**, *12*. [CrossRef]
46. Bergström, J. 2—Experimental Characterization Techniques. In *Mechanics of Solid Polymers*; Bergström, J., Ed.; William Andrew Publishing: San Diego, CA, USA, 2015; pp. 19–114. [CrossRef]
47. Wattanakornsiri, A.; Pachana, K.; Kaewpirom, S.; Traina, M.; Migliaresi, C. Preparation and properties of green composites based on tapioca starch and differently recycled paper cellulose fibers. *J. Polym. Environ.* **2012**, *20*, 801–809. [CrossRef]
48. Forssell, P.M.; Mikkilä, J.M.; Moates, G.K.; Parker, R. Phase and glass transition behaviour of concentrated barley starch-glycerol-water mixtures, a model for thermoplastic starch. *Carbohydr. Polym.* **1997**, *34*, 275–282. [CrossRef]
49. Karlberg, A.-T. Colophony: Rosin in unmodified and modified form. In *Kanerva's Occupational. Dermatology*, 2nd ed.; Springer-Verlag: Berlin/Heidelberg, Germany, 2012; Volume 1, pp. 467–479. [CrossRef]
50. Liu, C.; Yu, J.; Sun, X.; Zhang, J.; He, J. Thermal degradation studies of cyclic olefin copolymers. *Polym. Degrad. Stab.* **2003**, *81*, 197–205. [CrossRef]
51. Teixeira, E.d.M.; Pasquini, D.; Curvelo, A.A.; Corradini, E.; Belgacem, M.N.; Dufresne, A. Cassava bagasse cellulose nanofibrils reinforced thermoplastic cassava starch. *Carbohydr. Polym.* **2009**, *78*, 422–431. [CrossRef]
52. Sessini, V.; Arrieta, M.P.; Kenny, J.M.; Peponi, L. Processing of edible films based on nanoreinforced gelatinized starch. *Polym. Degrad. Stab.* **2016**, *132*, 157–168. [CrossRef]
53. Cerruti, P.; Santagata, G.; d'Ayala, G.G.; Ambrogi, V.; Carfagna, C.; Malinconico, M.; Persico, P. Effect of a natural polyphenolic extract on the properties of a biodegradable starch-based polymer. *Polym. Degrad. Stab.* **2011**, *96*, 839–846. [CrossRef]
54. Mendes, J.; Paschoalin, R.; Carmona, V.; Neto, A.R.S.; Marques, A.; Marconcini, J.; Mattoso, L.; Medeiros, E.; Oliveira, J. Biodegradable polymer blends based on corn starch and thermoplastic chitosan processed by extrusion. *Carbohydr. Polym.* **2016**, *137*, 452–458. [CrossRef]
55. Kizil, R.; Irudayaraj, J.; Seetharaman, K. Characterization of irradiated starches by using FT-Raman and FTIR spectroscopy. *J. Agric. Food Chem.* **2002**, *50*, 3912–3918. [CrossRef]
56. Musa, M.; Yoo, M.; Kang, T.; Kolawole, E.; Ishiaku, U.; Yakubu, M.; Whang, D. Characterization and thermomechanical properties of thermoplastic potato starch. *J. Eng. Technol* **2013**, *2*, 9–16.
57. Dang, K.M.; Yoksan, R. Development of thermoplastic starch blown film by incorporating plasticized chitosan. *Carbohydr. Polym.* **2015**, *115*, 575–581. [CrossRef]

58. El-Ghazawy, R.A.; El-Saeed, A.M.; Al-Shafey, H.I.; Abdul-Raheim, A.-R.M.; El-Sockary, M.A. Rosin based epoxy coating: Synthesis, identification and characterization. *Eur. Polym. J.* **2015**, *69*, 403–415. [CrossRef]
59. Campos, A.; Teodoro, K.; Teixeira, E.; Corrêa, A.; Marconcini, J.; Wood, D.; Williams, T.; Mattoso, L. Properties of thermoplastic starch and TPS/polycaprolactone blend reinforced with sisal whiskers using extrusion processing. *Polym. Eng. Sci.* **2013**, *53*, 800–808. [CrossRef]

applied sciences

MDPI

Article

Valorization of Rice Straw into Cellulose Microfibers for the Reinforcement of Thermoplastic Corn Starch Films

Pedro A. V. Freitas [1,*], Carla I. La Fuente Arias [1,2], Sergio Torres-Giner [1], Chelo González-Martínez [1] and Amparo Chiralt [1]

[1] Research Institute of Food Engineering for Development (IIAD), Universitat Politècnica de València (UPV), 46022 Valencia, Spain; carlalfa@usp.br (C.I.L.F.A.); storresginer@upv.es (S.T.-G.); cgonza@tal.upv.es (C.G.-M.); dchiralt@tal.upv.es (A.C.)

[2] Departament of Agri-Food Industry, Food and Nutrition (LAN), School of Agriculture Luiz de Queiroz (ESALQ), Universidade de Sao Paulo, Piracicaba 13418-900, Brazil

* Correspondence: pedvidef@doctor.upv.es

Citation: Freitas, P.A.V.; Arias, C.I.L.F.; Torres-Giner, S.; González-Martínez, C.; Chiralt, A. Valorization of Rice Straw into Cellulose Microfibers for the Reinforcement of Thermoplastic Corn Starch Films. *Appl. Sci.* **2021**, *11*, 8433. https://doi.org/10.3390/app11188433

Academic Editor: Giorgia Spigno

Received: 31 July 2021
Accepted: 9 September 2021
Published: 11 September 2021

Abstract: In the present study, agro-food waste derived rice straw (RS) was valorized into cellulose microfibers (CMFs) using a green process of combined ultrasound and heating treatments and were thereafter used to improve the physical properties of thermoplastic starch films (TPS). Mechanical defibrillation of the fibers gave rise to CMFs with cumulative frequencies of length and diameters below 200 and 5–15 μm, respectively. The resultant CMFs were successfully incorporated at, 1, 3, and 5 wt% into TPS by melt mixing and also starch was subjected to dry heating (DH) modification to yield TPS modified by dry heating (TPSDH). The resultant materials were finally shaped into films by thermo-compression and characterized. It was observed that both DH modification and fiber incorporation at 3 and 5 wt% loadings interfered with the starch gelatinization, leading to non-gelatinized starch granules in the biopolymer matrix. Thermo-compressed films prepared with both types of starches and reinforced with 3 wt% CMFs were more rigid (percentage increases of ~215% for TPS and ~207% for the TPSDH), more resistant to break (~100% for TPS and ~60% for TPSDH), but also less extensible (~53% for TPS and ~78% for TPSDH). The incorporation of CMFs into the TPS matrix at the highest contents also promoted a decrease in water vapor (~15%) and oxygen permeabilities (~30%). Finally, all the TPS composite films showed low changes in terms of optical properties and equilibrium moisture, being less soluble in water than the TPSDH films.

Keywords: waste valorization; rice straw; thermoplastic starch; cellulose; thermal modification; microcomposites

1. Introduction

Today, the proper management of agro-industrial wastes is an essential socioeconomic and environmental issue. Most of these residues or by-products, generated in large quantities globally, are currently discarded or burned in the harvest fields [1]. In this sense, the use and valorization of agro-industrial waste into added-value materials are an emerging challenge that seeks to minimize environmental problems and promote the correct residue disposal. Agro-food and industrial waste derived biomass is essentially composed of lignocellulosic fractions, which are constituent biopolymers, such as cellulose, hemicellulose, and lignin [2,3]. Due to its renewability and biodegradability, the use of lignocellulosic-rich wastes can yield environmentally friendly materials through cost-effective processes with low environmental impact [4,5]. Several authors have reported the use of plant residues to obtain lignocellulosic fractions with potential application for biodiesel production [6], catalysts for biodiesel synthesis [7], antioxidant and antimicrobial materials [8,9] as well as reinforcing agents [10,11]. Particularly, cellulose and its derivatives extracted from biomass, such as cellulose microfibers (CMFs) and cellulose nanofibers, are highlighted renewable materials at the micro- and nanoscale, respectively, with low density, good compatibility, and excellent mechanical properties [12]. Thus, these lignocellulosic fractions can be used as reinforcing fillers of biopolymers in green composites for food packaging applications.

After harvesting the rice grain (*Oryza sativa* L.), one of the most important primary crops, large quantities of rice straw (RS) are obtained worldwide. In fact, the annual rice production is estimated at 782 million tons and 1 kg of rice grain yields approximately 1.5 kg of RS [13,14]. Generally, RS is directly burned in the paddies, which intensifies air pollution and damages soil feasibility and population health. Therefore, constant efforts are currently being made to propose different management approaches for this biomass waste [14,15]. Since RS is a lignocellulosic material that is composed of approximately 39% cellulose, 20% lignin, 23% hemicellulose, and 15% ashes, this waste derived biomass can be used to obtain several renewable added-value materials with interesting properties [16]. Among the most practical alternatives, RS can be valorized into CMFs by different extraction methods. Typically, CMFs from RS has a strong crystalline structure and low density, which can be used as reinforcing agent in biodegradable polymers instead to petroleum derived plastics [2,17].

Starch is a semicrystalline biopolymer consisting of linear and branched chains of D-glucopyranose units known as amylose and amylopectin, respectively [18]. The proportion of both constituents varies according to the source, from which the most common are cassava, maize, sweet potato, rice, oats, and peas [19,20]. Starch is widely investigated for several applications due to its renewability, biodegradability, low cost, availability, and biocompatibility [21–23]. In its native state, starch can be processed with a plasticizer in an extruder at temperatures between 140 and 160 °C and high pressure, giving rise to an amorphous material with thermoplastic behavior [21]. Although thermoplastic starch (TPS) films are excellent barriers to lipids, oxygen, and carbon dioxide, their use in food packaging is still limited since the barrier and mechanical properties are highly dependent on moisture content [18,24,25]. To overcome these shortcomings, the preparation of biodegradable TPS films reinforced with CMFs obtained from agro-industrial wastes is an interesting alternative. Previous studies have reported that the incorporation of cellulosic materials in TPS-based green composite films exhibited enhanced mechanical properties [20,26], water vapor [27] and oxygen gas barrier [22], and thermal stability [28]. These recent advances indicate the significance and potential of green composites from starch and cellulose micro- and nanofibrils in the development of sustainable and cost-effective materials for food packaging [26,29] and other uses [30], from semistructural applications to biofoams or reinforcing adhesives. Likewise, another strategy to modify and/or improve the properties of TPS films is pre-treating the starch in its native state by the dry heating (DH) method. The so-called DH modification is a green physical pre-treatment that involves dry heating of native starch at temperatures between 110 and 150 °C for 1 to 4 h [31,32]. This method is very promising since it is simple, fast, does not produce effluents, and can modify the original physicochemical properties of native starch to produce materials with improved properties [32–34].

Thus, the objective of this work was to study the morphogeometric characteristics of CMFs obtained from waste derived RS, before and after fiber grinding, and its subsequent incorporation by melt mixing into TPS. Prior to melt mixing, native starch was also subjected to DH modification to yield thermoplastic dry heated corn starch (TPSDH). The resultant doughs of TPS and TPSDH filled with CMFs were thereafter shaped into films by thermo-compression and characterized in terms of their morphological characteristics, barrier properties, water solubility, water uptake, thermal behavior, and mechanical performance to evaluate their potential application in food packaging.

2. Materials and Methods

2.1. Materials

RS was obtained as waste of the rice industry that was collected in L'Albufera field (Valencia, Spain). It was supplied by the "Banco de Paja" (Valencia, Spain). The as-received fibers were dried under vacuum (50 ± 2 °C, 0.5 mmbar) for 16 h and milled (3 cycles of 90 s each, IKA, model M20, Staufen, Germany), and sieved to <0.5 mm particle size.

Corn starch was supplied by Roquette (Roquette Laisa, Benifaió, Spain). Glycerol, magnesium nitrate ($Mg(NO_3)_2$), phosphorus pentoxide (P_2O_5), dimethyl sulfoxide (DMSO), methanol, and acetic acid were all obtained from Panreac Quimica S.L.U. (Castellar del Vallés, Barcelona, Spain). Sodium acetate trihydrate was purchased by Fluka^{TM} (Steinheim, Germany). Sodium chlorite and heptane were provided by Sigma-Aldrich S.A. (Madrid, Spain).

2.2. Extraction of Cellulose Microfiber

A combination of ultrasound and heating treatments are known to lead to a certain degree of leaching out of hemicellulose and lignin present into the lignocellulosic RS matrix [8]. Based on this, CMFs were extracted according to our previous methodology [35]. Briefly, an aqueous dispersion of RS particles with 5% (*w*/*v*) solids was sonicated for 30 min at 25 °C (maintaining the temperature in an ice bath) using an ultrasonic homogenizer (750 W power, 20 kHz frequency, continuous mode) equipped with a high-intensity probe (Vibra CellTM VCX750, Sonics & Material, Newton, MA, USA). After sonication, the RS dispersion was heated in a reflux heating apparatus at 100 °C for 1 h. Then, the plant dispersion was filtered using a qualitative filter (Filter Lab, Vidra Foc, Barcelona, Spain) and the solid fraction was dried at 40 °C for 48 h. Afterwards, the bleaching step of the cellulosic material was carried out by mixing the dry powder residue and the bleaching solution at 5% (*w*/*v*). The bleaching solution consisted of equal parts of sodium hypochlorite solution (1.7%, *w*/*v*), acetate buffer solution (2 N), and distilled water. The dispersion was treated under reflux heating for 4 cycles of 4 h each. The supernatant was filtered and the residue was washed with distilled water several times to remove the residual bleach solution at each stage. Then, the bleached solid fraction was dried at 35 °C for 48 h and milled using a milling machine (pulses of 2 s for 20 min, model M20, IKA Werke GmbH & Co. KG, Staufen, Germany) to obtain the so-called CMFs.

2.3. Dry Heating Treatment

Prior to melt mixing, corn starch was submitted to a DH pre-treatment, according to Maniglia et al. [33]. To this end, 50 g of starch were spread (1 mm thick layer) and compacted in an aluminum foil envelope. Then, the packaged sample was heated in a hot air oven (J.P. Selecta, S.A., Barcelona, Spain) at 130 °C for 4 h. Thereafter, the resultant starch was placed in a dark bottle and stored in a desiccator containing P_2O_5 at 6 ± 2 °C until further use.

2.4. Film Preparation

TPS and TPSDH films were prepared by melt blending and subsequent compression molding using glycerol as plasticizer 30 wt% with respect to the total starch mass. Green composites were obtained by addition of CMFs in concentrations of 1, 3, and 5 wt% in relation to the starch content based on the optimization performed in our previous study [36]. Table 1 summarizes the different sample compositions used to produce the films. The film components were previously dry mixed in a beaker and then melt-blended using an internal mini-mixer (HAAKETM PolyLabTM QC, Thermo Fisher Scientific, Karlsruhe, Germany) at 130 °C and 50 rpm for 10 min. The obtained dough was cold milled in the IKA's model and pre-conditioned at 25 °C and 53% RH using an $Mg(NO_3)_2$ over-saturated solution for one week.

The obtained doughs were compression molded to yield films in a hydraulic press (Model LP20, Labtech Engineering, Samut Prakan, Thailand). To this end, approximately 4 g of each dough was placed between two Teflon sheets and compressed in the hot plates. Control films (TPS) and those incorporating 1 wt% (TPS1), 3 wt% (TPS3), and 5 wt% (TPS5) of CMFs were preheated for 3 min at 160 °C in the plate press, then compressed at 160 °C for 2 min at 50 bars, followed by a second compression at 160 °C for 4 min at 130 bars, and finally cooled down to 80 °C. The processing conditions of the films obtained using modified DH starch with CMFs (1 wt%: TPSDH1, 3 wt%: TPSDH3, and 5 wt%: TPSDH5) or without CMFs (TPSDH) were set as follows: preheating at 150 °C for 5 min, compression at 30 bars and 150 °C for 2 min, followed by 130 bars for 6 min, and then, final cooling to 80 °C. All films were conditioned at 25 °C and 53% RH ($Mg(NO_3)_2$) for one week before characterizations.

Table 1. Summary of compositions according to the mass fraction (g/g) of thermoplastic corn starch (TPS), thermoplastic dry heated corn starch (TPSDH), glycerol (Gly), and cellulose microfiber (CMF).

Sample	X_s	X_{Gly}	X_{CMF}
TPS	0.770	0.230	-
TPS1	0.763	0.229	0.008
TPS3	0.753	0.225	0.022
TPS5	0.741	0.222	0.037
TPSDH	0.770	0.230	-
TPSDH1	0.763	0.229	0.008
TPSDH3	0.753	0.225	0.022
TPSDH5	0.741	0.222	0.037

2.5. Material Characterization

2.5.1. Microscopy

An optical microscope (Optika Microscope B-350, OPTIKA S.r.l., Ponteranica, Italy) equipped with a camera (Optikam B2) was used to evaluate the morphology of the untreated RS and the resultant CMF samples before and after the milling process. To this end, the samples were extended with a water drop on the holder glass plate and the images were taken at 10× and 40× magnification. Morphogeometric properties of CMFs, expressed as cumulative distributions, were obtained by measuring the particle length (l) in different regions of the sample, as shown in the imaging procedure of Figure 1. The Optika Vision Lite program was used to obtain a minimum of 60 measurements.

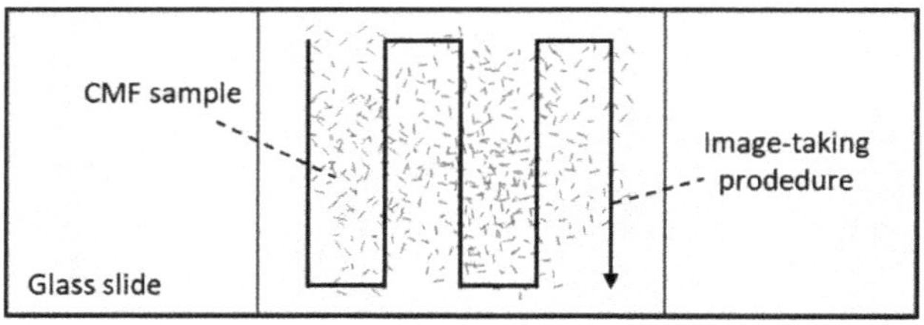

Figure 1. Image acquisition procedure to obtain the lengths of the cellulose microfibers (CMFs).

The morphologies of RS, CMFs, and the cross-sections of the cryo-fractured films obtained by immersion in liquid nitrogen were evaluated using a Field Emission Scanning Electron Microscope (FESEM) equipped with focused ion gun (AURIGA Compact, Zeiss, Oxford Instruments). The conditioned samples (P_2O_5 at 25 ± 2 °C for 1 week) were covered with a platinum layer (EM MED020 sputter coater, Leica Biosystems, Barcelona, Spain), and the images were taken at 2.0 kV acceleration voltage. The SmartTiff program (version 2, Zeiss, Oxford Instruments, Abingdon, UK) was used to measure the particle width (w) a minimum of 60 times.

2.5.2. Optical Evaluation

Optical properties of the films were obtained according the Kubelka–Munk theory of multiple scattering to determine the film reflection spectra (R) using the black (R_0) and white (R_g) backgrounds. The internal transmittance (T_i) of the films ranging from 400 to

700 nm, the film color coordinates L^* (lightness), a^* (redness-greenness), and b^* (yellowness-blueness) were determined using Equations (1)–(3), respectively. The colorimetric and transparency properties of the films were also evaluated in terms of hue angle ($h_{ab}{}^*$) (Equation (4)), chroma ($C_{ab}{}^*$) (Equation (5)), opacity (O) (Equation (6)), and total color difference (ΔE^*) (Equation (7)). The measurements were performed in triplicate.

$$T_i = \sqrt{(a + R_0)^2 - b^2} \tag{1}$$

$$a = \frac{1}{2}\left[R + \left(\frac{R_0 - R + R_g}{R_0 \times R_g}\right)\right] \tag{2}$$

$$b = \sqrt{a^2 - 1} \tag{3}$$

$$h_{ab}{}^* = arctg\left(\frac{b^*}{a^*}\right) \tag{4}$$

$$C_{ab}{}^* = \sqrt{a^{*2} + b^{*2}} \tag{5}$$

$$O = A_{500} x\ l \tag{6}$$

In which A_{500} and l are the absorbance at 500 nm and the film thickness, respectively.

$$\Delta E^* = \sqrt{(\Delta L^*)^2 + (\Delta a^*)^2 + (\Delta b^*)^2} \tag{7}$$

where $\Delta L^* = (L^* - L_0^*)$; $\Delta a^* = (a^* - a_0^*)$; $\Delta b^* = (b^* - b_0^*)$; and L_0^*, a_0^*, and b_0^* are the color coordinates of the control TPS film without fibers. The color differences were evaluated according to the following criteria [37]: $\Delta E^* < 1$ indicate unnoticeable color change; $1 \leq \Delta E^* \leq 2$ suggests that only an experienced observer can notice the difference; $2 \leq \Delta E^* \leq 3.5$ means that an inexperienced observer notices the difference; $3.5 \leq \Delta E^* \leq 5$ indicate clear noticeable difference; and $\Delta E^* \geq 5$ suggests that the observer notices different colors.

2.5.3. Equilibrium Moisture Content

The equilibrium moisture content of the films was determined by a gravimetric method. For this, film samples sizing 3 cm × 3 cm that were previously conditioned (25 °C and 53% RH for two weeks) were dried at 60 °C for 24 h using a drying oven. Thereafter, the dried samples were placed in a dissector containing P_2O_5 at 25 °C for two weeks to eliminate bonded water. Moisture content was determined from the total mass loss of the conditioned film during the drying process. The measurements were carried out in triplicate.

2.5.4. Water Solubility

Water solubility was determined based on the Talón et al. [38] methodology. A known mass of dry film sample (2 cm × 2 cm), conditioned in P_2O_5, were placed on a mesh, and immersed in a crucible with distilled water at 25 °C for 24 h. Afterwards, the sample-containing system was dried in an oven (J.P. Selecta, S.A.) at 60 °C for 48 h and, afterwards, conditioned in a desiccator with P_2O_5 at 25 °C for 1 week. The mass of residual film was weighed and compared with the initial mass of dry film. The measurements were carried out in triplicate and the results were expressed as g of solubilized film per 100 g film.

2.5.5. Barrier Measurements

Water vapor permeability (WVP) of the films was determined according to ASTM E96/E96M (ASTM, 2005) gravimetric methodology following the modifications proposed by McHUGH [39]. Film samples (Ø = 3.5 cm) were placed and sealed in Payne permeability cups filled with 5 mL of distilled water (100% RH). Then, the cups were placed into desiccators containing $Mg(NO_3)_2$ over-saturated solution (53% RH) and weighed periodically (ME36S, Sartorius, ± 0.00001 g, Fisher Scientific, Hampton, NH, USA) every

1.5 h for 25 h. For each treatment, the WVP was calculated considering the water vapor transmission rate (WVTR), which was determined from the slope of the weight loss vs. time. The measurements were performed in triplicate.

Oxygen permeability (OP) of the films was determined using an Ox-Tran equipment (Model 1/50, Mocon, Minneapolis, MN, USA) at 25 °C and 53% of RH, according to ASTM D3985-05 (ASTM, 2010). The used area of films was 50 cm^2 and the oxygen transmission rate (OTR) was obtained every 15 min until equilibrium was reached. The measurements were carried out in triplicate.

2.5.6. Mechanical Characterization

The tensile properties of the films were obtained according to ASTM D882 (ASTM, 2012). A universal testing machine (Stable Micro Systems, TA.XT plus, Stable Micro Systems, Godalming, UK) was used to determine elongation at break (EB), tensile strength at break (TS), and elastic modulus (EM). Film samples with dimensions of 25 mm × 10 mm were grabbed by two grips initially separated by 50 mm and stretched at a crosshead speed of 50 $mm.min^{-1}$. Eight samples were evaluated for each formulation. Before the analysis, the films were conditioned at 53% RH ($Mg(NO_3)_2$) and 25 °C for two weeks.

2.5.7. Thermal Analysis

The thermal stability of the films was determined by thermogravimetric analysis (TGA) using a thermogravimetric analyzer (TGA 1 Star^e System analyzer, Mettler-Toledo GmbH, Greifensee, Switzerland). Film samples, with a weight of 3–5 mg, were heated from 25 to 600 °C at a heating rate of 20 $°C.min^{-1}$ under nitrogen atmosphere (10 mL/min). For each thermal event, the initial (T_{onset}) and final (T_{final}) degradation temperatures, temperature of maximum degradation rate (T_{peak}), and the residual mass were determined by analyzing the TGA curves and their first derivative thermogravimetry (DTG) curves. The measurements were carried out in triplicate.

The phase transitions of the samples were investigated by differential scanning calorimeter (DSC) according to the method described by Collazo-Bigliardi et al. [40] with a DSC 1 Star^e System analyzer (Mettler-Toledo GmbH) operating under a nitrogen atmosphere (10 $mL.min^{-1}$). Samples of about 5–6 mg were weighted in aluminum pans and heated from 25 to 160 °C, cooled to 25 °C, and then heated (second heating step) to 160 °C at 10 $°C.min^{-1}$ using, in all cases, rates of 10 °C/min.

2.6. Statistical Analysis

The experimental data were submitted to multifactorial analysis of variance (ANOVA) at a confidence level of 95% using Minitab Statistical Program (version 17). Tukey's studentized range (HSD) test, considering the least significant difference of 5%, was applied to determine the influence of CMFs on the properties of the starch-based films.

3. Results

3.1. Morphological Characterization of CMFs

Figure 2 shows the FESEM micrographs of the untreated ground RS, the lignocellulosic residue after the combined ultrasound-heating treatment, the bleached cellulose fibers, and the resultant CMFs obtained after the milling step. It can be observed that the untreated RS (Figure 2a) presented heterogeneous particle sizes, with a predominance of rod-like shape, as seen in the magnified FESEM image. The RS particles showed mean sizes of width and length in the 20–60 μm and 16–270 μm ranges, respectively. Prior to the RS fiber extraction, the grinding step was applied to improve the effectiveness of the combined ultrasound-heating method and the bleaching treatment. According to Chen et al. [2], RS has a strong crystalline structure, making the extraction and purification of cellulose fibers difficult. Likewise, RS presents a high proportion of non-fibrous cells, such as parenchyma, epidermis, and vessel cells [41]. The traditional alkaline pre-treatment, which aims to eliminate hemicellulose and lignin [42], was replaced by a more environmentally friendly aqueous treatment based on the application of ultrasound, followed by reflux heating. Figure 2b reveals that the application of ultrasound and heating promoted a certain degree of fibrillar plane de-bonding and distorted the RS particles, as previously reported [8]. The acoustic cavitation produced by the ultrasound waves, combined with the thermal erosion generated by reflux heating, promoted the disruption of the primary structure of RS and improved the leaching out of non-cellulosic components from the cellulosic matrix [43–45].

Figure 2. Field emission scanning electron microscope (FESEM) micrographs of untreated rice straw (RS) (**a**), RS treated with combined ultrasound-heating method (**b**), bleached cellulose microfibers (CMFs) before (**c**) and after (**d**) the milling step. Images were taken at 300× with insets at 2000×.

As shown in Figure 2c, the bleached cellulose fibers exhibited a swollen, rough, and unregular appearance. The removal of non-fibrous components from the lignocellulosic matrix, such as hemicellulose, lignin, and waxes, gave rise to an entangled material, as observed by other authors [46]. The lignocellulosic matrix comprises cellulose fibrils organized in fibers arranged in a longitudinal orientation and inserted in a cementing matrix consisting of hemicellulose and lignin [2]. Thus, after removing the amorphous components, the hydroxyl (-OH) groups of cellulose could establish extensive hydrogen (-H) bonds among their chains, leading to a more compact and disordered arrangement [47]. This phenomenon can be observed in Figure 2c in the FESEM image taken at higher magnification. Thereafter, the cellulose fibers were ground to break up the agglomerates and obtain thinner CMFs (Figure 2d). Image analysis using both optical microscopy and FESEM techniques served to determine the cumulative distributions of CMF lengths and widths, respectively shown in Figure 3. Therefore, the mechanical defibrillation of the cellulose fiber bundles was evident, which was considered to occur by detaching CMFs into long fibrils with a major cumulative frequency of lengths below 200 μm (Figure 3a). Likewise, there was a marked reduction in the CMF thickness, showing predominant values between 5–15 μm. The magnified image of the ground material, shown in Figure 2d, revealed that CMFs were isolated with a smooth surface pattern, confirming that the milling step was an efficient method to disrupt the tangled microfiber arrangement. In this regard, Jiang et al. [48] observed that mechanical defibrillation was effective at detaching cellulose fibrils from RS treated with the 2,2,6,6-tetramethylpiperidine-1-oxyl (TEMPO)-

mediated oxidation method. The authors reported less uniform defibrillated particles, as also detected herein in Figure 2d. In this sense, the micrographs suggested that, due to the detachment of the fiber bundles into CMFs, the cellulosic material obtained could promote an efficient force transferring from the microfibers to polymer matrices, such as TPS [49].

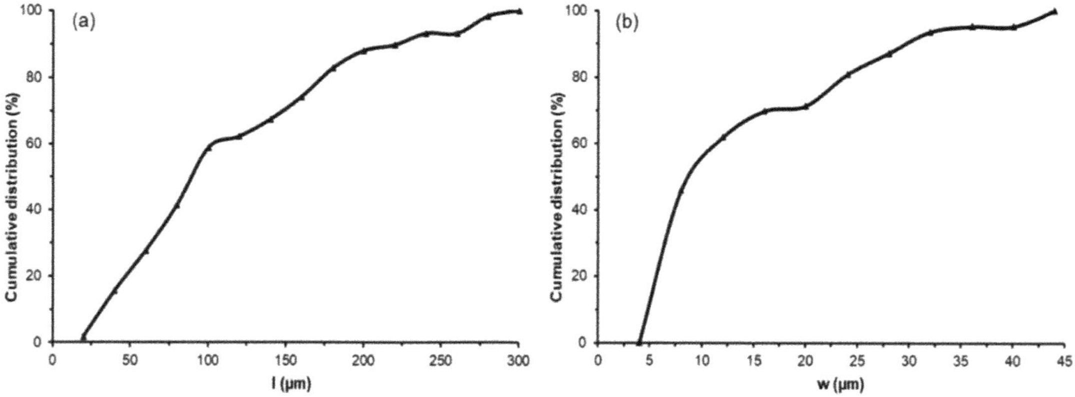

Figure 3. Cumulative distributions of cellulose microfiber (CMF) length (l) (**a**) and width (w) (**b**) obtained from image analyses of optical microscopy and field emission scanning electron microscopy (FESEM), respectively. A minimum of 60 measurements for each parameter was performed.

3.2. Microstructure of the Films

Figure 4 shows the FESEM micrographs of the cross-sections of the TPS-based films incorporating different CMF concentrations (1, 3, and 5 wt%) obtained by cryo-fracture and their corresponding samples using DH-modified starch, that is, TPSDH. The TPS control film samples revealed a smooth and homogeneous cross-section surface pattern, indicating a good starch thermoplasticization, as also reported by other authors for starch films obtained in similar conditions [50,51]. In contrast, the TPSDH films exhibited remaining starch granules, which suggests changes in the starch thermoplasticization behavior after the DH treatment. In this context, Chandanasree et al. [31] reported that the treatment of native starch by DH provokes partial damages in the semi-crystalline and crystalline regions of the granules, which prevent the granule swelling and delays its gelatinization. Furthermore, the DH treatment can oxidize and convert the -OH groups present in the starch chains to carbonyl (C=O) and carboxyl (-COOH) groups, which alters the gelatinization profile [52]. A progressive increase in the number of non-gelatinized starch granules was also observed in the cross-sections of the films of non-modified starch when the CMF concentration increased. However, this number increased in the samples prepared with DH-modified starch.

The starch thermoplasticization occurs during heating, in the extruder or melt-mixing equipment, in the presence of a plasticizer, such as glycerol, at high temperatures (130 °C) and pressures [21]. The microstructural features of the microcomposite films revealed that the CMF incorporation interfered the starch gelatinization with glycerol. This could be explained by the competitive interaction of the plasticizer with both CMFs and starch through -H intermolecular bonds with the cellulose chains, decreasing their availability and subsequently provoking gelatinization of the granular structure of starch. This effect was mainly observed in the TPSDH samples, where the thermal treatment involved changes in its granular structure that already limits its gelatinization.

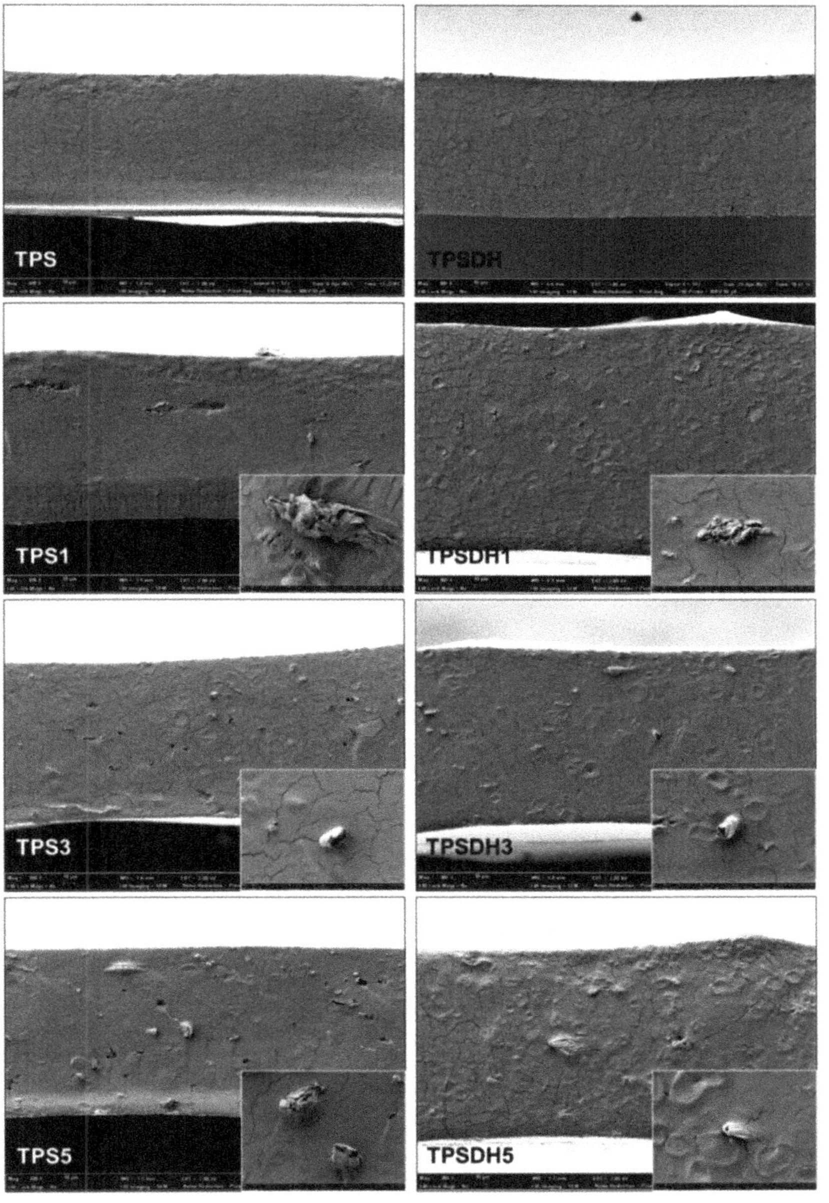

Figure 4. Field emission scanning electron microscope (FESEM) micrographs of the fracture surfaces of thermoplastic starch (TPS) and TPS modified by dry heating (TPSDH) films with contents of cellulose microfibers (CMFs) of 1 wt% (TPS1, TPSDH1), 3 wt% (TPS3, TPSDH3), and 5 wt% (TPS5, TPSDH5). Images were taken at 300× with insets at 2000×.

Figure 4, which shows the fracture surfaces of film samples, revealed that the CMFs incorporated into the films exhibited good compatibility with both types of TPS matrices, that is, non-modified and DH-modified starch. This observation can be deduced from the good interfacial adhesion observed between the fibers and starch matrix due to the absence of gap between the fiber surface and the TPS matrix. Nevertheless, the film cryo-fracture does not seem to promote the breakage of the fibers, but their separation from the matrix, resulting in hollows that arise from the outgoing fiber during the fracture of the film samples. Similar findings were observed by Collazo-Bigliardi et al. [40] and Kargarzadeh et al. [28] in TPS films reinforced with CMFs from rice husk. The apparently good interfacial adhesion properties of fibers in the starch matrices could derive from the obtaining process of CMFs, without alkaline pre-treatment, which promoted the presence of exposed -OH groups on the surface of the fibers [53], thus enhancing the -H bond interactions with the -OH groups of starch.

3.3. Optical Properties: Color and Transparency

Film color and transparency are not only important factors in food packaging designs, but also very useful for evaluating the filler-polymer matrix compatibility. Table 2 summarizes the color coordinates in terms of L^*, C_{ab}^*, and h_{ab}^* as well as the total color difference (ΔE^*) with respect to the neat TPS films without DH treatment and CMFs and the opacity coefficient of the different films. Likewise, the T_i spectra of the different films are shown in Figure 5. Both the CMF incorporation and starch DH modification significantly ($p < 0.05$) affected the evaluated optical properties. Comparison of TPS and TPSDH films without fibers permits to elucidate that DH gave rise to slightly darker films with more saturated, that is, higher C_{ab}^* values, and a yellower color, which suggest that Maillard or sugar caramelization reactions could occur during thermal treatment of native starch. Likewise, as the CMF concentration increased, both TPS and TPSDH films exhibited a slight decrease in their lightness (L^*) while color saturation (C_{ab}^*) increased and hue tended to be more yellowish. This was reflected on the ΔE^* values, in which the color coordinates L^*, a^*, and b^* of the TPS control film were used as a reference. In the TPS films without DH treatment, the ΔE^* values attained after the incorporation of low contents of CMFs were below 1 indicating unnoticeable color changes, whereas only in the microcomposite with 5 wt% CMFs, that is, TPS5, an unexperienced observer can already observe color differences ($2 \leq \Delta E^* \leq 3.5$). In the TPSDH films, as opposite, the combined effect of dispersed phases, that is, CMFs and non-gelatinized starch granules, and colored compounds formed during thermal modification of starch [31,54], gave rise to higher color variations with respect to the neat TPS films. Thus, the TPSDH and its microcomposite with the lowest CMF content showed clear noticeable differences ($3.5 \leq \Delta E^* \leq 5$), whereas the samples with higher fiber contents yielded films with a noticeable different color ($\Delta E^* \geq 5$).

Table 2. Optical properties in terms of lightness (L^*), chroma (C_{ab}^*), hue angle (h_{ab}^*), color difference (ΔE^*), and opacity (O) of thermoplastic starch (TPS) and TPS modified by dry heating (TPSDH) films with different contents of cellulose microfibers (CMFs).

Formulation	L^*	C_{ab}^*	h_{ab}^*	ΔE^*	O
TPS	$88.5 \pm 0.1^{a,1}$	$7.6 \pm 0.1^{f,2}$	$92.6 \pm 0.1^{a,1}$	-	$0.141 \pm 0.001^{b,2}$
TPS1	$88.5 \pm 0.1^{a,3}$	$7.5 \pm 0.2^{f,2}$	$92.8 \pm 0.1^{a,1}$	0.07 ± 0.01^{e}	$0.146 \pm 0.009^{b,2}$
TPS3	$88.1 \pm 0.2^{b,5}$	$8.2 \pm 0.3^{e,2}$	$92.6 \pm 0.2^{a,1}$	0.70 ± 0.02^{e}	$0.151 \pm 0.005^{ab,2}$
TPS5	$87.6 \pm 0.2^{c,7}$	$9.4 \pm 0.3^{d,2}$	$91.5 \pm 0.2^{b,1}$	2.03 ± 0.01^{d}	$0.140 \pm 0.005^{b,2}$
TPSDH	$86.0 \pm 0.1^{d,2}$	$11.9 \pm 0.5^{c,1}$	$89.2 \pm 0.2^{cd,2}$	4.76 ± 0.05^{c}	$0.163 \pm 0.001^{ab,1}$
TPSDH1	$86.2 \pm 0.2^{d,4}$	$11.3 \pm 0.2^{c,1}$	$89.5 \pm 0.3^{c,2}$	4.64 ± 0.10^{c}	$0.163 \pm 0.006^{ab,1}$
TPSDH3	$85.4 \pm 0.1^{e,6}$	$12.4 \pm 0.2^{b,1}$	$88.9 \pm 0.1^{d,2}$	6.24 ± 0.01^{b}	$0.171 \pm 0.007^{a,1}$
TPSDH5	$84.5 \pm 0.1^{f,8}$	$14.2 \pm 0.2^{a,1}$	$88.8 \pm 01^{d,2}$	7.88 ± 0.15^{a}	$0.156 \pm 0.007^{ab,1}$

Different subscript letters indicate significant differences between samples of the same group (TPS or TPSDH films). Different numbers indicated significant differences between TPS and TPSDH samples with the same ratio of CMF (Tukey test, $p < 0.05$).

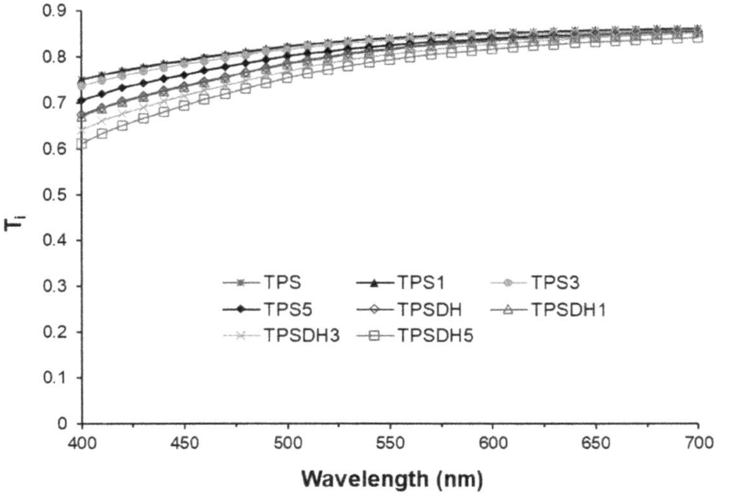

Figure 5. Internal transmittance (T_i) of thermoplastic starch (TPS) and TPS modified by dry heating (TPSDH) films with different contents of cellulose microfibers (CMFs).

The T_i spectra of the films within the wavelength range of 400–700 nm shown in Figure 5 revealed that the TPSDH films without CMFs exhibited lower transparency that the TPS films, with lower values of T_i, mainly at the lowest wavelength. This can be explained by the presence of colored compounds, formed during the starch DH modification process and the presence of dispersed starch granules in the films, which were non-completely gelatinized. These dispersed particles, having refractive indexes different from that of the starch continuous phase, promoted light scattering and film opacity. In both kinds of films, a slight and progressive decrease in light transmission was observed as the CMF concentration increased. This result agrees with the observations reported by other authors for CMF-containing TPS samples [36,55,56], where the filler incorporation as dispersed phase also caused changes in the refractive index of the polymer matrix, promoting light scattering and opacity in the films. Therefore, opacity increased due to the presence of CMFs and the non-gelatinized starch granules. In particular, the opacity parameter, also included in Table 2, increased from 0.141 for the neat TPS film up to 0.171 for the TPSDH3 film.

3.4. *Equilibrium Moisture Content and Water Solubility*

Table 3 shows the equilibrium moisture content and water solubility of the different films prepared with TPS and TPSDH. One can observe that the DH treatment of starch slightly promoted water sorption capacity and water solubility of the films. In this sense, Lim et al. [57] reported that DH treatment of starch can damage its granular structure, which culminates in the exposure of its -OH groups, promoting the ability to interact with water molecules and water sorption capacity. Oh et al. [32] also reported an increase in water solubility of rice starch modified with the DH treatment. Furthermore, the CMF incorporation slightly decreased the water sorption capacity of both TPS and TPSDH films, whereas it provided different effects on water solubility for each kind of starch film, reducing solubility for the TPS films but increasing it in the case of the TPSDH films. The small changes in the water sorption capacity provoked by CMFs can be explained in terms of the lower water affinity of the fibers in comparison with TPS [58], reflected on their low equilibrium water content (3%) at 53% RH [35]. In fact, the resulting water content of the microcomposites could be deduced from the mass balance considering the equilibrium water content of both starch matrix and fibers.

Table 3. Moisture, water solubility, water vapor permeability (WVP), and oxygen permeability (OP) of thermoplastic starch (TPS) and TPS modified by dry heating (TPSDH) films with different contents of cellulose microfibers (CMFs).

Formulation	Moisture (%)	Solubility (g Soluble Film·100 g^{-1} Film)	WVP (g.mm.kPa^{-1}.h^{-1}.m^{-2})	OP × 10^{14} (cm^3.m^{-1}.s^{-1}.Pa^{-1})
TPS	7.7 ± 0.1 ^{a,1}	42 ± 2 ^{a,2}	6.3 ± 0.2 ^{a,b,1}	9.0 ± 0.3 ^{a,1}
TPS1	7.5 ± 0.1 ^{a,1}	28 ± 1 ^{b,2}	6.5 ± 0.1 ^{a,1}	9.2 ± 0.4 ^{a,1}
TPS3	7.7 ± 0.1 ^{a,1}	37 ± 4 ^{ab,2}	5.9 ± 0.2 ^{b,1}	6.3 ± 0.6 ^{b,2}
TPS5	7.1 ± 0.2 ^{b,1}	36 ± 5 ^{ab,2}	5.3 ± 0.3 ^{c,1}	6.4 ± 0.1 ^{b,2}
TPSDH	8.1 ± 0.1 ^{a,2}	55 ± 8 ^{b,1}	0.25 ± 0.01 ^{a,2}	8.7 ± 0.5 ^{a,1}
TPSDH1	7.9 ± 0.1 ^{a,2}	80 ± 8 ^{a,1}	0.33 ± 0.05 ^{a,2}	8.8 ± 0.4 ^{a,1}
TPSDH3	7.7 ± 0.1 ^{b,1}	73 ± 12 ^{ab,1}	0.31 ± 0.02 ^{a,2}	7.7 ± 0.4 ^{a,1}
TPSDH5	7.9 ± 0.1 ^{a,2}	57 ± 6 ^{ab,1}	0.31 ± 0.01 ^{a,2}	8.4 ± 0.1 ^{a,1}

Different subscript letters indicate significant differences between samples of the same group (TPS or TPSDH films). Different numbers indicated significant differences between TPS and TPSDH samples with the same ratio of CMF (Tukey test, $p < 0.05$).

The different effect of the fibers on water solubility of TPS and TPSDH films can be explained by the different interactions of CMFs with both starch matrices. The -OH groups present in CMFs could interact with the hydrophilic groups in starch, making it unavailable to interact with the water molecules [59]. This effect was evident for the water solubility of the TPS films, which decreased significantly ($p < 0.05$) after the CMF incorporation, as previously reported by other authors [56,60,61]. Furthermore, the lower water solubility of these microcomposites due to the CMF presence could make starch less susceptible to hydrolysis, though this property mainly depends on its structural characteristics and modifications [62]. However, in the TPSDH films, the CMF incorporation promoted film water solubility, except for the samples containing 5 wt% CMFs. This different effect could be ascribed to a poorer integration and dispersion of the fibers in the starch matrix, having lower adhesion forces at the interface and, thus, favoring water penetration through the junction zones. The higher presence of non-gelatinized starch granules could also contribute to the higher solubility in the TPSDH films since this dispersed phase can reduce the cohesion forces of the matrix, which are related to the interchain bonds that depend on their interfacial adhesion in the continuous matrix.

3.5. Barrier Properties to Water Vapor and Oxygen Gas

The values of WVP and OP of the TPS films are also shown in Table 3. The non-modified starch films exhibited WVP values ranging from 5.3 to 6.3 g.mm/kPa·h·m^2, close to the range reported by other authors for corn starch films [34,40]. Films of TPS with 3 and 5 wt% CMFs exhibited slightly lower but still significant ($p < 0.05$) WVP values than the neat TPS film, reaching percentages increases in the barrier performance of 6% and 15%, respectively. Since WVP of packaging materials is mainly affected by the solubility coefficient in the polymer matrix and also, but to a lesser extent, by the vapor diffusion rate [63], the permeability decrease attained in the film samples containing CMFs can be ascribed to the chemical interaction by -H bonds of cellulose with hydrophilic groups of the starch-glycerol matrix described above [4,49,64]. Other authors also reported a decrease in WVP of TPS films from different sources incorporated with cellulosic fractions [61,65]. Furthermore, the CMF addition did not influence significantly ($p > 0.05$) the WVP values of the TPSDH films, which already presented values remarkably lower ($p < 0.05$) than the TPS samples. The latter lower permeability to water vapor of DH-modified starch can be ascribed to the large presence of intact starch granules, previously observed in the FESEM images, which would increase the tortuosity factor of the modified matrix, further limiting the diffusion of water molecules through the films. Other authors also reported lower values of VWP for DH-modified rice starch films [66].

As concerns the OP values, the incorporation of CMFs at contents of 3 and 5 wt% promoted a significant ($p < 0.05$) OP decrease of approximately 30% with respect to the neat TPS film. This barrier enhancement can be attributed to the increase of the tortuosity factor associated with the fiber distribution along the TPS matrix, with a good compatibility and dispersion, which would limit the diffusion rate of oxygen molecules through the films. Other studies have also reported a decrease in the OP of starch-based films with cellulosic fractions [36,65], which are in agreement with the results reported herein. However, non-significant differences ($p > 0.05$) in the OP values were observed for the TPSDH films after fiber incorporation. These films exhibited significantly ($p < 0.05$) higher OP values than TPS microcomposite films with 3 and 5 wt% CMFs and similar to those observed for TPS films without and with 1 wt% CMFs. The lack of influence of the incorporated fibers on the barrier properties (both WVP and OP) of the TPSDH films suggests that the expected fiber promotion of the tortuosity factor is probably compensated by lower adhesion forces between the fibers and biopolymer matrix that promoted mass transfer through the union points. Moreover, the larger content of C=O and -COOH groups produced by depolymerization of heat-treated starch chains could promoted the solubility of oxygen molecules in the films, enhancing the transmission rate [67].

Therefore, incorporating CMFs from RS in TPS films represents an adequate strategy to enhance the oxygen barrier capacity of the biopolymer films, being suitable to reduce or modulate oxidation reactions in packaged foods. In contrast, a very low effect of CMFs on the barrier properties of the DH-modified starch films was observed, though these biopolymer films already exhibited a better water vapor barrier capacity (90% reduction of WVP with respect to untreated TPS films).

3.6. Mechanical Properties

The mechanical properties of the films, namely EB, TS, and EM were determined by tensile tests and are summarized in Table 4. Both TPS and TPSDH films exhibited similar stiffness or elastic modulus, though the TPSDH films were slightly more rigid and remarkably more resistant to break but also less extensible. The TPS control film had an EB value of approximately 30%, in line with that observed by Hernández-García et al. [50] for TPS films. Furthermore, the TPSDH film without fibers showed an EB value of 23%, being lower than the neat TPS film. This mechanical embrittlement can be attributed to the changes occurred in the starch chains, such as short-chain amyloses produced by heat-induced hydrolysis of starch [54] or the increase in the content of C=O and -COOH groups during the DH of starch [67]. Other authors [60,68] also suggested that the conversion of native -OH groups to C=O and -COOH ones in oxidized starch may promote -H bonds between amylose and amylopectin molecules, increasing the integrity of the biopolymer matrix and improving the tensile resistance of the films.

Table 4. Mechanical properties in terms of elongation at break (EB%), tensile strength at break (TS), and elastic modulus (EM) of thermoplastic starch (TPS) and TPS modified by dry heating (TPSDH) films with different contents of cellulose microfibers (CMFs).

Formulation	EB (%)	TS (MPa)	EM (MPa)
TPS	$30 \pm 4^{a,1}$	$3.4 \pm 0.6^{c,2}$	$180 \pm 50^{c,1}$
TPS1	$30 \pm 7^{a,1}$	$4.2 \pm 1.0^{bc,2}$	$140 \pm 30^{c,2}$
TPS3	$14 \pm 5^{b,1}$	$6.8 \pm 1.2^{a,2}$	$550 \pm 180^{a,1}$
TPS5	$18 \pm 4^{b,1}$	$4.6 \pm 0.9^{b,2}$	$300 \pm 100^{b,1}$
TPSDH	$23 \pm 4^{a,2}$	$6.4 \pm 0.4^{b,1}$	$190 \pm 30^{f,1}$
TPSDH1	$9 \pm 3^{bc,2}$	$6.9 \pm 0.5^{b,1}$	$320 \pm 70^{e,1}$
TPSDH3	$5 \pm 1^{c,2}$	$10.2 \pm 1.5^{a,1}$	$590 \pm 70^{d,1}$
TPSDH5	$10 \pm 3^{b,2}$	$7.2 \pm 1.2^{b,1}$	$290 \pm 100^{ef,1}$

Different subscript letters indicate significant differences between samples of the same group (TPS or TPSDH films). Different numbers indicated significant differences between TPS and TPSDH samples with the same ratio of CMF (Tukey test, $p < 0.05$).

The incorporation of CMFs at 3 wt% increased notably film stiffness in both kinds of films, being this effect less noticeable at other ratios of fibers. Several studies have also reported higher EM values of starch films containing cellulosic fractions from different sources [36,56,69]. In general, all the here-developed films became more resistant and less stretchable as higher loading of CMFs were incorporated. Higher fiber contents, of 3 and 5 wt% CMFs, decreased the TPS film stretchability by about 50%, whereas a similar decrease was observed for the TPSDH film with only 1 wt% CMFs. The lack of stretchability limits the overall extensibility of the films without markedly reducing the resistance to break. As shown in FESEM micrographs, CMFs were strongly bonded to the polymer matrix, which restricted the movement of starch chains, decreasing their stretchability but also reinforcing the TPS matrix [70]. The highest reinforcing effect was observed for the 3 wt% CMFs in both TPS and TPSDH films. These films particularly exhibited the highest EM (about 550 and 590 MPa for TPS and TPSDH, respectively) and resistance to break (approximately 7 and 10 MPa for TPS and TPSDH, respectively), but also with reduced elongation capacity (14 and 5% for TPS and TPSDH, respectively). In this regard, Fazeli et al. [65] also observed a threshold concentration in cellulose nanofibers from henequen to reinforce starch films.

Therefore, the obtained results in tensile behavior of TPS and TPSDH films demonstrated the reinforcing capacity of CMFs for starch films, which indicates good compatibility and dispersion along the TPS matrix. However, contents higher than 3 wt% led to lower mechanical performances, which can be attributed to a loss of cohesion forces or a lower degree of dispersion and certain agglomeration of the fibers in the TPS matrix.

3.7. Thermal Properties

Figure 6 shows the TGA and first derivative (DTG) curves obtained for the TPS and TPSDH films and their microcomposites films prepared with CMFs. Table 5 gathers the thermal stability parameters, namely T_{onset}, T_{final}, and T_{deg} as well as the mass loss at T_{deg}. One can observe that fiber incorporation did not notably influence the thermal degradation properties of the TPS and TPSDH films. In particular, the TPS films showed the main degradation event ranging at approximately 150–455 °C, corresponding to a mass loss of nearly 80% with a T_{deg} value of 341 °C. This event is associated with the thermal decomposition of glycerol [71], starch [28,72], and CMFs. Figure 6a shows the thermal decomposition profile of the CMF fractions, showing the previously reported mass loss steps [35]: evaporation of bonded water (30–140 °C), cellulose and hemicellulose decomposition (177–350 °C) and lignin degradation (350–500 °C). The second thermal degradation step is referred as the "active pyrolysis zone" since the mass loss rate is high, whereas the third one is called "passive pyrolysis zone" since the percentage of mass loss is smaller, and the mass loss rate is also much lower compared to that in the second zone [73]. In relation to the degradation profile of the TPS films, the sample containing 5 wt% CMFs exhibited slightly higher mass loss from 180 to 280 °C, which can be attributed to the thermal degradation of the cellulose fractions, more noticeable due to its high content. However, this TPS5 sample was more stable than the other TPS films in the thermal range of 280–310 °C (see Figure 6a,c), suggesting that the CMF incorporation delayed the thermal degradation of the shortest chains of starch produced during its thermo-processing [74].

Regarding the TPSDH films, these samples exhibited a slight initial mass loss (3–5%) in the thermal range from approximately 50 to 150 °C, which is ascribed to evaporation of water adsorbed in the inner regions of the non-gelatinized starch granules and it may also derive from thermal decomposition of low-molecular weight (M_W) components produced during the DH treatment [54]. This mass loss can be observed in Figure 6b, where the mass loss curves of the TPS and TPSDH films were compared. As shown in Table 4 and also Figure 6d, TPSDH films exhibited slightly higher T_{deg} values than the TPS films. This thermal stability enhancement may be attributed to the presence of C=O and -COOH groups produced by the heat treatment of starch that allows for more -H bonds between the polymer chains [68]. Furthermore, the non-gelatinized starch granules, with more crystalline regions, could also contribute to the resultant thermal behavior. In the TPSDH films, the concentration increases in CMFs also implied a slight increase in the thermal stability, which could be due to the increasing presence of starch granules in the matrix.

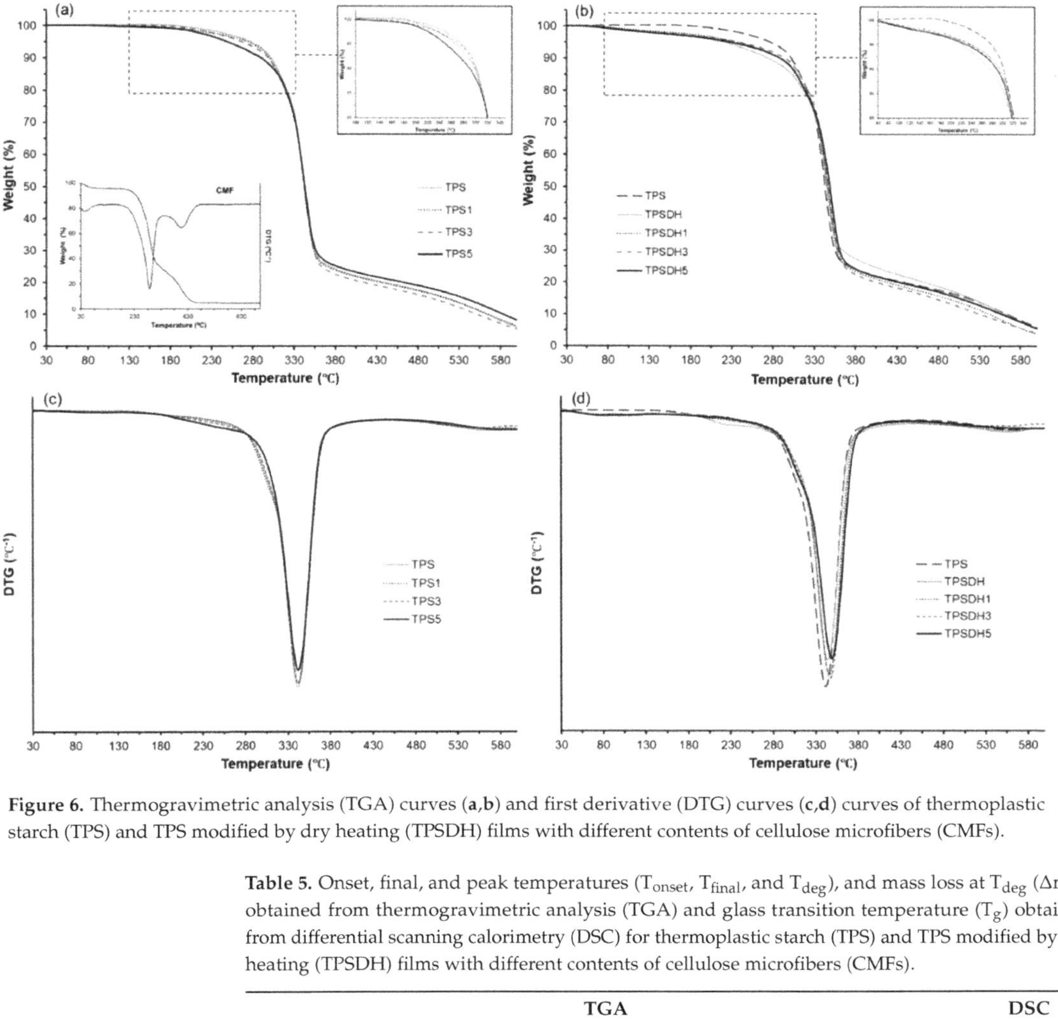

Figure 6. Thermogravimetric analysis (TGA) curves (**a**,**b**) and first derivative (DTG) curves (**c**,**d**) curves of thermoplastic starch (TPS) and TPS modified by dry heating (TPSDH) films with different contents of cellulose microfibers (CMFs).

Table 5. Onset, final, and peak temperatures (T_{onset}, T_{final}, and T_{deg}), and mass loss at T_{deg} (Δm%) obtained from thermogravimetric analysis (TGA) and glass transition temperature (T_g) obtained from differential scanning calorimetry (DSC) for thermoplastic starch (TPS) and TPS modified by dry heating (TPSDH) films with different contents of cellulose microfibers (CMFs).

	TGA				DSC
Formulation	T_{onset} (°C)	T_{final} (°C)	T_{peak} (°C)	Δm (%)	T_g
TPS	156 ± 1	455 ± 3	341 ± 1	80.7 ± 1.0	92 ± 4
TPS1	153 + 1	456 ± 1	343 ± 1	81.3 ± 0.7	86 ± 1
TPS3	152 ± 4	453 ± 1	342 ± 1	82 ± 1.0	85 ± 6
TPS5	154 ± 1	452 ± 3	343 ± 1	79.2 ± 0.2	82 ± 9
TPSDH	179 ± 2	455 ± 1	345 ± 1	80 ± 1	106 ± 1
TPSDH1	183 ± 1	454 ± 2	346 ± 2	78 ± 2	79 ± 2
TPSDH3	181 ± 1	449 ± 2	347 ± 1	80 ± 1	-
TPSDH5	182 ± 1	450 ± 1	349 ± 3	79 ± 3	124 ± 4

Finally, DSC analysis was also performed to evaluate the effect of the starch modification by DH and CMF incorporation on the thermal phase transitions of starch. The first heating scans, gathered in Figure 7, showed the typical endotherm of starch gelatinization associating peak temperatures in the 55–110 °C range for the film samples of TPSDH and TPS with the highest CMF loadings (TPS3 and TPS5). This agrees with the above-reported FESEM observations that revealed the presence of non-gelatinized starch granules in the starch matrix. It can also be observed that the gelatinization enthalpy increased with the CMF ratio due to the aforementioned higher presence of starch granules in the films. In the second heating scan, the TPS films also exhibited the glass transition (T_g) of the already gelatinized starch, being this transition less appreciable, that is, having a lower change in specific heat (ΔC_p) for TPSDH samples. The T_g values (midpoint) of the TPSDH sample was slightly higher than that of TPS, which is coherent with the stronger interactions promoted in the matrix when starch was DH modified. This result also agrees with the more resistance but less ductile behavior of the films attained during the mechanical characterization. As expected, the addition of CMFs did not induce notable changes in the starch's T_g, when compared with the corresponding control sample, since the fibers are not miscible with starch. The small variations of T_g in the TPS composites with CMFs can be attributed to potential interactions between fibers or their remaining water contents and plasticizer that could modify the plasticization level of starch [31]. In this context, Collazo-Bigliardi et al. [40] found similar results for corn starch films filled with fibers derived from rice husk.

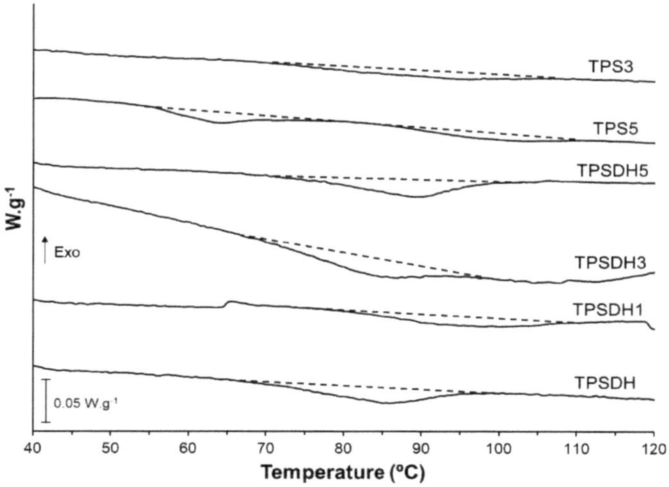

Figure 7. Differential scanning calorimetry (DSC) thermograms taken during first heating scan of thermoplastic starch (TPS) and TPS modified by dry heating (TPSDH) films with different contents of cellulose microfibers (CMFs).

4. Conclusions

CMFs with good reinforcing capacity for starch films were obtained from RS, a waste material derived from the agricultural and food industry, using a green process and mechanical defibrillation. Fibrils with a major cumulative frequency of lengths below 200 μm and 5–15 μm of thickness were obtained. Incorporation of CMFs into starch matrices at 3 wt%, the most optimal content, led to stiffer and more resistance to break films for both non-modified and DH-modified corn starch films. Likewise, fibers at 3 and 5 wt% improved the oxygen and water vapor barrier capacity of TPS films but did not modify barrier properties of TPSDH films. Fibers interfered starch gelatinization during the starch–glycerol thermo-processing, more markedly in DH-modified starch with modified granular structure, giving rise to non-gelatinized starch granules into the film matrix. The dispersed phase (starch granules or fibers) also modified the optical properties of the TPS films, making them slightly less transparent. Therefore, it can be concluded that RS is a good source of CMFs that can be used to improve properties of starch films with different purposes, such as their use as food packaging material. This contributes to the valorization of the agricultural and food waste into added-value materials, useful for improving the performance of biodegradable food packaging materials. Likewise, this study further confirms that the DH process of starch represents a green physical method to enhance the water barrier capacity and tensile strength of starch films. Future studies will deal with the hydrolysis resistance of these microcomposite films and their practical application as packaging materials in food preservation.

Author Contributions: P.A.V.F.: conceptualization, methodology, formal analysis, investigation, writing—original draft preparation, writing—review and editing; C.I.L.F.A.: conceptualization, methodology, formal analysis, investigation, data analysis; S.T.-G.: writing—original draft preparation, writing—review and editing; C.G.-M.: conceptualization, methodology, investigation, writing—original draft preparation, writing—review and editing; A.C.: conceptualization, methodology, investigation, writing—original draft preparation, writing—review and editing. All authors have read and agreed to the published version of the manuscript.

Funding: This research was funded by Spanish Ministry of Science and Innovation (MICI), grant number PID2019-105207RB-I00.

Institutional Review Board Statement: Not applicable.

Informed Consent Statement: Not applicable.

Acknowledgments: P.A.V.F. is grateful to Generalitat Valenciana (GVA) for the Grisolia P/2019/115 grant. C.L.F.A. acknowledges the São Paulo Research Foundation (FAPESP) for the Research Internships Abroad (BEPE) fellowship (2020/10498-0). S.T.-G. acknowledges MICI for his Ramón y Cajal contract (RYC2019-027784-I).

Conflicts of Interest: The authors declare no conflict of interest.

References

1. Hafemann, E.; Battisti, R.; Marangoni, C.; Machado, R.A.F. Valorization of Royal Palm Tree Agroindustrial Waste by Isolating Cellulose Nanocrystals. *Carbohydr. Polym.* **2019**, *218*, 188–198. [CrossRef]
2. Chen, X. Study on Structure and Thermal Stability Properties of Cellulose Fibers from Rice Straw. *Carbohydr. Polym.* **2011**, *85*, 245–250. [CrossRef]
3. Saini, J.K.; Saini, R.; Tewari, L. Lignocellulosic Agriculture Wastes as Biomass Feedstocks for Second-Generation Bioethanol Production: Concepts and Recent Developments. *3 Biotech.* **2015**, *5*, 337–353. [CrossRef] [PubMed]
4. Ng, H.-M.; Sin, L.T.; Tee, T.-T.; Bee, S.-T.; Hui, D.; Low, C.-Y.; Rahmat, A.R. Extraction of Cellulose Nanocrystals from Plant Sources for Application as Reinforcing Agent in Polymers. *Compos. Part B Eng.* **2015**, *75*, 176–200. [CrossRef]
5. Sharma, B.; Vaish, B.; Singh, U.K.; Singh, P.; Singh, R.P. Recycling of Organic Wastes in Agriculture: An Environmental Perspective. *Int. J. Env. Res.* **2019**, *13*, 409–429. [CrossRef]
6. Casabar, J.T.; Ramaraj, R.; Tipnee, S.; Unpaprom, Y. Enhancement of Hydrolysis with Trichoderma Harzianum for Bioethanol Production of Sonicated Pineapple Fruit Peel. *Fuel* **2020**, *279*, 118437. [CrossRef]
7. Basumatary, B.; Basumatary, S.; Das, B.; Nath, B.; Kalita, P. Waste *Musa Paradisiaca* Plant: An Efficient Heterogeneous Base Catalyst for Fast Production of Biodiesel. *J. Clean. Prod.* **2021**, *305*, 127089. [CrossRef]
8. Freitas, P.A.V.; González-Martínez, C.; Chiralt, A. Application of Ultrasound Pre-Treatment for Enhancing Extraction of Bioactive Compounds from Rice Straw. *Foods* **2020**, *9*, 1657. [CrossRef]
9. Prakash, A.; Vadivel, V.; Banu, S.F.; Nithyanand, P.; Lalitha, C.; Brindha, P. Evaluation of Antioxidant and Antimicrobial Properties of Solvent Extracts of Agro-Food by-Products (Cashew Nut Shell, Coconut Shell and Groundnut Hull). *Agric. Nat. Resour.* **2018**, *52*, 451–459. [CrossRef]
10. Ilyas, R.A.; Sapuan, S.M.; Ibrahim, R.; Abral, H.; Ishak, M.R.; Zainudin, E.S.; Atikah, M.S.N.; Mohd Nurazzi, N.; Atiqah, A.; Ansari, M.N.M.; et al. Effect of Sugar Palm Nanofibrillated Cellulose Concentrations on Morphological, Mechanical and Physical Properties of Biodegradable Films Based on Agro-Waste Sugar Palm (Arenga Pinnata (Wurmb.) Merr) Starch. *J. Mater. Res. Technol.* **2019**, *8*, 4819–4830. [CrossRef]
11. Kassab, Z. Cellulosic Materials from Pea (Pisum Sativum) and Broad Beans (Vicia Faba) Pods Agro-Industrial Residues. *Mater. Lett.* **2020**, *280*, 128–539. [CrossRef]
12. De Souza Lima, M.M.; Borsali, R. Rodlike Cellulose Microcrystals: Structure, Properties, and Applications. *Macromol. Rapid Commun.* **2004**, *25*, 771–787. [CrossRef]
13. FAOSTAT. Available online: http://www.fao.org/faostat/en/#data/QC/visualize (accessed on 4 November 2020).
14. Peanparkdee, M.; Iwamoto, S. Bioactive Compounds from By-Products of Rice Cultivation and Rice Processing: Extraction and Application in the Food and Pharmaceutical Industries. *Trends Food Sci. Technol.* **2019**, *86*, 109–117. [CrossRef]
15. Sarkar, N.; Ghosh, S.K.; Bannerjee, S.; Aikat, K. Bioethanol Production from Agricultural Wastes: An Overview. *Renew. Energy* **2012**, *37*, 19–27. [CrossRef]

16. El-Tayeb, T.S.; Abdelhafez, A.A.; Ali, S.H.; Ramadan, E.M. Effect of Acid Hydrolysis and Fungal Biotreatment on Agro-Industrial Wastes for Obtainment of Free Sugars for Bioethanol Production. *Braz. J. Microbiol.* **2012**, *43*, 1523–1535. [CrossRef] [PubMed]
17. Neto, W.P.F.; Silvério, H.A.; Vieira, J.G.; da Costa e Silva Alves, H.; Pasquini, D.; de Assunção, R.M.N.; Dantas, N.O. Preparation and Characterization of Nanocomposites of Carboxymethyl Cellulose Reinforced with Cellulose Nanocrystals. *Macromol. Symp.* **2012**, *319*, 93–98. [CrossRef]
18. Galdeano, M.C.; Mali, S.; Grossmann, M.V.E.; Yamashita, F.; García, M.A. Effects of Plasticizers on the Properties of Oat Starch Films. *Mater. Sci. Eng. C* **2009**, *29*, 532–538. [CrossRef]
19. Forssell, P.; Lahtinen, R.; Lahelin, M.; MyllaÈrinen, P. Oxygen Permeability of Amylose and Amylopectin Fims. *Carbohydr. Polym.* **2002**, *47*, 125–129. [CrossRef]
20. Zainuddin, S.Y.Z. Potential of Using Multiscale Kenaf Fibers as Reinforcing Filler in Cassava Starch-Kenaf Biocomposites. *Carbohydr. Polym.* **2013**, *92*, 2299–2305. [CrossRef]
21. Carvalho, A.J.F. Starch: Major Sources, Properties and Applications as Thermoplastic Materials. In *Monomers, Polymers and Composites from Renewable Resources*; Elsevier: Amsterdam, The Netherlands, 2008; pp. 321–342, ISBN 978-0-08-045316-3.
22. González, K.; Retegi, A.; González, A.; Eceiza, A.; Gabilondo, N. Starch and Cellulose Nanocrystals Together into Thermoplastic Starch Bionanocomposites. *Carbohydr. Polym.* **2015**, *117*, 83–90. [CrossRef]
23. Teixeira, E.d.M.; Pasquini, D.; Curvelo, A.A.S.; Corradini, E.; Belgacem, M.N.; Dufresne, A. Cassava Bagasse Cellulose Nanofibrils Reinforced Thermoplastic Cassava Starch. *Carbohydr. Polym.* **2009**, *78*, 422–431. [CrossRef]
24. Collazo-Bigliardi, S. Isolation and Characterisation of Microcrystalline Cellulose and Cellulose Nanocrystals from Coffee Husk and Comparative Study with Rice Husk. *Carbohydr. Polym.* **2018**, *191*, 205–215. [CrossRef]
25. Ghanbarzadeh, B.; Almasi, H.; Entezami, A.A. Improving the Barrier and Mechanical Properties of Corn Starch-Based Edible Films: Effect of Citric Acid and Carboxymethyl Cellulose. *Ind. Crop. Prod.* **2011**, *33*, 229–235. [CrossRef]
26. Othman, S.H.; Majid, N.A.; Tawakkal, I.S.M.A.; Basha, R.K.; Nordin, N.; Shapi'I, R.A. Tapioca Starch Films Reinforced with Microcrystalline Cellulose for Potential Food Packaging Application. *Food Sci. Technol.* **2019**, *39*, 605–612. [CrossRef]
27. Chen, J.; Wang, X.; Long, Z.; Wang, S.; Zhang, J.; Wang, L. Preparation and Performance of Thermoplastic Starch and Microcrystalline Cellulose for Packaging Composites: Extrusion and Hot Pressing. *Int. J. Biol. Macromol.* **2020**, *165*, 2295–2302. [CrossRef] [PubMed]
28. Kargarzadeh, H. Starch Biocomposite Film Reinforced by Multiscale Rice Husk Fiber. *Compos. Sci. Technol.* **2017**, *151*, 147–155. [CrossRef]
29. Syafri, E.; Wahono, S.; Irwan, A.; Asrofi, M.; Sari, N.H.; Fudholi, A. Characterization and Properties of Cellulose Microfibers from Water Hyacinth Filled Sago Starch Biocomposites. *Int. J. Biol. Macromol.* **2019**, *137*, 119–125. [CrossRef] [PubMed]
30. Abdul Khalil, H.P.S.; Bhat, A.H.; Ireana Yusra, A.F. Green Composites from Sustainable Cellulose Nanofibrils: A Review. *Carbohydr. Polym.* **2012**, *87*, 963–979. [CrossRef]
31. Chandanasree, D.; Gul, K.; Riar, C.S. Effect of Hydrocolloids and Dry Heat Modification on Physicochemical, Thermal, Pasting and Morphological Characteristics of Cassava (Manihot Esculenta) Starch. *Food Hydrocoll.* **2016**, *52*, 175–182. [CrossRef]
32. Oh, I.K.; Bae, I.Y.; Lee, H.G. Effect of Dry Heat Treatment on Physical Property and in Vitro Starch Digestibility of High Amylose Rice Starch. *Int. J. Biol. Macromol.* **2018**, *108*, 568–575. [CrossRef]
33. Maniglia, B.C.; Lima, D.C.; Matta Junior, M.D.; Le-Bail, P.; Le-Bail, A.; Augusto, P.E.D. Preparation of Cassava Starch Hydrogels for Application in 3D Printing Using Dry Heating Treatment (DHT): A Prospective Study on the Effects of DHT and Gelatinization Conditions. *Food Res. Int.* **2020**, *128*, 108803. [CrossRef] [PubMed]
34. Sun, Q.; Xu, Y.; Xiong, L. Effect of Microwave-Assisted Dry Heating with Xanthan on Normal and Waxy Corn Starches. *Int. J. Biol. Macromol.* **2014**, *68*, 86–91. [CrossRef] [PubMed]
35. Freitas, P.A.V.; González-Martínez, C.; Chiralt, A. Applying ultrasound-assisted processing to obtain cellulose fibers from rice straw to be used as reinforcing agents. *Innov. Food Sci. Emerg. Technol.* **2021**, in press.
36. Collazo-Bigliardi, S.; Ortega-Toro, R.; Chiralt Boix, A. Reinforcement of Thermoplastic Starch Films with Cellulose Fibres Obtained from Rice and Coffee Husks. *J. Renew. Mater.* **2018**, *6*, 599–610. [CrossRef]
37. Rojas-Lema, S.; Quiles-Carrillo, L.; Garcia-Garcia, D.; Melendez-Rodriguez, B.; Balart, R.; Torres-Giner, S. Tailoring the Properties of Thermo-Compressed Polylactide Films for Food Packaging Applications by Individual and Combined Additions of Lactic Acid Oligomer and Halloysite Nanotubes. *Molecules* **2020**, *25*, 1976. [CrossRef]
38. Talón, E.; Trifkovic, K.T.; Nedovic, V.A.; Bugarski, B.M.; Vargas, M.; Chiralt, A.; González-Martínez, C. Antioxidant Edible Films Based on Chitosan and Starch Containing Polyphenols from Thyme Extracts. *Carbohydr. Polym.* **2017**, *157*, 1153–1161. [CrossRef]
39. McHugh, T.H.; Avena-Bustillos, R.; Krochta, J.M. Hydrophilic Edible Films: Modified Procedure for Water Vapor Permeability and Explanation of Thickness Effects. *J. Food Sci.* **1993**, *58*, 899–903. [CrossRef]
40. Collazo-Bigliardi, S.; Ortega-Toro, R.; Chiralt, A. Improving Properties of Thermoplastic Starch Films by Incorporating Active Extracts and Cellulose Fibres Isolated from Rice or Coffee Husk. *Food Packag. Shelf Life* **2019**, *22*, 100383. [CrossRef]
41. Jin, S.; Chen, H. Structural Properties and Enzymatic Hydrolysis of Rice Straw. *Process Biochem.* **2006**, *41*, 1261–1264. [CrossRef]
42. Zhang, Z.; Smith, C.; Li, W. Extraction and Modification Technology of Arabinoxylans from Cereal By-Products: A Critical Review. *Food Res. Int.* **2014**, *65*, 423–436. [CrossRef]
43. Cheung, Y.-C.; Wu, J.-Y. Kinetic Models and Process Parameters for Ultrasound-Assisted Extraction of Water-Soluble Components and Polysaccharides from a Medicinal Fungus. *Biochem. Eng. J.* **2013**, *79*, 214–220. [CrossRef]

44. Hayat, K.; Abbas, S.; Hussain, S.; Shahzad, S.A.; Tahir, M.U. Effect of Microwave and Conventional Oven Heating on Phenolic Constituents, Fatty Acids, Minerals and Antioxidant Potential of Fennel Seed. *Ind. Crop. Prod.* **2019**, *140*, 111610. [CrossRef]
45. Machado, I. Characterization of the Effects Involved in Ultrasound-Assisted Extraction of Trace Elements from Artichoke Leaves and Soybean Seeds. *Ultrason. Sonochem.* **2019**, *59*, 104752. [CrossRef]
46. Moslemi, A. Addition of Cellulose Nanofibers Extracted from Rice Straw to Urea Formaldehyde Resin; Effect on the Adhesive Characteristics and Medium Density Fiberboard Properties. *Int. J. Adhes. Adhes.* **2020**, *99*, 102582. [CrossRef]
47. Bochek, A.M. Effect of Hydrogen Bonding on Cellulose Solubility in Aqueous and Nonaqueous Solvents. *Russ. J. Appl. Chem.* **2003**, *76*, 1711–1719. [CrossRef]
48. Jiang, F.; Han, S.; Hsieh, Y.-L. Controlled Defibrillation of Rice Straw Cellulose and Self-Assembly of Cellulose Nanofibrils into Highly Crystalline Fibrous Materials. *RSC Adv.* **2013**, *3*, 12366. [CrossRef]
49. Littunen, K.; Hippi, U.; Saarinen, T.; Seppälä, J. Network Formation of Nanofibrillated Cellulose in Solution Blended Poly(Methyl Methacrylate) Composites. *Carbohydr. Polym.* **2013**, *91*, 183–190. [CrossRef]
50. Hernández-García, E.; Vargas, M.; Chiralt, A. Thermoprocessed Starch-Polyester Bilayer Films as Affected by the Addition of Gellan or Xanthan Gum. *Food Hydrocoll.* **2021**, *113*, 106509. [CrossRef]
51. Menzel, C.; González-Martínez, C.; Vilaplana, F.; Diretto, G.; Chiralt, A. Incorporation of Natural Antioxidants from Rice Straw into Renewable Starch Films. *Int. J. Biol. Macromol.* **2020**, *146*, 976–986. [CrossRef] [PubMed]
52. Oladebeye, A.O.; Oshodi, A.A.; Amoo, I.A.; Karim, A.A. Functional, Thermal and Molecular Behaviours of Ozone-Oxidised Cocoyam and Yam Starches. *Food Chem.* **2013**, *141*, 1416–1423. [CrossRef] [PubMed]
53. Abraham, E. Extraction of Nanocellulose Fibrils from Lignocellulosic Fibres: A Novel Approach. *Carbohydr. Polym.* **2011**, *86*, 1468–1475. [CrossRef]
54. Hung, P.V.; My, N.T.H.; Phi, N.T.L. Impact of Acid and Heat-Moisture Treatment Combination on Physicochemical Characteristics and Resistant Starch Contents of Sweet Potato and Yam Starches. *Starch—Stärke* **2014**, *66*, 1013–1021. [CrossRef]
55. Benito-González, I.; López-Rubio, A.; Martínez-Sanz, M. High-Performance Starch Biocomposites with Celullose from Waste Biomass: Film Properties and Retrogradation Behaviour. *Carbohydr. Polym.* **2019**, *216*, 180–188. [CrossRef]
56. Fourati, Y.; Magnin, A.; Putaux, J.-L.; Boufi, S. One-Step Processing of Plasticized Starch/Cellulose Nanofibrils Nanocomposites via Twin-Screw Extrusion of Starch and Cellulose Fibers. *Carbohydr. Polym.* **2020**, *229*, 115554. [CrossRef]
57. Lim, H.S.; Han, J.-A.; BeMiller, J.N.; Lim, S.-T. Physical Modification of Waxy Maize Starch by Dry Heating with Ionic Gums. *J. Appl. Glycosci.* **2006**, *53*, 281–286. [CrossRef]
58. Ma, X.; Yu, J.; Kennedy, J.F. Studies on the Properties of Natural Fibers-Reinforced Thermoplastic Starch Composites. *Carbohydr. Polym.* **2005**, *62*, 19–24. [CrossRef]
59. Avérous, L.; Fringant, C.; Moro, L. Plasticized Starch–Cellulose Interactions in Polysaccharide Composites. *Polymer* **2001**, *42*, 6565–6572. [CrossRef]
60. El Halal, S.L.M.; Colussi, R.; Deon, V.G.; Pinto, V.Z.; Villanova, F.A.; Carreño, N.L.V.; Dias, A.R.G.; Zavareze, E.d.R. Films Based on Oxidized Starch and Cellulose from Barley. *Carbohydr. Polym.* **2015**, *133*, 644–653. [CrossRef] [PubMed]
61. Müller, C.M.O.; Laurindo, J.B.; Yamashita, F. Effect of Cellulose Fibers Addition on the Mechanical Properties and Water Vapor Barrier of Starch-Based Films. *Food Hydrocoll.* **2009**, *23*, 1328–1333. [CrossRef]
62. Cornejo-Ramírez, Y.I.; Martínez-Cruz, O.; Del Toro-Sánchez, C.L.; Wong-Corral, F.J.; Borboa-Flores, J.; Cinco-Moroyoqui, F.J. The Structural Characteristics of Starches and Their Functional Properties. *Cyta—J. Food* **2018**, *16*, 1003–1017. [CrossRef]
63. Talja, R.A.; Helén, H.; Roos, Y.H.; Jouppila, K. Effect of Various Polyols and Polyol Contents on Physical and Mechanical Properties of Potato Starch-Based Films. *Carbohydr. Polym.* **2007**, *67*, 288–295. [CrossRef]
64. Hedenqvist, M.S. Barrier Packaging Materials. In *Handbook of Environmental Degradation of Materials*; Elsevier: Amsterdam, The Netherlands, 2012; pp. 833–862, ISBN 978-1-4377-3455-3.
65. Fazeli, M.; Keley, M.; Biazar, E. Preparation and Characterization of Starch-Based Composite Films Reinforced by Cellulose Nanofibers. *Int. J. Biol. Macromol.* **2018**, *116*, 272–280. [CrossRef]
66. Li, Y.; Shoemaker, C.F.; Ma, J.; Shen, X.; Zhong, F. Paste Viscosity of Rice Starches of Different Amylose Content and Carboxymethylcellulose Formed by Dry Heating and the Physical Properties of Their Films. *Food Chem.* **2008**, *109*, 616–623. [CrossRef]
67. Reyes, I.; Hernandez-Jaimes, C.; Vernon-Carter, E.J.; Bello-Perez, L.A.; Alvarez-Ramirez, J. Air Oxidation of Corn Starch: Effect of Heating Temperature on Physicochemical Properties and In Vitro Digestibility. *Starch—Stärke* **2021**, *73*, 2000237. [CrossRef]
68. Zamudio-Flores, P.B.; Vargas-Torres, A.; Pérez-González, J.; Bosquez-Molina, E.; Bello-Pérez, L.A. Films Prepared with Oxidized Banana Starch: Mechanical and Barrier Properties. *Starch—Stärke* **2006**, *58*, 274–282. [CrossRef]
69. Adeyi, A.J.; Durowoju, M.O.; Adeyi, O.; Oke, E.O.; Olalere, O.A.; Ogunsola, A.D. Momordica Augustisepala L. Stem Fibre Reinforced Thermoplastic Starch: Mechanical Property Characterization and Fuzzy Logic Artificial Intelligent Modelling. *Results Eng.* **2021**, *10*, 100222. [CrossRef]
70. Lu, Y.; Weng, L.; Cao, X. Biocomposites of Plasticized Starch Reinforced with Cellulose Crystallites from Cottonseed Linter. *Macromol. Biosci.* **2005**, *5*, 1101–1107. [CrossRef]
71. Maria, V.D.; Bernal, C.; Francois, N.J. Development of Biodegradable Films Based on Chitosan/Glycerol Blends Suitable for Biomedical Applications. *J. Tissue Sci. Eng.* **2016**, *7*. [CrossRef]
72. Merci, A.; Marim, R.G.; Urbano, A.; Mali, S. Films Based on Cassava Starch Reinforced with Soybean Hulls or Microcrystalline Cellulose from Soybean Hulls. *Food Packag. Shelf Life* **2019**, *20*, 100321. [CrossRef]

73. Melendez-Rodriguez, B.; Torres-Giner, S.; Aldureid, A.; Cabedo, L.; Lagaron, J.M. Reactive Melt Mixing of Poly(3-Hydroxybutyrate)/Rice Husk Flour Composites with Purified Biosustainably Produced Poly(3-Hydroxybutyrate-*Co*-3-Hydroxyvalerate). *Materials* **2019**, *12*, 2152. [CrossRef]
74. Menzel, C.; González-Martínez, C.; Chiralt, A.; Vilaplana, F. Antioxidant Starch Films Containing Sunflower Hull Extracts. *Carbohydr. Polym.* **2019**, *214*, 142–151. [CrossRef] [PubMed]

applied sciences

MDPI

Article

Manufacturing and Characterization of Green Composites with Partially Biobased Epoxy Resin and Flaxseed Flour Wastes

Diego Lascano [1,2], Daniel Garcia-Garcia [1,*], Sandra Rojas-Lema [1,2], Luis Quiles-Carrillo [1], Rafael Balart [1] and Teodomiro Boronat [1]

1 Technological Institute of Materials (ITM), Universitat Politècnica de València (UPV), Plaza Ferrándiz y Carbonell 1, 03801 Alcoy, Spain; dielas@epsa.upv.es (D.L.); sanrole@epsa.upv.es (S.R.-L.); luiquic1@epsa.upv.es (L.Q.-C.); rbalart@mcm.upv.es (R.B.); tboronat@dimm.upv.es (T.B.)

2 Escuela Politécnica Nacional, Quito 17-01-2759, Ecuador

* Correspondence: dagarga4@epsa.upv.es; Tel.: +(34)96-652-84-34

Received: 29 April 2020; Accepted: 25 May 2020; Published: 26 May 2020

Featured Application: In the present work, green-composites have been developed from a partially biobased epoxy resin reinforced with flaxseed flour wastes. The attractive aesthetic appearance, similar to wood, and the balanced overall properties of the obtained composites may be interesting for use in sectors such as decoration, furniture or automotive industry.

Abstract: In the present work, green-composites from a partially biobased epoxy resin (BioEP) reinforced with lignocellulosic particles, obtained from flax industry by-products or wastes, have been manufactured by casting. In this study, the flaxseed has been crushed by two different mechanical milling processes to achieve different particle sizes, namely coarse size (CFF), and fine size (FFF) particle flaxseed flour, with a particle size ranging between 100–220 μm and 40–140 μm respectively. Subsequently, different loadings of each particle size (10, 20, 30, and 40 wt%) were mixed with the BioEP resin and poured into a mold and subjected to a curing cycle to obtain solid samples for mechanical, thermal, water absorption, and morphological characterization. The main aim of this research was to study the effect of the particle size and its content on the overall properties of composites with BioEP. The results show that the best mechanical properties were obtained for composites with a low reinforcement content (10 wt%) and with the finest particle size (FFF) due to a better dispersion into the matrix, and a better polymer-particle interaction too. This also resulted in a lower water absorption capacity due to the presence of fewer voids in the developed composites. Therefore, this study shows the feasibility of using flax wastes from the seeds as a filler in highly environmentally friendly composites with a wood-like appearance with potential use in furniture or automotive sectors.

Keywords: Flax seed; biobased epoxy; green-composite; waste valorization; size particle

1. Introduction

During the last years, there has been a significant increase in social concern for the environmental problem generated by petrochemical polymeric materials [1–3]. For this reason, one of the main objectives of the scientific community is the research and development of new highly environmentally friendly materials which could be suitable to replace petroleum-based polymers to reduce their carbon footprint [4]. Many of these researches focus on the field of polymer composites reinforced with lignocellulosic particles giving rise to the so-called wood plastic composites (WPC). A series of advantages make the use of lignocellulosic reinforcements very attractive for their use as reinforcement, such as their low cost, low density, non-abrasive properties, non-toxic, biodegradable, and their environmentally friendly nature [5,6]. Besides, the lignocellulosic particles usually provide an aesthetic wood-like surface finish, which makes them very interesting for use in sectors where aesthetics is an essential factor, such as the furniture or automotive sectors. Furthermore, WPCs have several advantages over wood, such as low maintenance, high dimensional stability, and high resistance to biological attack [7,8]. There are many research works that have focused on the effect of different types of lignocellulosic particles on the properties of both thermoplastic and thermoset matrices. For example, interesting works have been developed with rice husks [9–11], peanut shells [7,12,13], almond shells [14–16], hazelnut shells [17,18], date palm seeds [19], lemon peel [20], *Posidonia oceanica* [21,22], olive stones [23], among others. Most of these lignocellulosic particles come from agricultural by-products or wastes that are currently used for animal feeding and just left on controlled landfills. Therefore, the use of these wastes, widely available as a source of fillers for composite materials, can be a new economic opportunity for the agricultural sector and thereby contributing to generate sustainable circular economies [24–26].

Flax (*Linum usitatissimum L.*) is a worldwide cultivated plant, mainly for obtaining fibers and oil-rich seeds [27]. It is estimated that the world production of flaxseed, also known as linseed, was approximately 3.2 million tons during 2018 [28]. Traditionally flaxseeds have been used as an oil source, due to its high triglyceride content (between 30 and 41 wt%), for use in paints and coatings, linoleum, inks, varnishes, cosmetics, soap production, among others [29,30]. However, over the past few years, the flaxseed oil has also gained popularity as a nutritional supplement due to its high content of α-linolenic acid (ALA), an omega-3 fatty acid beneficial in preventing cardiovascular disease or hypertension [31]. In addition, many studies have shown that flaxseed oil has a positive effect on diseases such as hyperlipidaemia, colon and breast cancer, or atherosclerosis [32]. The main by-product generated during oil extraction is flaxseed cake, which is the solid mass left after the seeds are pressed during the oil extraction process. For this reason, flaxseed cake is widely available and, a cost-effective and environmentally friendly material to be used in composites [33]. The flaxseed cake is rich in cyanogenic glycosides, which may be degraded to toxic hydrogen cyanide (HCN) upon ingestion and may represent a risk to human health if used in food applications without prior detoxification treatment [31]. For this reason, currently, a part of this by-product is spray-dried to obtain flour (flaxseed flour), which is used as a low-value by-product for obtaining livestock feed or fertilizers [34,35]. In other cases, this waste is used for composting or simply incinerated [36]. Therefore, the flaxseed flour can potentially be a candidate for use in composites.

Furthermore, a significant part of the plastic matrices used in WPCs is thermoplastic petroleum-derived polymers such as polypropylene (PP), polyethylene (PE), polystyrene (PS), polyvinylchloride (PVC), among others. However, there has been a tendency in recent years to replace these matrices by biobased and biodegradable (actually soil compostable) polymers such as polylactic acid (PLA) or polyhydroxyalkanoates (PHA) [37], to achieve fully biodegradable WPCs. Natural fiber reinforced plastics (NFRP) represent a wider group that includes WPCs and thermosetting-based composites, as well. For thermosetting resins, this trend is focused on the use of fully or partially biobased resins [38–42]. With this, it is possible to reduce the dependence on fossil fuels. Besides this, these fully/partially thermosetting resins, positively contribute to reducing the carbon footprint generated by their petrochemical counterparts such as phenolics (PF), epoxies (EP), or unsaturated polyesters (UP). A promising source for biobased epoxy resins and plasticizers are epoxidized vegetable oils, which are obtained by epoxidizing the C-C double bonds of unsaturated fatty acids contained in triglycerides, the main component of vegetable oils [43–45]. However, due to the long aliphatic chains in the triglycerides and their low cross-linking density, epoxy resins obtained from vegetable oils tend to have low glass transition temperatures (T_g) and lower mechanical properties than traditional epoxy resins [46]. Therefore, to obtain environmentally friendly materials with balanced mechanical and thermal properties, one of the most efficient approaches is copolymerization of epoxidized vegetable oils with petroleum-derived epoxy resins, thus giving rise to a partially biobased epoxy resin with high mechanical properties [47]. Niedermann et al. [48] investigated the effect of the epoxidized soybean oil (ESO) content (0, 25, 50, 75, and 100 wt%) on bisphenol-A based aromatic epoxy resin (DGEBA). The results showed a decrease in T_g as the ESO content increased but indicated that T_g of the DGEBA/ESO system (75/25) was very similar (138 °C) to the base DGEBA resin (140 °C). Obviously, this was remarkably higher than the T_g of neat crosslinked ESO (75 °C). Similar trends were also observed for mechanical properties such as tensile strength and impact energy, where the addition of 25 wt% ESO decreased the mechanical properties of DGEBA resin but made them more resistant than ESO resin. Regarding the manufacturing of materials with these cast resins, Wu et al. [49] reported impregnation bamboo with epoxy with previous delignification processes to obtain almost transparent bamboo goods. Salasinska et al. [50], reported new environmentally friendly composites with epoxy resins and *Pinus sibirica* lignocellulosic fillers at a constant wt% of 20%. These composites were manufactured at a laboratory scale by mixing the components and finally, a conventional cast method was used to pour the liquid mixture into a mold. Casting is the most widely used method to manufacture composites with thermosetting resins and lignocellulosic particles. Kumar et al. [51] reported manufacturing by casting and characterization of epoxy composites with up to 12.5 wt% wood particles. Centrifugal casting is a way to obtain gradation of the filler due to centrifugal forces as reported by Stabik et al. [52]. They report that as well as the particle gradation occurs, gradation of properties occurs too which could be interesting from different standpoints (decorative, percolation thresholds in some parts, and so on.

The main objective of the present work was to obtain green composites from a partially biobased epoxy resin reinforced with flaxseed flour waste using the casting method. Specifically, the effect of particle size and the content of flaxseed flour on the mechanical, morphological, thermal, and water absorption properties of the bioepoxy/flaxseed flour composites, has been investigated. In the present work, four compositions (10, 20, 30, and 40 wt%) and two-particle sizes (CFF and FFF) were investigated.

2. Materials and Methods

2.1. Materials

The matrix material used in this study was a commercial epoxy resin Resoltech® 1070 ECO (viscosity of 1750 mPa s and a density of 1.18 g cm^{-3} at 23 °C). The hardener was an amine-based system Resoltech® 1074 ECO (viscosity of 50 mPa s and density of 0.96 g cm^{-3} at 23 °C). Both were supplied by Castro Composites (Pontevedra, Spain). The epoxy resin was based on a mixture of a diglycidyl ether of bisphenol A (DGEBA) and a plant-based epoxy reactive diluent from vegetable oil epoxidation. The resin to hardener weight ratio was 100:35 (parts by weight), and as indicated by the supplier the cured resin contained 31 wt% biobased content (according to ASTM D6866-12).

Flaxseed (FS) used in this work was supplied by Sorribas S.A. (Polinyà, Spain). The raw FS was crushed in a grinder (Moulinex, Allençon, France) to obtain flaxseed flour (FF) with an average particle size of 157 μm (CFF–coarse flaxseed flour). Then, a part of the obtained CFF was ground using an ultra-centrifugal mill from Retsch Gmbh model Mill ZM 1000 (Haan, Germany) with a sieve of 250 μm and a rotating speed of 10,000 rpm, obtaining FF with an average particle size of 91 μm (FFF—fine flaxseed flour). Figure 1 shows an image of the raw FS and the obtained FF after each of the grinding processes.

Figure 1. Images corresponding to (**a**) raw flaxseeds (FS); (**b**) flaxseed flour (FF) obtained by grinder (CFF); (**c**) flaxseed flour (FF) obtained by ultracentrifuge (FFF).

2.2. BioEP/FF Composites' Manufacturing

Different contents (10, 20, 30, and 40 wt%) of CFF and FFF were added to the liquid epoxy resin Resoltech® 1070 ECO and were mixed at room temperature in a planetary mixer KAPL 5KPM5 from KitchenAid (Benton Harbor, MI, USA) with a total volume of 4.8 L. First, the FF was added to the mixer with the biobased resin and was subjected to initial homogenization at 40 rpm for 5 min. Then, Resoltech® 1074 ECO hardener was added in the stoichiometric ratio (100:35 wt/wt) to the BioEP/FF mixture and subjected to a second mixing cycle at 60 rpm for 2.5 min. After this two-stage mixing cycle, the resin was subjected to vacuum to remove air bubbles in a vacuum chamber MCP 00ILC from HEK-GmbH (Lubeck, Germany) for 5 min. A maximum vacuum of −1 bar was applied. The resulting resin-filler mixture was poured into a silicone mold designed with standardized cavities for mechanical characterization and then subjected to a curing cycle in an oven at 80 °C for 1 h. Cured samples were post-cured at 150 °C for 30 min. The optimal curing and post-curing conditions of the partially biobased epoxy resin were selected according to a previous study [53]. Finally, cured samples were demolded from the silicone molds and used for different characterizations.

The nomenclature of the samples is denoted with the acronym BioEP_iCFF for samples with FF obtained by a simple grinding process, and BioEP_iFFF for samples reinforced with FF obtained by using an ultra-centrifugal mill, where *i* represents the filler content (10, 20, 30 and 40 wt%).

2.3. BioEP/FF Composites' Characterization

2.3.1. Mechanical Properties

Flexural properties of BioEP resin and BioEP/FF composites were obtained at room temperature in a universal test machine Ibertest ELIB 30 (S.A.E. Ibertest, Madrid, Spain) equipped with a 5 kN load cell following the guidelines of the ISO 178. Rectangular samples with dimensions 80 × 40 × 4 mm^3 were subjected to a three-point bending flexural test with a crosshead speed of 5 mm min^{-1}. At least five specimens of each composition were tested, and characteristic average values were calculated.

Impact-absorbed energy of different samples was obtained in a 1-J Charpy's impact pendulum from Metrotec S.A. (San Sebastián, Spain) as indicated in ISO 179:1993. The values of the impact-absorbed energy of each sample were calculated as the average of the energies obtained for five different specimens.

Shore D hardness values of BioEP resin and BioEP/FF composites were obtained with a Shore D hardness durometer model 676-D from J. Bot Instruments S.A. (Barcelona, Spain) according to ISO 868. At least five different measurements were taken at room temperature, and average values were calculated.

2.3.2. Thermal Properties

Thermal stability at elevated temperatures of BioEP and BioEP/FF composites was studied by thermogravimetric analysis (TGA) using a TGA/SDT 851 thermobalance from Mettler-Toledo Inc. (Schwerzenbach, Switzerland). Samples with an average weight ranging from 7 to 9 mg were heated from 30 to 700 °C at a constant heating rate of 10 °C min^{-1}. All samples were tested in triplicate in a nitrogen atmosphere with a constant nitrogen flow rate of 66 mL min^{-1}. The onset degradation temperature (T_0) was assumed at a weight loss of 5 wt%, and the maximum degradation rate temperature (T_{max}) was obtained as the corresponding peak in the first derivative from TGA curves (DTG).

2.3.3. Thermo-Mechanical Properties

Dynamic mechanical thermal analysis (DMTA) of BioEP and BioEP/FF composites was carried out in torsion mode in an oscillatory rheometer AR G2 by TA Instruments (New Castle, DE, USA) equipped with a special clamp system for solid samples working in a combination of torsion and shear. Rectangular samples (40 × 10 × 4 mm^3) were subjected to a temperature sweep program from 30 °C to 140 °C at a constant heating rate of 2 °C min^{-1}. The frequency of the dynamic stress was set to 1 Hz, and a maximum shear strain (γ) of 0.1% was used in all tests. The evolution of the storage modulus (G′) and the dynamic damping factor (*tan* δ) were recorded as a function of increasing temperature.

The effect of temperature on the dimensional stability of BioEP/FF composites was studied by thermomechanical analysis (TMA) using a Q400 TMA analyzer from TA Instruments (New Castle, DE, USA). Samples with dimensions of 4 × 10 × 10 mm^3 were subjected to a heating ramp from 0 °C to 140 °C, at a constant heating rate of 2 °C min^{-1}, with an applied load of 20 mN. The coefficient of linear thermal expansion (CLTE) was calculated as the slope of the linear relationship between the expansion and temperature, both below and above T_g. All measurements were done in triplicate to obtain reliable values.

2.3.4. Morphological Properties

The morphology of fractured surfaces from an impact test of BioEP resin, different BioEP/FF composites, and flaxseed flour particles was observed using a field emission scanning electron microscope (FESEM) ZEISS model ULTRA55 (Eindhoven, The Netherlands) working at an acceleration voltage of 2 kV. Before the morphological characterization, all samples were surface coated with a thin layer of platinum in a high vacuum sputter coater EM MED20 from Leica Microsystems (Milton Keynes, UK) to provide electrical conductivity to samples.

2.3.5. Water Uptake

Water absorption of samples was carried out in triplicate by immersion of samples (80 × 10 × 4 mm^3) in distilled water at room temperature following ISO 62:2008. Samples were extracted at different times and appropriately weighed using an analytical balance with an accuracy of ± 0.001 g, after removing the residual water with a dry cloth. Before the initial water immersion, samples were dried at 60 °C for 24 h to remove residual moisture. Water absorption percentage was calculated by using the following expression:

$$Water\ uptake\ (\%) = \frac{(W_t - W_0)}{W_0} \times 100 \quad (1)$$

where W_t is the dry weight of the sample after the corresponding time t, and W_0 is the initial weight of the sample before water immersion. The evolution of water uptake was followed in a total period of 12 weeks.

The diffusion coefficient (D) for all samples was calculated by the application of the first Fick's law using the following equation [54]:

$$D = \pi\left[\frac{mh}{4}\right]^2 \tag{2}$$

where m is a slope value, that can be calculated from the plot of W_t/W_s (dry weight after the corresponding time/saturation weight of the sample) versus $t^{1/2}$, and h stands for the initial thickness of the sample.

The previous equation for the calculation of D is only valid for a one-dimensional shape. To obtain the accurately corrected diffusion coefficient (D_c) for three-dimensional shapes, the Stefan approximation was applied, which assumes that the diffusion rates are the same for all directions [18]:

$$D_c = D\left[1 + \frac{h}{L} + \frac{h}{w}\right]^{-2} \tag{3}$$

where L and w are the length and width of each sample, respectively.

2.3.6. Color Properties

The influence of the FF content and size in the color of BioEP/FF composites were studied in a colorimeter model KONICA CM-3600d Colorflex-DIFF2 from Hunter Associates Laboratory (Virginia, EEUU). The CIELab color scale was used to measure the degree of L^* (lightness), a^* (color coordinate from red to green) and b^* (color coordinate from yellow to blue). The total color difference (ΔE) was calculated using the following equation:

$$\Delta E = \sqrt{(\Delta L^*)^2 + (\Delta a^*)^2 + (\Delta b^*)^2} \tag{4}$$

where ΔL^*, Δa^*, and Δb^* are the differences between the corresponding color parameter of the composites and the color parameter values of the reference material, i.e., BioEP matrix. Measurements were done in triplicate.

3. Results and Discussion

3.1. Morphology of FF Particles

Figure 2 shows the FESEM images of the flax flour powder, as well as the particle size distribution after each grinding process. Figure 2 includes representative images of each particle size at different magnifications while the particle size distribution plots were obtained by taking at least 50 measurements on different FESEM images corresponding to each particle size, namely fine (F) or coarse (C), using the software analysis included in the FESEM microscope. The following parameters were obtained (area, angle, and length), and the histogram plots included the length.

As can be seen in the FESEM images at lower magnifications (Figure 2a,d), coarse particles obtained by simple grinding (CFF) as well as fine flax flour particles with smaller dimensions (FFF), tend to form aggregates due to their high hydrophilicity. Quantitatively we can see from the size distribution made by measuring randomly chosen particles (Figure 2c,f), that CFF offers a particle size of 100–220 μm with an average particle size around 157 μm and remarkably higher particle content in the 140–160 μm range. On the other hand, FFF particle distribution shows that their size changes in the 40–140 μm range with smaller average particle size, of about 91 μm, and the most abundant content of particles is in the range of 80–100 μm. Therefore, it is evident that with the grinding of FF by ultra-centrifugation, finer particle sizes were obtained, which resulted in a better filler dispersion in the matrix, as well as improved polymer-filler interactions, with a positive effect on the general properties of the developed composites.

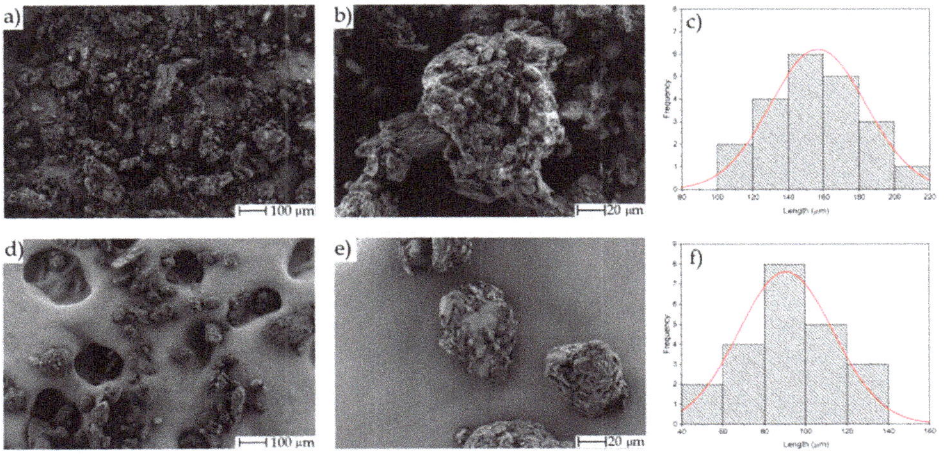

Figure 2. Field emission scanning electron microscope (FESEM) images corresponding to (**a**) CFF at 100× with a scale marker of 100 µm; (**b**) CFF at 500× with a scale marker of 20 µm; (**c**) particle size histogram of CFF; (**d**) FFF at 100× with a scale marker of 100 µm; (**e**) FFF at 500× with a scale marker of 20 µm; (**f**) particle size histogram of FFF.

The grinding process has a relevant effect on both particle aggregate and geometry. Despite this, these differences can be seen in Figure 2 since in general, both CFF and FFF particles had an irregular morphology with a rough surface and the presence of granular fractures (typical morphology of hard lignocellulosic particles after being subjected to crushing processes [54–56]). Figure 3 shows in a more detailed way the above-mentioned effects. Figure 3a shows the morphology of directly ground flax particles after the cold press process. One can see these particles show very irregular shapes. Moreover, it is possible to find particles with high size and very small particles. This is because no particle size separation has been carried out after this grinding process which led to coarse flax flour (CFF). In addition, aggregate formation is evident (in fact, the adhesive carbon tape cannot be seen).

Regarding fine flax flour (FFF), as shown in Figure 3b, it is evident that the shape is more likely spherical, the particle size range is narrower, and the aggregation phenomenon is less pronounced. The sieving process after ultra-centrifugation gives more homogeneity on the obtained morphologies and less aggregation.

Figure 3. FESEM images corresponding to (**a**) CFF at 100× with a scale marker of 100 µm; (**b**) FFF at 100× with a scale marker of 100 µm.

3.2. Mechanical Properties

Table 1 shows the mechanical behavior of unfilled BioEP resin and BioEP/FF composites reinforced with different contents of CFF and FFF. Regarding the flexural properties, it can be seen that FF addition to the BioEP matrix results in a significant decrease in flexural strength compared to the unfilled BioEP resin. This decrease is much more pronounced as the FF content (both CFF and FFF) increases. As expected, the lowest flexural strength is obtained in composites reinforced with 40 wt% of CFF and FFF, obtaining a flexural strength of 22.5 and 22.9 MPa, respectively, which represents a decrease of 76.4% and 75.9% in comparison to the flexural strength of the unfilled BioEP resin (95.2 MPa). This decrease in flexural strength is due to the lack of adhesion between the lignocellulosic filler and the surrounding matrix, which causes stress concentration phenomena that promote breakage [20]. Another issue with a high influence on the mechanical properties of composites is the filler aspect ratio. High filler aspect ratios (above 6) result in a better stress transfer from the matrix to the filler/reinforcement, thus improving their mechanical properties. However, this stresses transfer is poorer in the case of low aspect ratio filler, such as the FF used, with an aspect ratio comprised between 1–2 [57]. If the two types

FF, i.e., coarse and fine FF (CFF and FFF, respectively) are compared, it can be seen that the flexural strength is always lower for all composites containing CFF. This is because the bigger particle size causes a larger weak interfacial area between the BioEP resin and the reinforcement, i.e., it decreases the adhesion between the hydrophobic matrix and the hydrophilic reinforcement which generates an increase in the stress concentration phenomena that negatively affect the mechanical properties of composites [58]. The biggest difference in flexural strength between composites with CFF and FFF is obtained for samples with 10 wt% of FF. For this composition, the flexural strength was 24.2% higher for FFF-reinforced composite compared to CFF-reinforced composite.

Table 1. Summary of the main mechanical properties of BioEP and BioEP/FF composites reinforced with different content of coarse (CFF) and fine (FFF) flaxseed flour particles obtained by flexural, impact, and hardness tests.

Code	Flexural Properties		Impact Energy (kJ m^{-2})	Hardness (Shore D)
	FS (MPa)	E_f (MPa)		
BioEP	95.2 ± 3.4	2985 ± 115	21.8 ± 3.4	83.2 ± 1.0
BioEP_10CFF	40.4 ± 1.3	2840 ± 242	2.7 ± 0.9	83.3 ± 0.6
BioEP_20CFF	31.5 ± 0.5	2878 ± 86	2.6 ± 0.6	82.3 ± 1.0
BioEP_30CFF	28.0 ± 2.0	2930 ± 93	2.4 ± 0.2	82.2 ± 1.0
BioEP_40CFF	22.5 ± 0.8	2290 ± 127	2.2 ± 0.3	81.0 ± 1.0
BioEP_10FFF	50.2 ± 1.4	3579 ± 53	3.6 ± 0.5	83.3 ± 1.0
BioEP_20FFF	38.7 ± 7.0	2975 ± 84	3.4 ± 0.1	83.7 ± 0.5
BioEP_30FFF	31.8 ± 1.5	2696 ± 20	2.7 ± 0.4	83.0 ± 1.1
BioEP_40FFF	22.9 ± 2.4	2108 ± 204	2.4 ± 0.2	81.6 ± 0.8

Regarding the flexural modulus, different trends can be observed for each of the fillers. The flexural modulus of CFF-reinforced composites is hardly affected by 10, 20, and 30 wt% filler and the modulus values are similar to those obtained for BioEP resin (2840–2985 MPa). However, the flexural modulus of the composite reinforced with 40 wt% CFF decreases significantly to 2290 MPa, representing a decrease of 23.3% compared to the unfilled BioEP resin. This is directly related to an embrittlement process which is much more evident at higher filler loadings. It is important to bear in mind that the flexural modulus is directly related to the supported stress and inversely related to the flexural deformation. BioEP is intrinsically brittle, therefore, the change in the flexural deflection before fracture is very low. This is even reduced in BioEP/FF composites, but the most important parameter is a clear decrease of the flexural strength from 95.2 MPa down to 22.9 MPa. For this reason, at 40 wt% CFF, the modulus decreases in a remarkable way since the flexural stress is remarkably reduced while the elongation is almost identical to neat BioEP, which is an intrinsically brittle material.

In the case of FFF-reinforced composites, the addition of 10 wt% FFF to the BioEP matrix significantly increases the flexural modulus from 2985 MPa (unreinforced BioEP) resin up to 3579 MPa, which is a % increase of nearly 20%. This increase in the flexural modulus compared to the 10 wt% CFF reinforced composite may be due to the better dispersion of fine particles into the matrix and improved polymer-particle interaction due to its smaller size, which results in comparatively higher flexural strength values, and consequently higher modulus and stiffness. It should be noted that the sample reinforced with 10 wt% FFF has similar and even higher flexural mechanical properties than those obtained in commercial WPCs currently used in furniture applications [59,60]. For higher FFF contents, it is observed that the flexural modulus decreases as the reinforcement content increases, obtaining the lowest modulus for the sample reinforced with 40 wt% FFF, 2108 MPa, which shows the same behavior above-mentioned.

Regarding the impact energy (Charpy test), it can be seen in Table 1, that the incorporation of both CFF and FFF in the BioEP resin results in a remarkable decrease in impact-absorbed energy from 21.8 kJ m^{-2} corresponding to the unfilled BioEP resin to values around 3 kJ m^{-2} for FF reinforced composites. This behavior is typical of polymeric composites filled with lignocellulosic particles due to the lack of (or very poor) interfacial adhesion between the reinforcement and the matrix, which gives rise to stress concentration points, promoting the formation of microcracks at the interface when impact conditions are applied that easily induce crack propagation, thus decreasing their impact resistance [61,62]. Impact energy absorption is also influenced by the filler content in the matrix. As can be seen, the impact energy absorption becomes lower as the reinforcement content increases, obtaining the lowest impact-absorbed energy for composites reinforced with 40 wt% CFF and FFF, with an impact energy of 2.2 and 2.4 kJ m^{-2} respectively, which represents a decrease of nearly 90% in both cases with respect to the BioEP resin. This decrease in impact energy with the higher filler content is due to the greater lack of interaction between the filler and the matrix, which results in a higher void content, thus increasing the stress concentration phenomena. Particle size is another factor affecting the impact of energy absorption. When comparing the composites, the energy absorption values for the different percentages of reinforcement are slightly lower in the case of CFF-reinforced composites. This is because the coarse particles have less dispersion in the matrix as well as a larger surface area that leaves a greater amount of surface-exposed between the filler and the matrix, negatively affecting the mechanical properties [63]. It is worthy to note that the better results regarding mechanical properties, obtained with FFF were expected as the morphology of FFF is rounded (almost spherical in most cases) while CFF show very irregular shapes with angular geometries that contribute to micro crack formation and subsequent growth.

Regarding the Shore D hardness, it can be seen in Table 1 that the incorporation of the FF filler into the BioEP matrix hardly affects hardness. Only a slight decrease is seen in samples reinforced with 40 wt% CFF and FFF, in which a hardness of 81.0 and 81.6 Shore D, respectively, was obtained, which means a decrease of 2.6% and 1.9%, respectively, in comparison to the BioEP resin hardness (83.2 Shore D). The slight decrease in hardness in composites reinforced with a high amount of FF may be due to the lower lignocellulosic reinforcement hardness compared to the thermosetting matrix used [58]. By considering these mechanical properties, it seems that composites with 10 wt% FFF offer the best-balanced properties. Despite this, other compositions must not be discarded as they offer a higher biobased content and wood-like surface finish. Obviously, these high wt% FF composites would not be suitable for technical applications since there brittle and with low tensile strength, but they can find interesting applications in the decorative sector and leather goods (buckles, buttons, among others).

3.3. Thermal Properties

The thermal stability at high temperatures of FF, BioEP resin, and different BioEP/FF composites was obtained by thermogravimetric analysis (TGA). The temperature effect on the mass of each sample is shown in Table 2, while Figure 4 shows the corresponding TGA and DTG curves for FF, BioEP resin and BioEP/FF with varying CFF and FFF content from 10 to 40 wt%.

Table 2. Thermal parameters of BioEP and BioEP/FF composites reinforced with different contents of coarse (CFF) and fine (FFF) flaxseed flour, obtained by thermogravimetry (TGA).

Code	T_0 [1] (°C)	T_{max} (°C)	wt% Residual Mass
BioEP	306.0 ± 2.1	327.5 ± 1.7	8.5 ± 0.3
BioEP_10CFF	292.7 ± 2.8	322.2 ± 2.5	12.9 ± 0.4
BioEP_20CFF	273.7 ± 1.9	322.2 ± 1.9	13.5 ± 0.5
BioEP_30CFF	268.0 ± 3.1	322.0 ± 2.5	15.0 ± 0.6
BioEP_40CFF	249.2 ± 2.0	323.2 ± 1.4	18.7 ± 0.6
BioEP_10FFF	279.2 ± 1.0	322.2 ± 1.8	10.1 ± 0.5
BioEP_20FFF	270.1 ± 2.4	323.0 ± 3.2	12.7 ± 0.4
BioEP_30FFF	264.0 ± 1.6	321.1 ± 1.8	15.7 ± 0.3
BioEP_40FFF	244.2 ± 2.2	319.0 ± 2.1	20.3 ± 0.4

[1] T_0, calculated at 5% mass loss.

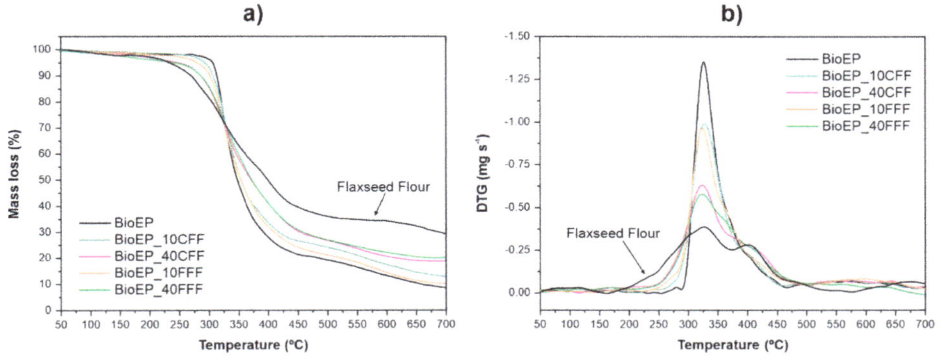

Figure 4. Thermal degradation of BioEP, FF, and BioEP/FF composites reinforced with a different content of coarse (CFF) and fine (FFF) flaxseed flour (**a**) thermogravimetry (TG) weight loss and (**b**) differential thermogravimetry (DTG) first derivative curves.

As shown in Figure 4a, the FF showed four typical degradation stages of lignocellulosic materials. In the first stage, produced at 50–150 °C, the moisture contained in the material evaporated, which was reflected by a mass loss of around 2.3 wt% [64]. In the second stage thermal depolymerization of hemicelluloses took place in the temperature range from 150 to 375 °C with a weight loss of about 41.4% [65]. In the third stage, which was located at 375–450 °C, cellulose degradation occurred, this stage being observed by a weight loss of 18.9% [61]. Finally, the lignin degradation was detected, which begins at around 250 °C, but due to its complex structure, it degraded more slowly, producing a progressive weight loss up to 500 °C. Its decomposition being overlapped with that of other compounds [14]. As shown in the TGA graph, the residual mass of FF was high, which can be due to its high mineral content [66]. On the other hand, as can be seen in Table 2, the thermal degradation of the BioEP resin occurred in a single step, starting its degradation (T_0) around 306 °C and with a maximum degradation temperature (T_{max}) of 327.5 °C. Regarding BioEP/FF composites degradation, it can be seen that the FF filler addition to the matrix results in a slight decrease of the composites' thermal stability, which is reflected by a T_0 decrease.

This is due to the low thermal stability of the lignocellulosic reinforcement, whose degradation onset temperature begins around 234 °C, which affects the overall thermal stability of composites negatively. Besides, the increased filler content in composites results in a reduced weight fraction of the BioEP resin, causing a more significant decrease in T_0 as the filler content increases [20]. Comparing the two types of composites obtained according to the filler size, it can be seen that the thermal stability was slightly lower, i.e., lower T_0, for composites with different FFF contents. This lower thermal stability was more evident for composites with 10 wt% FFF, where a decrease in T_0 of 27 °C compared to the T_0 of the BioEP resin can be observed, while this decrease is only 13 °C for the same composite containing 10 wt% CFF. By observing the DTG curves (Figure 4b) it can be seen that the thermal degradation of BioEP/FF composites takes place in two stages. These stages were more evident in the reinforced samples with high filler contents (40 wt%). In the first stage, BioEP resin was thermally degraded along with the low molecular weight components of FF such as hemicelluloses, while in the second stage the thermal degradation of cellulose and lignin occurred. For this reason, this stage was more evident in composites with high filler content. Regarding the maximum degradation temperature (T_{max}), obtained from the peak of the first degradation stage of the DTG curves, it can be seen that the incorporation of the reinforcement into the BioEP matrix slightly decreases this temperature (it almost remains constant). In this case, the T_{max} of composites did not vary significantly with the particle size or content, obtaining a T_{max} of around 322 °C for all developed composites, except for the composite reinforced with 40 wt% of FFF, which has a T_{max} slightly lower of about 319 °C. Therefore, after thermogravimetric analysis, the results suggest that as the lignocellulosic reinforcement content increases, the thermal stability of the composites decreases, while the residual mass at high temperatures (700 °C) increases. Nevertheless, the overall thermal stability of these composites is not compromised by incorporating FF in both coarse and fine particle size.

3.4. Morphological Properties

The particle dispersion and its interaction with the matrix are two of the main aspects that influence the composite's mechanical properties. To study these phenomena, a morphological study was carried out using FESEM on impact fractured surfaces of BioEP resin and BioEP/FF composites filled with 10 wt% and 40 wt% CFF and FFF. Figure 5a shows the BioEP resin fracture surface, which is characterized by a smooth surface with the presence of cleavage planes characteristic of a brittle fracture.

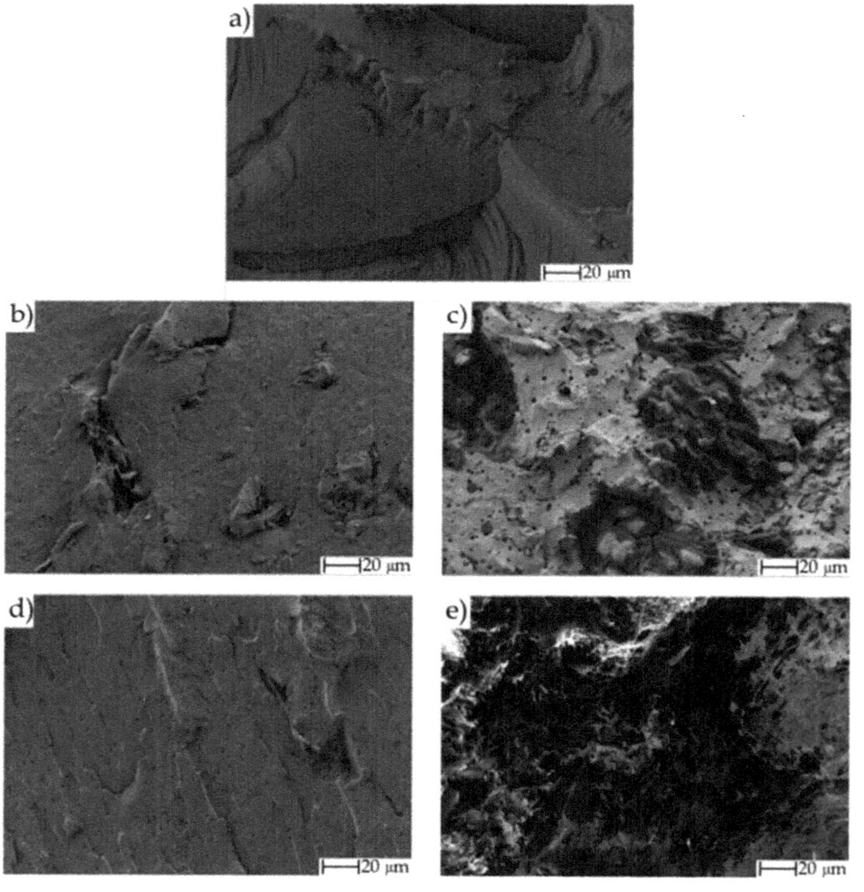

Figure 5. FESEM images at 500× of impact-fracture surface of: (**a**) BioEP; (**b**) BioEP_10CFF; (**c**) BioEP_40CFF; (**d**) BioEP_10FFF and (**e**) BioEP_40FFF.

After the addition of 10 wt% of both CFF and FFF filler (Figure 5b,d), it can be seen that the fracture surface acquires a noticeable roughness, and small holes appear randomly located on the surface, which corresponds to the particles pulled out after the impact test. These voids/holes are representative of poor adhesion between the polymeric matrix and the lignocellulosic filler, which results, as above-mentioned, in low stresses transfer from the matrix to the filler, thus negatively affecting the overall mechanical properties of the obtained composites. Comparing the two types of composites reinforced with 10 wt% FF, it can be seen that fine particles (FFF) reinforced composites, a better particle dispersion in the matrix is achieved due to the smaller particle size, which is reflected in the absence of aggregates in the fracture surface. Such aggregates are observed on the fracture surface of the coarse particles (CFF) reinforced composites, as the larger particle size promotes their aggregation. The higher lack of adhesion between these aggregates and the matrix due to the increase of the exposed surface area intensifies the stress concentration phenomena that negatively affects the mechanical properties, as it has been evidenced by a lower impact energy absorption and lower flexural strength than the composite reinforced with 10 wt% FFF. Higher filler contents (Figure 5c,e) lead to an increase in the fracture surface roughness with the presence of particles embedded in the matrix. An increase in the size of the voids/holes was also observed, possibly because the higher filler content promoted aggregate formation. This increase in the void size results in reduced interfacial adhesion between the reinforcement and the matrix for composites reinforced with high FF contents, which is reflected in poor mechanical properties. Comparing both composites reinforced with 40 wt% of FF, the presence of large aggregates in the CFF reinforced sample can be seen more clearly, while in the FFF reinforced sample, the dispersion of particles was more homogeneous. Therefore, the increase of the filler content in the matrix results in a significant decrease in ductile and resistant mechanical properties, as shown by the evolution of impact energy absorption and flexural strength, due to a reduction in interfacial adhesion between filler and matrix by a decrease in the resin's ability to fully embed the particles. This also occurs with coarse particles (CFF), which have a higher surface area, exposing more surface area to weak bonding with the matrix, resulting in lower flexural strength and higher water absorption capacity.

3.5. Thermo-Mechanical Properties

Figure 6 shows the evolution of the storage modulus (G') and the dynamic damping factor (*tan* δ) as a function of the temperature of BioEP resin and BioEP composites reinforced with 10 wt% and 40 wt% of CFF and FFF. On the other hand, Table 3 shows the storage modulus at 40 °C and 110 °C, as well as the glass transition temperature (T_g), obtained from the peak maximum of the *tan* δ curve for all considered composites.

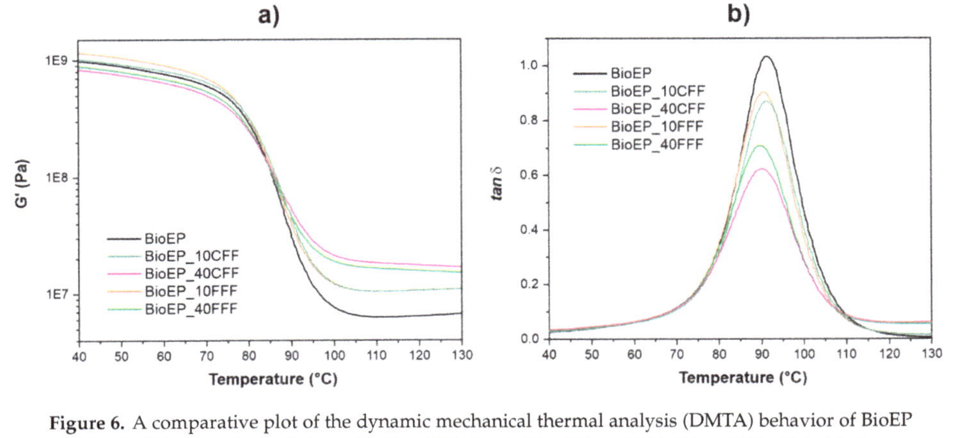

Figure 6. A comparative plot of the dynamic mechanical thermal analysis (DMTA) behavior of BioEP and BioEP/FF composites reinforced with CFF (10 and 40 wt%) and FFF (10 and 40 wt%): (**a**) storage modulus (G'), and (**b**) dynamic damping factor (*tan* δ).

Table 3. Values of dynamic mechanical, thermal analysis (DMTA), and thermomechanical analysis (TMA) of BioEP and BioEP composites reinforced with different CFF and FFF content.

Code	DMTA			TMA	
	G' at 40 °C (MPa)	G' at 110 °C (MPa)	T_g (°C)	*CLTE* below T_g (μm m^{-1} K^{-1})	*CLTE* above T_g (μm m^{-1} K^{-1})
BioEP	981 ± 25	6.3 ± 0.3	91.8 ± 2.1	0.356 ± 0.008	0.565 ± 0.005
BioEP_10CFF	1015 ±36	10.4 ± 0.2	90.8 ± 2.7	0.395 ± 0.015	0.938 ± 0.005
BioEP_20CFF	1041 ± 58	12.5 ± 0.3	90.7 ± 1.7	0.385 ± 0.002	0.621 ± 0.005
BioEP_30CFF	1091 ± 49	18.4 ± 0.9	89.8 ± 2.5	0.570 ± 0.006	0.813 ± 0.005
BioEP_40CFF	826 ± 30	18.5 ± 0.8	90.2 ± 2.5	0.404 ± 0.010	0.727 ± 0.007
BioEP_10FFF	1147 ± 13	10.4 ± 0.2	89.8 ± 2.1	0.529 ± 0.013	0.824 ± 0.003
BioEP_20FFF	1064 ± 52	11.5 ± 0.3	89.1 ± 1.8	0.380 ± 0.009	0.687 ± 0.006
BioEP_30FFF	935 ± 40	14.1 ± 0.4	89.2 ± 1.6	0.436 ± 0.007	0.635 ± 0.002
BioEP_40FFF	878 ± 24	16.3 ± 0.4	89.9 ± 2.2	0.554 ± 0.002	0.770 ± 0.001

The storage modulus (G') (Figure 6a) shows different behavior depending on the filler content. As shown in Table 3, the addition of a lignocellulosic filler into the BioEP resin hardly affects the storage modulus at low temperatures (40 °C), resulting in very similar G′ values to the unfilled BioEP resin. However, at high temperatures (110 °C), it can be seen more clearly how the filler content leads to an increase in G'. As observed in Figure 6a, G' increases as the filler content increases, resulting in more rigid materials as the FF content increases. This is because the particles give rise to a high degree of mechanical restriction since they act as interlock points that reduce the mobility of the polymeric 3D-thermosetting net and their deformation ability. This phenomenon is much more pronounced at high temperatures when the chain motion or vibration is greater [22,61]. Comparing the composites according to the type of reinforcement, it can be seen that the reinforced composites with coarse particles (CFF) have a higher modulus at high temperatures, thus showing their higher capacity to maintain the mechanical load with recoverable viscoelastic deformation at high temperatures compared to the FFF-reinforced composites [56]. Figure 6b shows the dynamic damping factor (*tan δ*) evolution regarding temperature. As can be seen, the BioEP resin curve has the highest value of *tan δ*; however, this value decreases as the FF content in the matrix increases.

This decrease is due to the attenuation that the addition of the filler stiff domains causes in the resin since filler particles act as a steric hindrance [56]. As shown in Table 3, the T_g of the BioEP resin, obtained from the peak maximum of the dynamic damping factor, is around 92 °C. After the addition of the lignocellulosic filler, the T_g remains almost constant, with slight changes due to the restriction that the rigid particles randomly dispersed in the epoxy matrix cause [51]. Comparing both composites, it can be seen that the T_g decrease is slightly higher for FFF-reinforced composites, with a T_g of around 89 °C for all of them, on the other hand, it can be seen that T_g obtained for CFF-reinforced composites is about 90 °C. This slight difference in T_g between the two composites may be because the smaller size of the particles and the greater dispersion of them in the matrix results in greater interaction between the filler and the matrix and, therefore, the mobility of the polymer chains in these regions are more restricted [67]. Despite this hypothesis, the changes in T_g are so slight that it is not possible to hypothesize a remarkable effect of the filler on T_g. In addition, it is worthy to note that FF has been obtained after cold-pressing flaxseed and some residual oil could be present in the flour, thus explaining this slight decrease in T_g [68].

To analyze the dimensional stability of the obtained composites, the coefficient of linear thermal expansion (*CLTE*) was determined by thermomechanical analysis (TMA). Table 3 shows the *CLTE* of BioEP resin and FF-reinforced BioEP composites obtained from the slope of the thermal expansion curves in the rubbery (above T_g) and glassy (below T_g) regions. As shown in Table 3, the BioEP resin shows a low *CLTE*, both below and above the T_g, characteristic of thermosetting resins, which usually exhibit excellent dimensional stability [69]. As can be seen, the *CLTE* is lower for all the samples obtained at temperatures below T_g. This is due to the lower mobility of the polymeric chains in the glassy state, which results in low values of linear expansion [70]. Concerning *CLTE* below T_g, it was observed that the addition of FF filler to BioEP resin increased its value, which evidenced lower dimensional stability of composites with respect to BioEP resin; nevertheless, the dimensional stability of these composites was not compromised since the *CLTE* values obtained for all composites were still very low. In Table 3, it can be seen how the *CLTE* values below T_g in the developed composites did not show any clear trend regarding the filler content. This was also observed for *CLTE* above T_g, where there was an increase in this value after the addition of FF filler into the BioEP resin, but there was no trend with respect to the filler content used in the matrix. In this case, it can be seen how the lowest dimensional stability of the developed composites was obtained for composites filled with 10 wt% for both types of filler size, CFF and FFF, with *CLTE* values of 0.938 and 0.824 $\mu m\ m^{-1}\ K^{-1}$ respectively. Despite this slight increase, the dimensional stability of the developed composites was very high compared to conventional WPCs with *CLTE* values higher than 50 $\mu m\ m^{-1}\ K^{-1}$ [71–73].

3.6. Water Uptake Properties

Figure 7 shows the evolution of the water absorption over time for BioEP resin and BioEP/FF composites filled with 10 wt% and 40 wt% of CFF and FFF. The water diffusion in the wood plastic composites is based on three different mechanisms, the first one involves the water molecules diffusion inside the micro-voids of the polymeric chains, the second mechanism is based on the water absorption by capillarity inside the voids and defects present at the matrix-filler interface, finally, the third mechanism involves the transport through the micro cracks that appear from the swelling of the fillers, namely, the swelling of the reinforcement produced by the contact with the water gives rise to the appearance of microcracks in the fragile thermosetting resin, which facilitates its penetration in the interface between the filler and the matrix [74]. As can be seen in Figure 7, the addition of FF into the BioEP resin significantly increased its water absorption capacity, which rose as the reinforcement content in the matrix increased. This could be due to two possible reasons, one of them is to the hydrophilic nature of the lignocellulosic filler, since cellulose and hemicellulose contain hydroxyl (-OH) groups in their structure that can easily interact with water molecules through hydrogen bonding, thus allowing a path for water entering [61,75]. The other reason is the presence of small voids, pores and microcracks in the internal structure due to the lack of interfacial interaction between the filler and matrix and to the filler swelling. that facilitates the water accumulation in the composite by capillarity [24,76]. By observing the water absorption curves of BioEP/FF composites, two stages can be clearly differentiated. We can see an initial stage with rapid water absorption, followed by a second stage where the curve stabilized into an asymptotic value, thus indicating saturation. Therefore, the water absorption of BioEP/FF composites followed Fickian's diffusion behavior. However, at low immersion times, Figure 7a, it can be seen how the weight gain in the curves of the composites stops being gradual and a rapid increase in weight appears between 10 and 12 h of immersion in all of them. This can be due to the appearance of deformations or damage to the matrix for that immersion time, such as the appearance of microcracks due to the swelling of the filler or the fiber/matrix debonding [77]. As shown in Table 4, the water saturation of composites increases as the filler content does. In this case, it can be seen how the water saturation of composites reinforced with 10 wt% of FF (both CFF and FFF) reaches values of 4.8 wt%. Obviously, composites with 40 wt% of FF water saturation are located at about 12.5 wt%, which means an increase in water absorption compared to neat BioEP resin of 114.3% and 458% respectively. A comparison of the two types of composites obtained shows that CFF-reinforced composites have a slightly higher water absorption compared to their FFF counterparts. This may be due to the larger particle size of the CFF filler, which results in less interaction with the matrix, producing a higher number of voids within the structure of composites. These voids allow the accumulation of water at the interface between the particle and the matrix thus increasing the absorption capacity of the composites [78]. In addition, the increased contact between FF particles as their size and content in the matrix increases contributes to water absorption by capillarity due to the creation of a percolating path through the filler network [79]. It should be noted that the water absorption of FF waste is lower than that obtained from other agroforestry waste that has been used in similar amounts as reinforcements in thermosetting polymeric matrices, such as peanut shells [63] or palm kernel shell [80]. As can be seen in Figure 7, there are some slight differences between the sort-term behavior (first 24 h, Figure 7a) and the long-term behavior (90 days immersion, Figure 7b). Despite this visual difference, it is worthy to note the scale. For the short-term graph, some specimens do not follow the expected behavior (e.g., BioEP/CFF 10 wt%), but the changes in the water absorption are very low, of about 0.2–0.3 wt%, which could be even included in the measurement error. In addition, this initial stage (24 h), is very sensitive to the surface, i.e., the presence of particles not fully embedded and directly exposed to water, which could cause these slight changes. The stationary water absorption after 90 days, represents the actual water uptake behavior as all these initial phenomena disappear, and this agrees with the expected behavior in terms of particle size and loading.

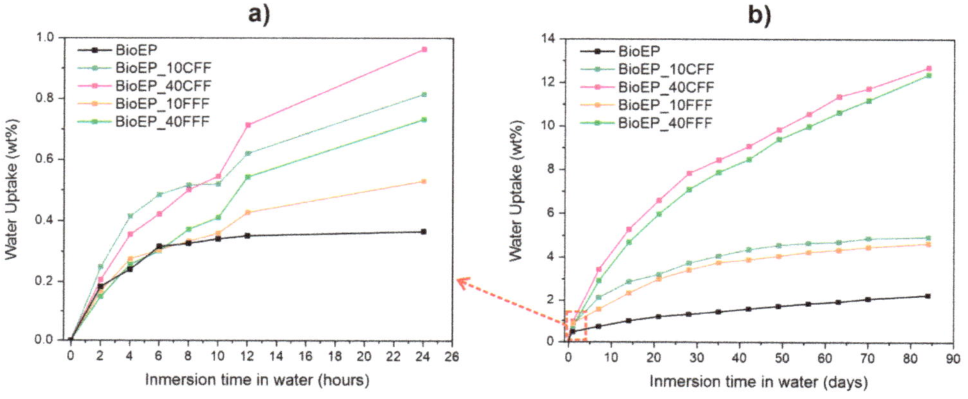

Figure 7. Evolution of water absorption over time of BioEP resin and BioEP composites filled with CFF (10 and 40 wt%) and FFF (10 and 40 wt%): (**a**) water absorption during the first 24 h; (**b**) water absorption during a period of 12 weeks.

Table 4. Values of water saturation (W_s), diffusion coefficient (D) and the corrected diffusion coefficient (D_c) for BioEP resin and BioEP composites reinforced with different CFF and FFF content.

Code	W_s (wt%)	$D \times 10^{-9}$ (cm^2 s^{-1})	$D_c \times 10^{-9}$ (cm^2 s^{-1})
BioEP	2.2 ± 0.0	0.72 ± 0.90	0.36 ± 0.42
BioEP_10CFF	4.9 ± 0.1	1.51 ± 0.40	4.99 ± 2.03
BioEP_20CFF	7.5 ± 0.1	2.07 ± 0.06	6.84 ± 2.9
BioEP_30CFF	8.4 ± 0.2	3.35 ± 0.10	11.1 ± 2.5
BioEP_40CFF	12.7 ± 0.1	4.27 ± 0.05	14.2 ± 0.9
BioEP_10FFF	4.6 ± 0.1	1.38 ± 0.01	4.56 ± 0.2
BioEP_20FFF	6.5 ± 0.1	2.23 ± 0.05	7.37 ± 0.6
BioEP_30FFF	7.7 ± 0.1	2.39 ± 0.04	7.9 ± 0.8
BioEP_40FFF	12.4 ± 0.1	3.39 ± 0.05	11.2 ± 0.9

Table 4 shows the values of the saturation by weight (W_s), as well as the diffusion coefficient (D) and the corrected diffusion coefficient (D_c) of BioEP resin and the obtained BioEP/FF composites. The diffusion coefficient is one of the most important parameters of the Fick's model, which is related to the initial diffusion of water molecules into the matrix surface by entering through external micro-voids towards the internal structure of the composites [81]. As can be seen, the BioEP resin is characterized by a W_s of 2.2 wt%, a D of 0.72 × 10^{-9} cm^2 s^{-1}, and a D_c of 0.36 × 10^{-9} cm^2 s^{-1}. After the incorporation of the highly hydrophilic lignocellulosic residue, it is observed, as expected, that D and D_c increase as the CFF and FFF content does, following the behavior shown by other composites reinforced with natural fillers [18,82,83]. Therefore, it can be concluded that the increase in the amount of lignocellulosic filler increases the ability of water molecules to enter through the composite. Comparing both composites (with CFF and FFF), it can be seen how CFF-filled materials tend to have a higher diffusion coefficient than FFF-filled composites, mainly due to the larger particle size and the more significant presence of voids in the internal structure of the composite due to the lack of matrix-filler interaction [84].

3.7. Color Properties

Figure 8 shows the resulting visual aspect of BioEP resin and BioEP/FF composites filled with coarse particles-CFF (Figure 8a) and fine particles-FFF (Figure 8b) after the curing/post-curing cycle.

Figure 8. Visual aspect of BioEP and BioEP composites reinforced with different FF content; (**a**) (left to right) BioEP resin and BioEP composites reinforced with 10, 20, 30 and 40 wt% of CFF; (**b**) (left to right) BioEP resin and BioEP composites reinforced with 10, 20, 30 and 40 wt% of FFF.

Table 5 gathers the color parameters for the BioEP resin and BioEP/FF composites. As can be seen in Figure 8, BioEP/FF composites obtained by casting show dimensional uniformity as well as a uniform and defect-free surface appearance. This situation is even more evident in composites with FFF. As can be seen, the unfilled BioEP resin sample has some degree of transparency, which is reflected by the low a^* and b^* values. After the addition of the FF filler, composites become brown due to the natural color of the lignocellulosic filler (see Figure 1), which produces an increase in the values in the a^* and b^* coordinates (Table 5). As it can be seen in Figure 8, and quantitatively analyzed by the CIElab coordinates, the color of composites filled with FF is very similar in all of them except that with 10 wt% of FFF, which presents a lighter brown color, obtaining a lower ΔE with respect to the BioEP than the rest of the composites. Comparing both types of composites according to the filler size (CFF or FFF), it can be seen how the surface appearance of FFF-filled composites is much more homogeneous than CFF-filled composites. In the case of CFF-reinforced composites, random whitish spots are present on the surface, which may be due to the formation of large filler aggregates, which negatively affect both aesthetics and mechanical properties. The brown wood color acquired by the samples after the filler incorporation, and the excellent surface appearance acquired mainly in the samples reinforced with FFF can be attractive from the aesthetic point of view for applications in sectors such as furniture, construction, or automotive [85,86].

Table 5. Color parameters from the CIELab space of BioEP resin and BioEP composites reinforced with different CFF and FFF content.

Code	L^*	a^*	b^*	ΔE
BioEP	39.50	−0.48	1.60	-
BioEP_10CFF	39.51	8.97	19.02	19.82
BioEP_20CFF	40.77	9.51	19.36	20.42
BioEP_30CFF	40.93	9.59	19.54	20.62
BioEP_40CFF	41.48	9.68	19.83	20.97
BioEP_10FFF	35.82	9.61	15.80	17.80
BioEP_20FFF	36.27	9.52	16.83	18.50
BioEP_30FFF	37.77	9.44	17.11	18.49
BioEP_40FFF	39.55	9.63	18.42	19.62

4. Conclusions

The main objective of the present work was to evaluate the influence of FF particle size, i.e., coarse (CFF) and fine (FFF), and the filler content (10, 20, 30 and 40 wt%) on the mechanical, thermal, water uptake and morphological properties of composites of a partially biobased epoxy (BioEP) resin processed by casting. The addition of this lignocellulosic filler into the BioEP matrix resulted in a decrease in flexural strength and impact absorption energy with increasing filler content. This is because of the lack of (or very poor) polymer matrix-particle filler interactions. However, this decrease is lower for composites with finer particles, due to their better dispersion and better interaction with the matrix. In this work, it has been observed that the composite filled with 10 wt% of FFF presents a flexural strength and impact energy absorption 24.2% and 33.3% higher than its counterpart with CFF. Particle size has also a high influence on water absorption. It has been observed that composites with fine particles (FFF) filler offer less water absorption due to the presence of fewer voids in their structure as a result of better dispersion and fewer aggregates. In the case of CFF-filled composites, the larger size increases the lack of adhesion with the matrix generating voids and aggregates that allow the water entering, thus leading to a higher water absorption and diffusion coefficient.

Therefore, the present work has revealed that the finer filler particle size results in better mechanical and water absorption properties in these natural fiber reinforced plastics (NFRPs). Besides, a partially biobased, high environmentally friendly material can be obtained. On one hand, the partial biobased epoxy content (31 wt%) can be increased up to almost 70 wt% renewable origin in composites with 40 wt% FF (both CFF or FFF). In addition, this work has revealed an alternative to upgrading wastes from the flaxseed industry. The developed BioEP/FF composites offer an attractive wood-like aesthetic appearance, and therefore, they can be used in sectors such as decoration, furniture, or the automotive industry.

Once the particle size and the amount of FF have been optimized to obtain more balanced mechanical, thermal and water absorption properties in the BioEP/FF composite, further studies will focus on the use of highly reactive coupling agents, i.e., silanes such as (3-glycidyloxypropyl) trimethoxy silane and (3-aminopropyl) trimethoxy silane, to provide increased interaction between the epoxy matrix and the embedded FF particles. All these steps, including an industrial scalation and cost study will assess the viability of these materials to contribute to the circular economy in the flax industry.

Author Contributions: Conceptualization, R.B. and T.B.; Data curation, D.L., D.G.-G. and T.B.; Formal analysis, D.L. and L.Q.-C.; Investigation, D.L. and S.R.-L.; Methodology, D.L., D.G.-G. and L.Q.-C.; Project administration, R.B. and T.B.; Supervision, D.G.-G. and R.B.; Validation, D.G.-G., R.B. and T.B.; Writing–original draft, D.L. and S.R.-L.; Writing–review & editing, D.G.-G. and R.B. All authors have read and agreed to the published version of the manuscript.

Funding: This research was funded by Spanish Ministry of Science, Innovation, and Universities (MICIU), project numbers MAT2017-84909-C2-2-R. This work was supported by the POLISABIO program grant number (2019-A02).

Acknowledgments: D. Lascano thanks Universitat Politècnica de València (UPV) for the grant received through the PAID-01-18 program. D. Garcia-Garcia wants to thank Generalitat Valenciana (GVA) for their financial support through a post-doctoral grant (APOSTD/2019/201). S. Rojas-Lema is a recipient of a Santiago Grisolía contract (GRISOLIAP/2019/132) from GVA. L. Quiles-Carrillo wants to thank GV for his FPI grant (ACIF/2016/182) and MECD for his FPU grant (FPU15/03812). Microscopy services at UPV are acknowledged for their help in collecting and analyzing FESEM images.

Conflicts of Interest: The authors declare no conflict of interest.

References

1. Antonio, C.; Newson, W.; Olsson, R.; Hedenquist, M.; Johansson, E. Advances in the use of protein-based materials: Towards sustainable naturally sourced absorbent materials. *ACS Sustain. Chem. Eng.* **2019**, *7*. [CrossRef]
2. Babu, R.P.; O'connor, K.; Seeram, R. Current progress on bio-based polymers and their future trends. *Prog. Biomater.* **2013**, *2*, 8. [CrossRef]
3. Kim, J.-Y.; Lee, H.W.; Lee, S.M.; Jae, J.; Park, Y.-K. Overview of the recent advances in lignocellulose liquefaction for producing biofuels, bio-based materials and chemicals. *Bioresour. Technol.* **2019**, *279*, 373–384. [CrossRef]
4. Boronat, T.; Fombuena, V.; Garcia-Sanoguera, D.; Sanchez-Nacher, L.; Balart, R. Development of a biocomposite based on green polyethylene biopolymer and eggshell. *Mater. Des.* **2015**, *68*, 177–185. [CrossRef]
5. Naghmouchi, I.; Mutjé, P.; Boufi, S. Olive stones flour as reinforcement in polypropylene composites: A step forward in the valorization of the solid waste from the olive oil industry. *Ind. Crops Prod.* **2015**, *72*, 183–191. [CrossRef]
6. Zaaba, N.F.; Ismail, H. Thermoplastic/Natural Filler Composites: A Short Review. *J. Phys. Sci.* **2019**, *30*, 81–99. [CrossRef]
7. Garcia-Garcia, D.; Carbonell-Verdu, A.; Jordá-Vilaplana, A.; Balart, R.; Garcia-Sanoguera, D. Development and characterization of green composites from bio-based polyethylene and peanut shell. *J. Appl. Polym. Sci.* **2016**, *133*. [CrossRef]
8. Torres-Giner, S.; Montanes, N.; Fenollar, O.; García-Sanoguera, D.; Balart, R. Development and optimization of renewable vinyl plastisol/wood flour composites exposed to ultraviolet radiation. *Mater. Des.* **2016**, *108*, 648–658. [CrossRef]
9. Ghofrani, M.; Pishan, S.; Mohammadi, M.R.; Omidi, H. A study on rice-husk/recycled high density polyethylene composites–their physical and mechanical properties. *Environ. Sci.* **2011**, *9*, 99–112.
10. Dimzoski, B.; Bogoeva-Gaceva, G.; Gentile, G.; Avella, M.; Grozdanov, A. Polypropylene-based eco-composites filled with agricultural rice hulls waste. *Chem. Biochem. Eng. Q.* **2009**, *23*, 225–230.
11. Prabu, V.A.; Johnson, R.D.J.; Amuthakkannan, P.; Manikandan, V. Usage of industrial wastes as particulate composite for environment management: Hardness, tensile and impact studies. *J. Environ. Chem. Eng.* **2017**, *5*, 1289–1301. [CrossRef]
12. Prabhakar, M.; Shah, A.U.R.; Rao, K.C.; Song, J.-I. Mechanical and thermal properties of epoxy composites reinforced with waste peanut shell powder as a bio-filler. *Fibers Polym.* **2015**, *16*, 1119–1124. [CrossRef]
13. Ikladious, N.; Shukry, N.; El-Kalyoubi, S.; Asaad, J.; Mansour, S.; Tawfik, S.; Abou-Zeid, R. Eco-friendly composites based on peanut shell powder/unsaturated polyester resin. *Proc. Inst. Mech. Eng. Part L J. Mater. Des. Appl.* **2019**, *233*, 955–964. [CrossRef]
14. Quiles-Carrillo, L.; Montanes, N.; Garcia-Garcia, D.; Carbonell-Verdu, A.; Balart, R.; Torres-Giner, S. Effect of different compatibilizers on injection-molded green composite pieces based on polylactide filled with almond shell flour. *Compos. Part B Eng.* **2018**, *147*, 76–85. [CrossRef]
15. Singh, V.; Bansal, G.; Agarwal, M.; Negi, P. Experimental determination of mechanical and physical properties of almond shell particles filled biocomposite in modified epoxy resin. *J. Mater. Sci. Eng.* **2016**, *5*. [CrossRef]
16. Liminana, P.; Garcia-Sanoguera, D.; Quiles-Carrillo, L.; Balart, R.; Montanes, N. Development and characterization of environmentally friendly composites from poly (butylene succinate)(PBS) and almond shell flour with different compatibilizers. *Compos. Part B Eng.* **2018**, *144*, 153–162. [CrossRef]

17. Barczewski, M.; Sałasińska, K.; Szulc, J. Application of sunflower husk, hazelnut shell and walnut shell as waste agricultural fillers for epoxy-based composites: A study into mechanical behavior related to structural and rheological properties. *Polym. Test.* **2019**, *75*, 1–11. [CrossRef]
18. Balart, J.; Montanes, N.; Fombuena, V.; Boronat, T.; Sánchez-Nacher, L. Disintegration in compost conditions and water uptake of green composites from poly (lactic acid) and hazelnut shell flour. *J. Polym. Environ.* **2018**, *26*, 701–715. [CrossRef]
19. Ameh, A.O.; Isa, M.T.; Sanusi, I. Effect of particle size and concentration on the mechanical properties of polyester/date palm seed particulate composites. *Leonardo Electron. J. Pract. Technol.* **2015**, *26*, 65–78.
20. Sharma, H.; Singh, I.; Misra, J.P. Mechanical and thermal behaviour of food waste (Citrus limetta peel) fillers–based novel epoxy composites. *Polym. Polym. Compos.* **2019**, *27*, 527–535. [CrossRef]
21. Garcia-Garcia, D.; Quiles-Carrillo, L.; Montanes, N.; Fombuena, V.; Balart, R. Manufacturing and Characterization of Composite Fibreboards with Posidonia oceanica Wastes with an Environmentally-Friendly Binder from Epoxy Resin. *Materials* **2018**, *11*, 35. [CrossRef] [PubMed]
22. Ferrero, B.; Fombuena, V.; Fenollar, O.; Boronat, T.; Balart, R. Development of natural fiber-reinforced plastics (NFRP) based on biobased polyethylene and waste fibers from Posidonia oceanica seaweed. *Polym. Compos.* **2015**, *36*, 1378–1385. [CrossRef]
23. Koutsomitopoulou, A.; Bénézet, J.; Bergeret, A.; Papanicolaou, G. Preparation and characterization of olive pit powder as a filler to PLA-matrix bio-composites. *Powder Technol.* **2014**, *255*, 10–16. [CrossRef]
24. Naghmouchi, I.; Espinach, F.X.; Mutjé, P.; Boufi, S. Polypropylene composites based on lignocellulosic fillers: How the filler morphology affects the composite properties. *Mater. Des. (1980–2015)* **2015**, *65*, 454–461. [CrossRef]
25. D'Amato, D.; Veijonaho, S.; Toppinen, A. Towards sustainability? Forest-based circular bioeconomy business models in Finnish SMEs. *For. Policy Econ.* **2020**, *110*, 101848. [CrossRef]
26. Capezza, A.J.; Lundman, M.; Olsson, R.T.; Newson, W.R.; Hedenqvist, M.S.; Johansson, E. Carboxylated wheat gluten proteins—A green solution for production of sustainable superabsorbent materials. *Biomacromolecules* **2020**, *21*, 1709–1719. [CrossRef]
27. Barczewski, M.; Mysiukiewicz, O.; Kloziński, A. Complex modification effect of linseed cake as an agricultural waste filler used in high density polyethylene composites. *Iran. Polym. J.* **2018**, *27*, 677–688. [CrossRef]
28. Food and Agriculture Organization. Available online: http://www.fao.org (accessed on 28 April 2020).
29. Zhang, Z.-S.; Wang, L.-J.; Li, D.; Jiao, S.-S.; Chen, X.D.; Mao, Z.-H. Ultrasound-assisted extraction of oil from flaxseed. *Sep. Purif. Technol.* **2008**, *62*, 192–198. [CrossRef]
30. Yasmeen, M.; Nisar, S.; Tavallali, V.; Khalid, T. A review of phytochemicals and uses of flaxseed. *Int. J. Chem. Biochem. Sci.* **2018**, *13*, 70–75.
31. Bekhit, A.E.-D.A.; Shavandi, A.; Jodjaja, T.; Birch, J.; Teh, S.; Ahmed, I.A.M.; Al-Juhaimi, F.Y.; Saeedi, P.; Bekhit, A.A. Flaxseed: Composition, detoxification, utilization, and opportunities. *Biocatal. Agric. Biotechnol.* **2018**, *13*, 129–152. [CrossRef]
32. Zhang, Z.-S.; Wang, L.-J.; Li, D.; Li, S.-J.; Özkan, N. Characteristics of flaxseed oil from two different flax plants. *Int. J. Food Prop.* **2011**, *14*, 1286–1296. [CrossRef]
33. Mannucci, A.; Castagna, A.; Santin, M.; Serra, A.; Mele, M.; Ranieri, A. Quality of flaxseed oil cake under different storage conditions. *LWT* **2019**, *104*, 84–90. [CrossRef]
34. Brison, L. Evaluation of the Effect of Nitrogen Fertilization and Tillage on the Yield and the Nutritional Profile of Flaxseed. Master's Thesis, Université catholique de Louvain, Ottignies-Louvain-la-Neuve, Belgium, 2019.
35. Wirkijowska, A.; Zarzycki, P.; Sobota, A.; Nawrocka, A.; Blicharz-Kania, A.; Andrejko, D. The possibility of using by-products from the flaxseed industry for functional bread production. *LWT* **2020**, *118*, 108860. [CrossRef]
36. Kolodziejczyk, P.; Ozimek, L.; Kozłowska, J. The application of flax and hemp seeds in food, animal feed and cosmetics production. In *Handbook of Natural Fibres*; Elsevier: Amsterdam, The Netherlands, 2012; pp. 329–366.
37. Chan, C.M.; Vandi, L.-J.; Pratt, S.; Halley, P.; Richardson, D.; Werker, A.; Laycock, B. Composites of wood and biodegradable thermoplastics: A review. *Polym. Rev.* **2018**, *58*, 444–494. [CrossRef]
38. España, J.; Samper, M.; Fages, E.; Sánchez-Nácher, L.; Balart, R. Investigation of the effect of different silane coupling agents on mechanical performance of basalt fiber composite laminates with biobased epoxy matrices. *Polym. Compos.* **2013**, *34*, 376–381. [CrossRef]

39. Bertomeu, D.; García-Sanoguera, D.; Fenollar, O.; Boronat, T.; Balart, R. Use of eco-friendly epoxy resins from renewable resources as potential substitutes of petrochemical epoxy resins for ambient cured composites with flax reinforcements. *Polym. Compos.* **2012**, *33*, 683–692. [CrossRef]
40. Samper, M.; Fombuena, V.; Boronat, T.; García-Sanoguera, D.; Balart, R. Thermal and mechanical characterization of epoxy resins (ELO and ESO) cured with anhydrides. *J. Am. Oil Chem. Soc.* **2012**, *89*, 1521–1528. [CrossRef]
41. Fombuena, V.; Samper, M.; Sanchez-Nacher, L. Study of the properties of thermoset materials derived from epoxidized soybean oil and protein fillers. *J. Am. Oil Chem. Soc.* **2013**, *90*, 449–457. [CrossRef]
42. Carbonell-Verdu, A.; Bernardi, L.; Garcia-Garcia, D.; Sanchez-Nacher, L.; Balart, R. Development of environmentally friendly composite matrices from epoxidized cottonseed oil. *Eur. Polym. J.* **2015**, *63*, 1–10. [CrossRef]
43. Anusic, A.; Resch-Fauster, K.; Mahendran, A.R.; Wuzella, G. Anhydride Cured Bio-Based Epoxy Resin: Effect of Moisture on Thermal and Mechanical Properties. *Macromol. Mater. Eng.* **2019**, *304*, 1900031. [CrossRef]
44. Quiles-Carrillo, L.; Duart, S.; Montanes, N.; Torres-Giner, S.; Balart, R. Enhancement of the mechanical and thermal properties of injection-molded polylactide parts by the addition of acrylated epoxidized soybean oil. *Mater. Des.* **2018**, *140*, 54–63. [CrossRef]
45. Fenollar, O.; Sanchez-Nacher, L.; Garcia-Sanoguera, D.; López, J.; Balart, R. The effect of the curing time and temperature on final properties of flexible PVC with an epoxidized fatty acid ester as natural-based plasticizer. *J. Mater. Sci.* **2009**, *44*, 3702–3711. [CrossRef]
46. Liu, X.; Huang, W.; Jiang, Y.; Zhu, J.; Zhang, C. Preparation of a bio-based epoxy with comparable properties to those of petroleum-based counterparts. *Express Polym. Lett.* **2012**, *6*, 293–298. [CrossRef]
47. Niedermann, P.; Szebényi, G.; Toldy, A. Characterization of high glass transition temperature sugar-based epoxy resin composites with jute and carbon fibre reinforcement. *Compos. Sci. Technol.* **2015**, *117*, 62–68. [CrossRef]
48. Niedermann, P.; Szebényi, G.; Toldy, A. Effect of epoxidized soybean oil on curing, rheological, mechanical and thermal properties of aromatic and aliphatic epoxy resins. *J. Polym. Environ.* **2014**, *22*, 525–536. [CrossRef]
49. Wu, Y.; Wang, Y.; Yang, F.; Wang, J.; Wang, X. Study on the Properties of Transparent Bamboo Prepared by Epoxy Resin Impregnation. *Polymers* **2020**, *12*, 863. [CrossRef]
50. Salasinska, K.; Mizera, K.; Barczewski, M.; Borucka, M.; Gloc, M.; Celiński, M.; Gajek, A. The influence of degree of fragmentation of Pinus sibirica on flammability, thermal and thermomechanical behavior of the epoxy-composites. *Polym. Test.* **2019**, *79*, 106036. [CrossRef]
51. Kumar, R.; Kumar, K.; Bhowmik, S. Mechanical characterization and quantification of tensile, fracture and viscoelastic characteristics of wood filler reinforced epoxy composite. *Wood Sci. Technol.* **2018**, *52*, 677–699. [CrossRef]
52. Stabik, J.; Chomiak, M. Graded epoxy-hard coal composites: Analysis of filler particle distribution in the epoxy matrix. *J. Compos. Mater.* **2016**, *50*, 3663–3677. [CrossRef]
53. Lascano, D.; Quiles-Carrillo, L.; Torres-Giner, S.; Boronat, T.; Montanes, N. Optimization of the curing and post-curing conditions for the manufacturing of partially bio-based epoxy resins with improved toughness. *Polymers* **2019**, *11*, 1354. [CrossRef]
54. Agüero, Á.; Lascano, D.; Garcia-Sanoguera, D.; Fenollar, O.; Torres-Giner, S. Valorization of Linen Processing By-Products for the Development of Injection-Molded Green Composite Pieces of Polylactide with Improved Performance. *Sustainability* **2020**, *12*, 652. [CrossRef]
55. Bledzki, A.K.; Mamun, A.A.; Volk, J. Barley husk and coconut shell reinforced polypropylene composites: The effect of fibre physical, chemical and surface properties. *Compos. Sci. Technol.* **2010**, *70*, 840–846. [CrossRef]
56. Salasinska, K.; Barczewski, M.; Górny, R.; Kloziński, A. Evaluation of highly filled epoxy composites modified with walnut shell waste filler. *Polym. Bull.* **2018**, *75*, 2511–2528. [CrossRef]
57. Kwon, H.-J.; Sunthornvarabhas, J.; Park, J.-W.; Lee, J.-H.; Kim, H.-J.; Piyachomkwan, K.; Sriroth, K.; Cho, D. Tensile properties of kenaf fiber and corn husk flour reinforced poly (lactic acid) hybrid bio-composites: Role of aspect ratio of natural fibers. *Compos. Part B Eng.* **2014**, *56*, 232–237. [CrossRef]
58. Bisht, N.; Gope, P.C. Mechanical properties of rice husk flour reinforced epoxy bio-composite. *Int. J. Eng. Res. Appl.* **2015**, *5*, 123128.
59. Novowood. Available online: https://www.novowood.it/en/download-technical-sheet-wpc_39c7.html (accessed on 18 May 2020).

60. Jeluplast. Available online: https://www.jeluplast.com/wp-content/uploads/2013/09/WPC-PE-H70-800-03.pdf (accessed on 18 May 2020).
61. García-García, D.; Carbonell, A.; Samper, M.; García-Sanoguera, D.; Balart, R. Green composites based on polypropylene matrix and hydrophobized spend coffee ground (SCG) powder. *Compos. Part B Eng.* **2015**, *78*, 256–265. [CrossRef]
62. Yussuf, A.; Massoumi, I.; Hassan, A. Comparison of polylactic acid/kenaf and polylactic acid/rise husk composites: The influence of the natural fibers on the mechanical, thermal and biodegradability properties. *J. Polym. Environ.* **2010**, *18*, 422–429. [CrossRef]
63. Raju, G.; Kumarappa, S. Experimental study on mechanical and thermal properties of epoxy composites filled with agricultural residue. *Polym. Renew. Resour.* **2012**, *3*, 117–138. [CrossRef]
64. Chen, W.-H.; Kuo, P.-C. A study on torrefaction of various biomass materials and its impact on lignocellulosic structure simulated by a thermogravimetry. *Energy* **2010**, *35*, 2580–2586. [CrossRef]
65. Moriana, R.; Vilaplana, F.; Ek, M. Forest residues as renewable resources for bio-based polymeric materials and bioenergy: Chemical composition, structure and thermal properties. *Cellulose* **2015**, *22*, 3409–3423. [CrossRef]
66. Hussain, S.; Anjum, F.; Butt, M.; Sheikh, M. Chemical composition and functional properties of flaxseed (*Linum usitatissimum*) flour. *Sarhad J. Agric.* **2008**, *24*, 649–653.
67. Sengupta, S.; Maity, P.; Ray, D.; Mukhopadhyay, A. Stearic acid as coupling agent in fly ash reinforced recycled polypropylene matrix composites: Structural, mechanical, and thermal characterizations. *J. Appl. Polym. Sci.* **2013**, *130*, 1996–2004. [CrossRef]
68. Mysiukiewicz, O.; Barczewski, M. Utilization of linseed cake as a postagricultural functional filler for poly (lactic acid) green composites. *J. Appl. Polym. Sci.* **2019**, *136*, 47152. [CrossRef]
69. Liu, R.; Wang, J.; He, Q.; Zong, L.; Jian, X. Interaction and properties of epoxy-amine system modified with poly (phthalazinone ether nitrile ketone). *J. Appl. Polym. Sci.* **2016**, *133*. [CrossRef]
70. Balart, J.; García-Sanoguera, D.; Balart, R.; Boronat, T.; Sánchez-Nacher, L. Manufacturing and properties of biobased thermoplastic composites from poly (lactid acid) and hazelnut shell wastes. *Polym. Compos.* **2018**, *39*, 848–857. [CrossRef]
71. Balart, J.; Fombuena, V.; Fenollar, O.; Boronat, T.; Sánchez-Nacher, L. Processing and characterization of high environmental efficiency composites based on PLA and hazelnut shell flour (HSF) with biobased plasticizers derived from epoxidized linseed oil (ELO). *Compos. Part B Eng.* **2016**, *86*, 168–177. [CrossRef]
72. Liminana, P.; Quiles-Carrillo, L.; Boronat, T.; Balart, R.; Montanes, N. The Effect of Varying Almond Shell Flour (ASF) Loading in Composites with Poly (Butylene Succinate (PBS) Matrix Compatibilized with Maleinized Linseed Oil (MLO). *Materials* **2018**, *11*, 2179. [CrossRef]
73. Reixach, R.; Puig, J.; Méndez, J.A.; Gironès, J.; Espinach, F.X.; Arbat, G.; Mutjé, P. Orange wood fiber reinforced polypropylene composites: Thermal properties. *Bioresources* **2015**, *10*. [CrossRef]
74. Akil, H.M.; Cheng, L.W.; Ishak, Z.M.; Bakar, A.A.; Rahman, M.A. Water absorption study on pultruded jute fibre reinforced unsaturated polyester composites. *Compos. Sci. Technol.* **2009**, *69*, 1942–1948. [CrossRef]
75. Ferrero, B.; Boronat, T.; Moriana, R.; Fenollar, O.; Balart, R. Green composites based on wheat gluten matrix and posidonia oceanica waste fibers as reinforcements. *Polym. Compos.* **2013**, *34*, 1663–1669. [CrossRef]
76. Alander, B.; Capezza, A.; Wu, Q.; Johansson, E.; Olsson, R.T.; Hedenqvist, M. A facile way of making inexpensive rigid and soft protein biofoams with rapid liquid absorption. *Ind. Crops Prod.* **2018**, *119*, 41–48. [CrossRef]
77. Ben Daly, H.; Ben Brahim, H.; Hfaied, N.; Harchay, M.; Boukhili, R. Investigation of water absorption in pultruded composites containing fillers and low profile additives. *Polym. Compos.* **2007**, *28*, 355–364. [CrossRef]
78. Udhayasankar, R.; Kathikeyan, B. Preparation and properties of cashew nut shell liquid-Based composite reinforced by coconut shell particles. *Surf. Rev. Lett.* **2019**, *26*, 1850174. [CrossRef]
79. Naghmouchi, I.; Mutjé, P.; Boufi, S. Polyvinyl chloride composites filled with olive stone flour: Mechanical, thermal, and water absorption properties. *J. Appl. Polym. Sci.* **2014**, *131*. [CrossRef]
80. Shehu, U.; Aponbiede, O.; Ause, T.; Obiodunukwe, E. Effect of particle size on the properties of polyester/palm kernel shell (PKS) particulate composites. *J. Mater. Environ. Sci.* **2014**, *5*, 366–373.

81. Thirmizir, M.A.; Ishak, Z.M.; Taib, R.M.; Sudin, R.; Leong, Y. Mechanical, water absorption and dimensional stability studies of kenaf bast fibre-filled poly (butylene succinate) composites. *Polym. Plast. Technol. Eng.* **2011**, *50*, 339–348. [CrossRef]
82. Zabihzadeh, S.M.; Omidvar, A.; Marandi, M.A.B.; Dastoorian, F.; Mirmehdi, S.M. Effect of filler loading on physical and flexural properties of rapeseed stem/PP composites. *BioResources* **2011**, *6*, 1475–1483.
83. Ishak, Z.M.; Yow, B.; Ng, B.; Khalil, H.A.; Rozman, H. Hygrothermal aging and tensile behavior of injection-molded rice husk-filled polypropylene composites. *J. Appl. Polym. Sci.* **2001**, *81*, 742–753. [CrossRef]
84. Najafi, A.; Khademi-Eslam, H. Lignocellulosic filler/recycled HDPE composites: Effect of filler type on physical and flexural properties. *BioResources* **2011**, *6*, 2411–2424.
85. Sapuan, S.; Harun, N.; Abbas, K. Design and fabrication of a multipurpose table using a composite of epoxy and banana pseudostem fibres. *J. Trop. Agric.* **2008**, *45*, 66–68.
86. Sapuan, S.; Maleque, M. Design and fabrication of natural woven fabric reinforced epoxy composite for household telephone stand. *Mater. Des.* **2005**, *26*, 65–71. [CrossRef]

applied sciences

MDPI

Article

Torrefaction of Coffee Husk Flour for the Development of Injection-Molded Green Composite Pieces of Polylactide with High Sustainability

Diana L. Ortiz-Barajas [1], Johan A. Arévalo-Prada [1,*], Octavio Fenollar [2], Yesid J. Rueda-Ordóñez [1] and Sergio Torres-Giner [3,*]

1 Grupo de Investigación en Energía y Medio Ambiente (GIEMA), Universidad Industrial de Santander (UIS), 680002 Bucaramanga, Colombia; diana.ortiz10@correo.uis.edu.co (D.L.O.-B.); yjruedao@uis.edu.co (Y.J.R.-O.)

2 Technological Institute of Materials (ITM), Universitat Politècnica de València (UPV), Plaza Ferrándiz y Carbonell 1, 03801 Alcoy, Spain; ocfegi@epsa.upv.es

3 Novel Materials and Nanotechnology Group, Institute of Agrochemistry and Food Technology (IATA), Spanish National Research Council (CSIC), Calle Catedrático Agustín Escardino Benlloch 7, 46980 Paterna, Spain

* Correspondence: johan.arevalo@correo.uis.edu.co (J.A.A.-P.); storresginer@iata.csic.es (S.T.-G.); Tel.: +57-315-443-3621 (J.A.A.-P.); +34-963-900-022 (S.T.-G.)

Received: 29 August 2020; Accepted: 15 September 2020; Published: 17 September 2020

Featured Application: Green composite pieces of polylactide (PLA) containing 20 wt% of coffee waste derived filler can be applied for rigid packaging, such as food trays and containers, beverage cups, and food contact disposables, such as cutlery and plates.

Abstract: Coffee husk, a major lignocellulosic waste derived from the coffee industry, was first ground into flour of fine particles of approximately 90 μm and then torrefied at 250 °C to make it more thermally stable and compatible with biopolymers. The resultant torrefied coffee husk flour (TCHF) was thereafter melt-compounded with polylactide (PLA) in contents from 20 to 50 wt% and the extruded green composite pellets were shaped by injection molding into pieces and characterized. Although the incorporation of TCHF reduced the ductility and toughness of PLA, filler contents of 20 wt% successfully yielded pieces with balanced mechanical properties in both tensile and flexural conditions and improved hardness. Contents of up to 30 wt% of TCHF also induced a nucleating effect that favored the formation of crystals of PLA, whereas the thermal degradation of the biopolyester was delayed by more than 7 °C. Furthermore, the PLA/TCHF pieces showed higher thermomechanical resistance and their softening point increased up to nearly 60 °C. Therefore, highly sustainable pieces were developed through the valorization of large amounts of coffee waste subjected to torrefaction. In the Circular Bioeconomy framework, these novel green composites can be used in the design of compostable rigid packaging and food contact disposables.

Keywords: PLA; coffee husk; torrefaction; green composites; waste valorization; Circular Bioeconomy

1. Introduction

Valorization of food processing by-products and agricultural waste is acquiring increasing importance due to the high interest in the development of novel sustainable materials as part of the Circular Bioeconomy framework [1,2]. Agro-food wastes, such as flax fibers [3,4], rice husk [5], jute fibers [6,7], almond husk [8,9], walnut husk [10], orange peel [11], coconut fibers [12] or sisal fibers [13], represent examples of recently explored fillers to be incorporated into biopolymer matrices. The resultant green composites can successfully show technological advantages over conventional petroleum derived polymer composites, including cost reduction, lower density, no toxicity, balanced mechanical properties and, more importantly, lower environmental impact [14,15]. However, natural fillers also have some disadvantages, such as lower thermal stability and higher moisture absorption [15,16]. Therefore, the optimal amount of filler content for a given polymer matrix and the specific target application must be carefully determined. In addition, different compatibilization strategies, such as fillers modifications and/or the use of multi-functional additives, can be used [17].

Coffee is considered by many to be the most popular hot beverage in the world, being consumed for its refreshing and stimulating properties. It is the most consumed drink after water and tea, the second largest traded commodity after petroleum, and probably the most essential agricultural commodity [18]. Coffee beans are one of the most important tropical products, being cultivated primarily in the "the coffee belt" area between 25° N and 25° S latitude [19]. Brazil, Vietnam, Indonesia, Colombia, and India are the global leaders in production of green coffee beans. Annual production of coffee increased from 140 to 152 million bags of 60 kg in 2010 [20]. However, the 'fruit-to-cup' transformation of coffee beans generates huge quantities of by-products and wastes, such as leaf, flower, cherry as well as coffee silver skin (CSS), coffee husks, and spent coffee ground (SCG) [21,22]. In particular, approximately 68 kg of green coffee waste are produced per 100 kg of fresh coffee, in which 50 and 18 kg correspond to coffee pulp and husk, respectively [23]. The CSS, which is the main by-product of the roasting process, includes dietary fiber (50–60 wt%) protein (16–19 wt%), fat (1.56–3.28 wt%), and ash (7 wt%) [24]. The main current value-adding applications include biofuel [25] or energy purposes [26], fertilizer production [27], dietary fiber [28], and bioactive compound extraction [29]. However, animal feeding is limited by the presence of caffeine and tannins, phenolic compounds that show anti-nutritional properties [30]. Furthermore, chlorogenic acid is phytotoxic when used as a plant fertilizer [31]. Therefore, the valorization of coffee by-products is a technological challenge and new added value applications are pursued.

In a Circular Bioeconomy context, the use of coffee husk in the form of flour as possible reinforcing filler in polymer composites can represent a sustainable method of valorization [32]. Some previous studies have incorporated different fillers derived from by-products of the coffee industry into biopolymer materials. Some examples include the production of biodegradable compounds for food packaging applications using poly(butylene adipate-*co*-terephthalate) (PBAT) and coffee grounds [33], the reinforcement of high-density polyethylene (HDPE) by coffee hull (CH) [34] or sustainable composites of polypropylene (PP) and ground coffee powder for use as wood plastic composites (WPCs) [35]. However, given the origin of such solid residue, the typical content of total carbohydrates (58–85 wt%) and moisture (up to 15 wt%) in coffee husk is high, with reported contents of cellulose, hemi-cellulose, and lignin of 43, 7, and 9 wt% on a dry basis, respectively [36]. Although such levels are low or at most similar to other lignocellulosic residues [37], this fact limits its application for polymer composites. In this regard, torrefaction of biomass can represent a novel strategy to not only enhance its performance for renewable energy applications in terms of grindability, higher calorific value, long-term storage, and handling, but also improve its affinity with polymer matrices [38]. Torrefaction is a mild form of pyrolysis performed at temperatures ranging from 200 to 300 °C under atmospheric pressure and in the absence of oxygen [39]. During this thermochemical treatment, the water contained in cellulose, hemicellulose, and lignin, as well as superfluous volatiles, are released, typically around 20 wt%, resulting in a relatively dry product that reduces or eliminates its potential for organic decomposition and is also more hydrophobic [40].

Among the biopolymers, polylactide (PLA) currently has a key position in the bioplastic market for both technical and commodity products [41]. PLA is a thermoplastic biopolyester that can be obtained from natural and renewable sources such as cornstarch or cassava and its articles can be compostable in industrial conditions [42]. It is mainly used in compostable food packaging and disposables, biomedical devices and, more recently, 3D printing applications [43–45]. In this study, coffee husk flour (CHF) was first torrefied and then incorporated into PLA by twin-screw extrusion (TSE) and the resultant compounded pellets were shaped into pieces by injection molding. Different torrefied coffee husk flour (TCHF) contents were incorporated into PLA and the mechanical and thermal properties were determined to analyze the influence of the coffee waste derived lignocellulosic filler on the final performance of the green composite pieces and ascertain their potential for sustainable packaging and disposables.

2. Materials and Methods

2.1. Materials

PLA biopolymer commercial grade Ingeo™ 3001D was supplied in pellet form by NatureWorks LLC (Minnesota, MN, USA). This grade corresponds to a poly(D,L-lactide) (PDLLA) of high optical purity in L-lactic acid isomer, with 1.4 ± 0.2% of D-lactic acid fraction [46]. The number- and weight-average molecular weights were found to be M_n = 90.5 kD and M_W = 158.3 kD by gel permeation chromatography (GPC), which gives a polydispersity index (PDI = M_W/M_n) of ~1.75 [47]. It is also characterized by a density of 1.24 g/cm^3 following the American Society for Testing and Materials (ASTM) method D792 and a melt flow rate (MFR) of 22 g/10 min, measured at 210 °C and 2.16 kg, according to the ASTM method D1238, which makes it feasible for injection molding.

Coffee husks were separated from the previously natural dried grains by a threshing process. The husks were obtained from a coffee crop located at 1390 m above sea level in the San José de Motoso district of Girón in Santander, Colombia, and were dried in a laboratory oven at 105 °C for 24 h to remove moisture. Afterwards, the dried husks were milled at a speed of 8000 rpm using a ZM 200 centrifugal mill from Retsch (Düsseldorf, Germany) and finally sieved with a 250-µm mesh filter to the size of 110 microns to obtain the so-called CHF (see Figure 1).

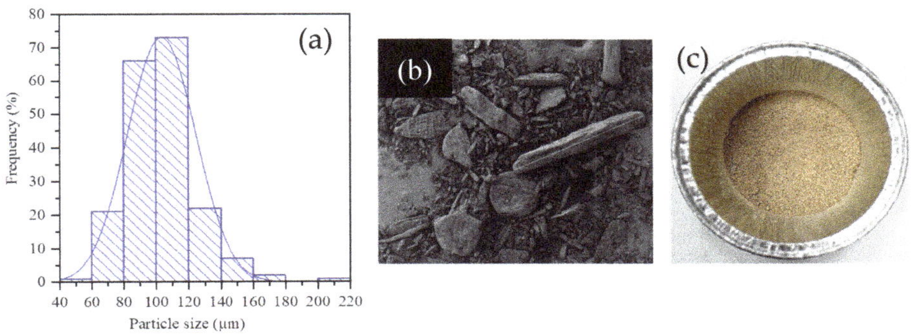

Figure 1. Detail of the coffee husk flour (CHF): (**a**) Particle size histogram; (**b**) Field emission scanning electron microscopy (FESEM) image taken at 55× with scale marker of 100 µm; (**c**) Visual aspect.

2.2. Torrefaction of CHF

The resultant CHF was torrefied in a laboratory oven Carbolite mod. 2416CG from Keison Products (Barcelona, Spain). The flour was first dried at 105 °C for 30 min, then heated up under nitrogen atmosphere to the torrefaction temperature of 250 °C at a heating rate of 5 °C/min, and finally this temperature was kept constant for 1 h. The resulting torrefied husk was milled and sieved to the size of approximately 90 µm using the same procedure applied in CHF to obtain the so-called TCHF (see Figure 2).

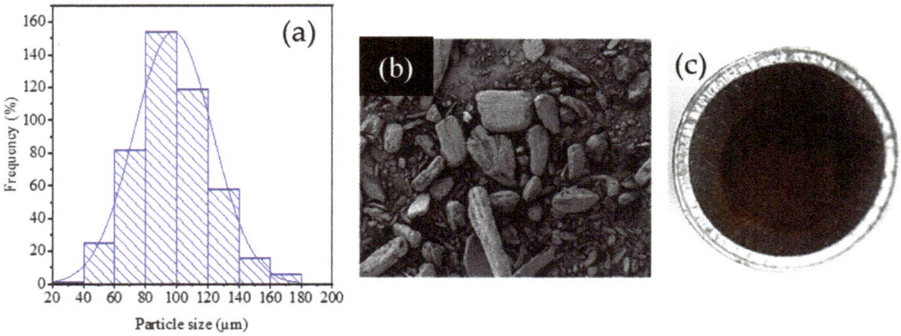

Figure 2. Detail of the torrefied coffee husk flour (TCHF): (**a**) Particle size histogram; (**b**) Field emission scanning electron microscopy (FESEM) image taken at 55× with scale marker of 100 μm; (**c**) Visual aspect.

2.3. Preparation of the Green Composite Pieces

The biopolymer pellets and the fillers were dried separately for 24 h at 60 °C prior to processing. PLA and TCHF were mixed in a zipped bag according to the contents indicated in Table 1. These composition ranges were selected as previous studies with lignocellulosic materials in PLA indicated that excessive loss of properties occurs with higher amounts of lignocellulosic fillers [13,48,49]. The mixtures were melt-compounded in a co-rotating twin-screw extruder (D = 25 mm, L/D = 24) from Dupra SL (Alicante, Spain), an equipment described in previous work [17]. The profile temperature was set, from the hopper to the die, to 165–170–175–180 °C and the rotating speed was 40 rpm. The extruded filaments were pelletized using an air-knife unit after being cooled down in an air stream. A green composite of PLA with 20 wt% of CHF was also prepared under the same conditions for comparison purposes.

Table 1. Set of samples prepared according to the weight percentage (wt%) of polylactide (PLA), coffee husk flour (CHF), and torrefied coffee husk flour (TCHF).

Sample	PLA (wt%)	CHF (wt%)	TCHF (wt%)
PLA	100	-	-
PLA/20CHF	80	20	-
PLA/20TCHF	80	-	20
PLA/30TCHF	70	-	30
PLA/40TCHF	60	-	40
PLA/50TCHF	50	-	50

The compounded pellets were thereafter injection molded to obtain pieces in a Meteor 270/75 from Mateu & Solé (Barcelona, Spain). The profile of temperatures in the barrel was set, from the hopper to the injection nozzle, to 165–170–175–180 °C and the mold temperature at 60 °C. A low injection time of 1 s was employed, whereas the applied clamping force was 20 tons and the pieces were cooled for 20 s in the mold. As a result, 4-mm pieces were produced that, prior to characterization, were stored for at least 48 h.

2.4. Mechanical Tests

Tensile, flexural, hardness, and impact standard tests were carried out for mechanical characterization of the green composites. Pieces with a geometry of 80 mm × 10 mm × 4 mm were tested in a universal test machine ELIB 30 from S.A.E Ibertest (Madrid, Spain) equipped with a 5-kN load cell in both tensile and flexural conditions. The guidelines of the International Organization for Standardization (ISO) 527 and ISO 178, were followed for the tensile and flexural tests using cross-head speeds of 10 and 5 mm/min, respectively. A Shore D durometer 673-D from Instrumentos J. Bot S.A. (Barcelona, Spain) was used to determine Shore D hardness according to ISO 868. The impact absorbed energy was determined following the ISO 179 in a Charpy pendulum of 6 J from Metrotec S.A. (San Sebastián, Spain) using unnotched samples of 80 mm × 10 mm × 4 mm. For each mechanical test, at least six different samples were tested and the values were averaged.

2.5. Microscopy

The fracture surfaces of the pieces obtained from the impact tests were coated in a vacuum with an ultrathin aurum-palladium layer in a high vacuum sputter Emitech mod. SC7620 (Quorum Technologies Ltd., East Sussex, UK). Then, field emission scanning electron microscopy (FESEM) was performed to analyze their morphologies in high-resolution Zeiss Ultra55 (Oxford Instruments, Abingdon, UK) operating at 2 kV.

2.6. Thermal Tests

Sample pieces with a mass of 6.5 ± 0.5 mg were placed in crucibles and analyzed by differential scanning calorimetry (DSC) in a DSC calorimeter from Mettler-Toledo Inc. (Schwerzenbach, Switzerland) under a nitrogen atmosphere (66 mL/min). The program consisted of a heating ramp from 30 to 250 °C to remove the thermal history, a cooling ramp down to 30 °C, and a second heating ramp to 350 °C. A heating rate of 10 °C/min were used in all cases. The maximum degree of crystallinity (X_{C_max}) was determined following Equation (1) [50]:

$$X_{C\_max} = \left[\frac{\Delta H_m}{\Delta H_m^0 \cdot (1-w)}\right] \cdot 100\ (\%) \quad (1)$$

where the melting enthalpy of PLA is ΔH_m (J/g), while the theoretical value of fully crystalline PLA is $\Delta H_m{}^\circ$ (J/g), taken as 93.7 J/g [51], and the weight fraction of PLA in the sample is $1 - w$.

Sample pieces with a mass of 15.5 ± 0.5 mg were placed in crucibles and subjected to thermogravimetric analysis (TGA) in a TGA/SDTA 851 thermobalance from Mettler-Toledo Inc. (Schwerzenbach, Switzerland). The samples were heated from 30 to 700 °C with a heating rate of 10 °C/min in nitrogen (66 mL/min). All the thermal tests were performed in triplicate.

2.7. Thermomechanical Tests

The Vicat softening point (VST) and heat deflection temperature (HDT) of the green composite pieces were determined using a VICAT/HDT station mod. VHDT 20 (Metrotec S.A., San Sebastián, Spain). VST tests were carried out following the ISO 306 with a load of 5 kg and a heating rate of 50 °C/h, whereas HDT measurements were performed according to the ISO 75 with a load of 320 g and a heating rate of 120 °C/h.

Rectangular samples of 40 mm × 10 mm × 4 mm were subjected to dynamic mechanical thermal analysis (DMTA) in torsion mode using an AR-G2 oscillatory rheometer from TA Instruments (New Castle, DE, USA) equipped with a special clamp system for solid samples. The temperature program consisted of a thermal ramp from 30 to 140 °C at a heating rate of 2 °C/min. The maximum deformation percentage (γ) was set to 0.1% at a frequency of 1 Hz. All the thermomechanical tests were performed in triplicate.

3. Results

3.1. Mechanical Performance of the Green Composite Pieces

Table 2 includes the mechanical properties of PLA and its green composite pieces to ascertain the effect of the TCHF content on PLA. One can notice that neat PLA resulted in a brittle piece with an elongation at break (ε_b) of nearly 9%. The tensile strength at yield (σ_y) was approximately 50 MPa and the tensile modulus (E_t) was 3.58 GPa. One can observe that the incorporation of both CHF and TCHF induced a progressive reduction in all the σ_y and ε_b values and an increase in E_t. For instance, with the addition of 20 wt% of CHF, the σ_y values diminished to 39.16 MPa and with the addition of 20 wt% of TCHF, the σ_y values diminished to 43.29 MPa, which represent a percentage decrease of 21% and 13%, respectively. Further reductions in the mechanical strength were observed at higher TCHF contents. For instance, the addition of 50 wt% of TCHF induced a percentage decrease of ~44%. Alternatively, the incorporation of CHF and TCHF improved the mechanical elasticity of the pieces. Thus, E_t increased to 3.48 GPa for the PLA/CHF composite and it was in the 3.59–4.45 GPa range for the PLA/TCHF composite pieces. This result suggests than an efficient dispersion and wettability of the torrefied filler in the biopolymer matrix was achieved. With regard to ductility, the addition of the lignocellulosic fillers led to a percentage decrease from 8.91%, for the neat PLA piece, to 6.98% for the PLA/20CH piece, 7.10% for the PLA/20TCH piece, and finally to the lowest value of 5.18% for the PLA/50TCH piece. It can be thus considered that the coffee waste derived fillers acted as a stress concentrator rather than a reinforcing element in the biopolymer matrix. Furthermore, this result highlights the lack of miscibility between both composite components and also the low particle–polymer interfacial interactions. This effect was, however, more noticeable in the piece containing the untreated CHF and, most notably, for the pieces filled with the highest TCHF contents. The latter observation may suggest that TCHF poorly disperses in the PLA matrix at high content [52,53]. Therefore, as also demonstrated by Lessa et al. [54] for chitosan films containing waste coffee grounds (WCG), high amounts of TCHF cannot be efficiently distributed in the biopolymer matrix.

Table 2. Mechanical properties of the injection-molded pieces of polylactide (PLA) with coffee husk flour (CHF) and torrefied coffee husk flour (TCHF) in terms of tensile modulus (E_t), tensile stress at yield (σ_y), elongation at break (ε_b), maximum flexural stress (σ_f), flexural modulus (E_f), Shore D hardness, and impact strength.

Piece	E_t (GPa)	σ_y (MPa)	ε_b (%)	E_f (GPa)	σ_f (MPa)	Shore D Hardness	Impact Strength (kJ/m^2)
PLA	3.58 ± 0.37	49.51 ± 1.92	8.91 ± 0.71	3.17 ± 0.19	98.88 ± 1.18	78.65 ± 2.64	25.10 ± 2.01
PLA/20CHF	3.48 ± 0.78	39.16 ± 0.69	6.98 ± 1.10	4.02 ± 0.20	84.88 ± 1.35	81.04 ± 1.38	12.00 ± 0.39
PLA/20TCHF	3.59 ± 0.53	43.29 ± 1.90	7.10 ± 0.47	4.33 ± 0.19	90.96 ± 2.18	81.10 ± 2.02	12.44 ± 2.17
PLA/30TCHF	3.66 ± 0.87	36.23 ± 1.83	5.92 ± 0.36	4.48 ± 0.27	73.58 ± 1.71	82.05 ± 1.80	8.85 ± 1.67
PLA/40TCHF	4.22 ± 1.07	30.33 ± 2.16	5.59 ± 0.45	5.10 ± 0.34	71.66 ± 2.21	82.10 ± 2.16	7.19 ± 0.31
PLA/50TCHF	4.45 ± 1.04	27.81 ± 1.65	5.18 ± 0.91	5.53 ± 0.40	61.42 ± 2.04	85.05 ± 1.19	7.19 ± 0.97

Nevertheless, one should also consider that this mechanical reduction was relatively low in terms of the high amount of lignocellulosic filler when compared with those attained in previously prepared green composites of PLA and coffee waste derived fillers. For instance, in the work reported by Suaduang et al. [55], the tensile strength at break (σ_b) of PLA films decreased from 79.20 MPa (transverse direction, TD) and 72.79 MPa (machine direction, MD) to 20.77 MPa (TD) and 18.31 MPa (MD) after the incorporation of 10 wt% of SCG. In another study, the σ_b value of PLA films decreased from 44.3 MPa to approximately 20 MPa after the incorporation of 40 wt% of SCG, which was attributed to the poor filler dispersion in the biopolymer matrix due to its low compatibility [56]. The improvement attained here may be attributed to the removal of impurities and other volatile TCHF components during torrefaction that would otherwise decrease the strength of interfacial interactions and adhesion between the filler and PLA matrix. Furthermore, the resultant increase in hydrophobicity of the lignocellulosic filler after torrefaction can also be a meaningful explanation for the improved mechanical properties attained in the green composite pieces with TCHF in comparison with the pieces based on untreated CHF. In this regard, one can expect that the higher presence of moisture in CHF can promote a larger decrease in the PLA's M_W during processing at high temperature due to hydrolysis of the ester groups, which can result in impairment of the final mechanical properties [57].

The flexural modulus (E_f) and strength (σ_f) of the unfilled PLA piece were 3.17 GPa and nearly 99 MPa, respectively. Similar to the tensile properties, the incorporation of TCHF yielded a progressive increase in E_f and a reduction in σ_f. In particular, the green composite pieces filled with 50 wt% of TCHF showed a E_f value of 5.53 GPa, which is approximately 1.7 times higher than that of the unfilled PLA, whereas σ_f was reduced to 61.42 MPa. Some previous studies have reported similar mechanical properties of green composites based on coffee waste derived fillers. For instance, Dominici et al. [58] showed a worsening of tensile strength and ductility when CSS was incorporated into bio-based high-density polyethylene (bio-HDPE). The described decrease of the mechanical properties was ascribed to the low aspect ratio of the CSS particles as well as their poor adhesion to the biopolymer matrix. In this regard, the use of coupling agents has improved both the tensile strength and ductility of PLA composites prepared with SCG, though the improvements achieved were relatively low and their overall and specific migration limits (OML and SML) should also be carefully determined [59].

One can also observe that, as the TCHF content increased, hardness also increased. The Shore D hardness value increased from 78.65, for the unfilled PLA piece, to 85.05 for the PLA/50TCH piece. This change in the mechanical behavior corresponds to the typical effect of a relatively stiff lignocellulosic filler on a polymer matrix [60]. In contrast to hardness, the impact absorbed energy of PLA progressively decreased after the TCHF incorporation. The impact energy of the neat PLA piece was 25.10 kJ/m^2 and it was reduced down to 7.19 kJ/m^2 for the green composite piece filled with 50 wt% of TCHF. This notable decrease in toughness can be related to the poor matrix–particle interactions, which could promote microcrack formations and stress concentrations during the impact test [60]. Therefore, the addition of the coffee waste derived lignocellulosic filler led to a more elastic and stiffer material, but the matrix-filler continuity was not achieved, and an early fracture was attained. Thus, in terms of mechanical performance, PLA pieces containing 20 wt% of TCHF were slightly more resistant but less ductile, showing characteristics comparable to those of neat PLA.

3.2. Aspect and Morphology of the Green Composite Pieces

Figure 3 shows the visual aspect of the injection-molded pieces of PLA, PLA filled with 20 wt% of CHF and TCHF, that is, PLA/20CHF and PLA/20TCHF, respectively. The increase in the characteristic dark brown color produced in the pieces of the green composites is due to the Maillard reaction of coffee husk, which took place during the torrefaction and its subsequent melt processing with PLA, leading to formation of melanoidins [35].

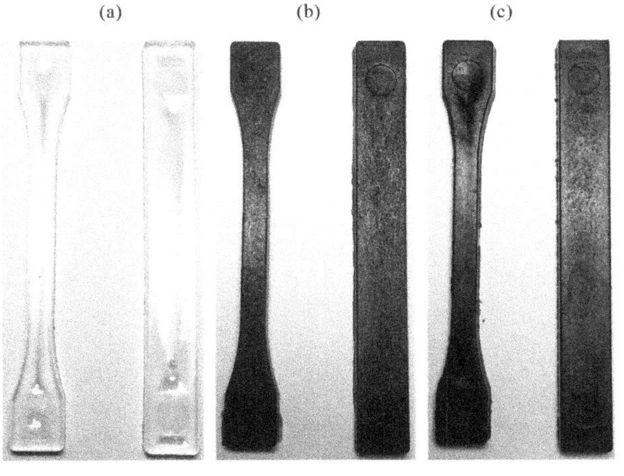

Figure 3. Visual aspect of the injection-molded pieces of: (**a**) Neat polylactide (PLA); (**b**) PLA filled with 20 wt% of coffee husk flour (CHF); (**c**) PLA filled with 20 wt% of torrefied coffee husk flour (TCHF).

The fracture surfaces after the impact tests of the neat PLA piece and the PLA/CHF and PLA/TCHF pieces at different filler contents are shown in Figure 4. The unfilled PLA piece, shown in Figure 4a, presented a typical brittle fracture with low plastic deformation [61]. In Figure 4b, one can see that the incorporation of untreated CHF led to a similar fracture surface but yet softer, indicating that the toughness of the piece was reduced. One can also observe the formation of a large gap between the coffee waste derived lignocellulosic fillers and the biopolymer PLA matrix, which points to their weak interactions [62,63]. The morphology of the green composite based on 20 wt% of TCHF, shown in Figure 4c, changed to a rougher surface, also showing a reduction in the filler-to-matrix gaps. However, no enhanced yielding was detected, which is in agreement with the ductility decrease reported in Table 2. In the case of the green composites prepared at higher TCHF contents, shown in Figure 4d–f, one can see that the fracture surface roughness increased with the filler addition. Moreover, the FESEM micrographs displayed some large voids, microcracks, and gaps in the interface between both components, that is, PLA and TCHF. This phenomenon led to a stiffer material, but with reduced impact absorbed energy, as also shown during the mechanical analysis. One can also observe in the FESEM micrographs that the PLA/TCHF pieces prepared at the highest filler contents, that is, 40 wt% and 50 wt%, showed the co-presence of single particles and irregular aggregates, being distributed randomly within the PLA matrix. Similar observations were described for pectin/coffee ground composites in which the formation of aggregates was ascribed to the prevalence of particle–particle interactions over the forces between the composite components [64].

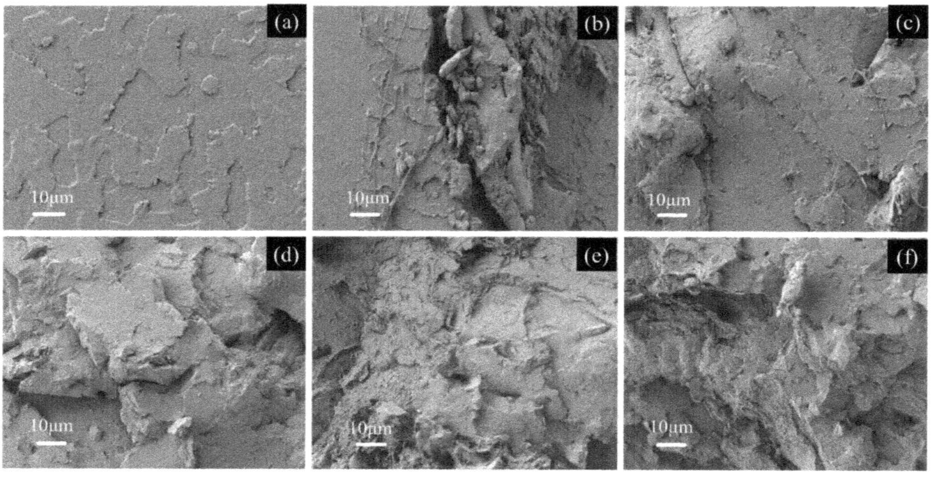

Figure 4. Field emission scanning electron microscopy (FESEM) images of the fracture surfaces of the injection-molded pieces of polylactide (PLA) with coffee husk flour (CHF) and torrefied coffee husk flour (TCHF): (**a**) Neat PLA; (**b**) PLA/20CHF; (**c**) PLA/20TCHF; (**d**) PLA/30TCH; (**e**) PLA/40TCH; (**f**) PLA/50TCH. Images were taken at 500× with scale markers of 10 μm.

3.3. *Thermal Performance of the Green Composite Pieces*

Figure 5 shows the DSC thermograms of PLA and its green composites with CHF and TCHF. Table 3 gathers the main thermal properties obtained from the cooling and the second heating scans, once the thermal history was removed from the samples. The neat PLA sample showed a glass transition temperature (T_g) of 62.3 °C and the incorporation of TCHF led to a slight decrease in the values up to 60.7 °C for the green composite filled at 50 wt%. In all curves, a low-intense endothermic peak was observed, which can be ascribed to the aging process of the biopolyester [42]. It can also be observed in Figure 5a that, during cooling, the PLA samples did not crystallize from the melt.

In contrast, in Figure 5b, one can see that the neat PLA samples showed an exothermic peak centered at 122.7 °C for neat PLA, which is ascribed to the biopolyester chain rearrangement during cold crystallization. The green composites showed slightly higher values of both cold crystallization temperature (T_{cc}) and cold crystallization enthalpy (ΔH_{cc}), showing the highest values for TCHF contents of 20 wt% and 30 wt%, respectively. The cold crystallization process is an exothermic process that takes place when the biopolymer chains become flexible, above PLA's T_g, which is caused by a primary nucleation, by which an ordered state is achieved. Since more PLA mass could cold crystallize, one can consider that intermediate filler amounts favored crystallization of PLA. A second endothermic peak, corresponding to the melting temperature (T_m), was observed at 151.3 °C for neat PLA, which slightly increased for filler contents of 20 wt% and then decreased for higher contents. An anticipation of the melting phenomena was also observed in HDPE by Arrigo et al. [65] after the addition of biochar particles derived from waste coffee grounds. This behavior was explained by the presence of specific interactions occurring between the biopolymer chains and the particle interfaces. In fact, the lower T_m values attained at the highest contents, that is, 40 wt% and 50 wt% of TCHF, in combination with the higher ΔH_{cc} values, suggest that the biopolymer chains were immobilized onto the surface of the lignocellulosic fillers. The maximum degree of crystallinity, that is, X_{C_max}, which does not consider the crystals that are formed during cold crystallization [50], increased progressively up to a TCHF content of 30 wt% and then slightly lower values were observed. This result indicates that the lignocellulosic fillers induced a nucleating effect that favored the formation of crystals in the green composites. In the case of the lower X_{C_max} values for the green composites containing 40 wt% and 50 wt% of TCHF, one can consider that the fillers presence at these contents also supposed an obstacle for the biopolymer chains to form stable crystals, generating a negative effect on crystallinity and leading to the formation of a more amorphous structure [66–68]. Furthermore, one can also observe that the torrefaction process did not yield any significant effect on the thermal properties of PLA at the studied filler loading, suggesting that nucleation was based on a physical phenomenon.

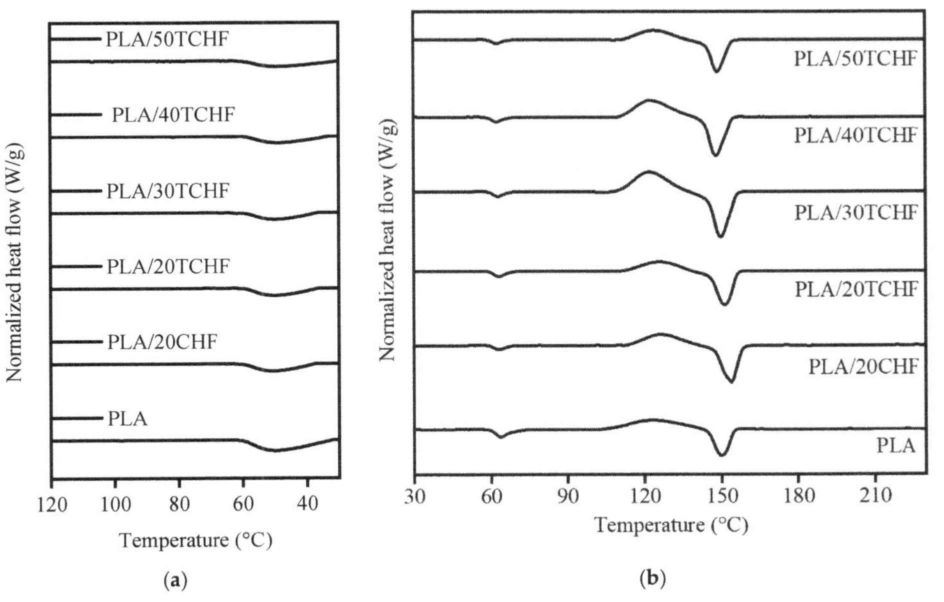

Figure 5. Differential scanning calorimetry (DSC) thermograms taken during (**a**) cooling and (**b**) second heating of the injection-molded pieces of polylactide (PLA) with coffee husk flour (CHF) and torrefied coffee husk flour (TCHF).

Table 3. Thermal properties of the injection-molded pieces of polylactide (PLA) with coffee husk flour (CHF) and torrefied coffee husk flour (TCHF) in terms of glass transition temperature (T_g), cold crystallization temperature (T_{cc}), melting temperature (T_m), cold crystallization enthalpy (ΔH_{cc}), melting enthalpy (ΔH_m), and maximum degree of crystallinity (X_{C_max}).

Piece	T_g (°C)	T_{cc} (°C)	ΔH_{cc} (J/g)	T_m (°C)	ΔH_m (J/g)	X_{C_max} (%)
PLA	62.3 ± 0.4	122.7 ± 1.9	4.30 ± 0.35	151.3 ± 0.4	11.32 ± 0.99	12.08 ± 1.80
PLA/20CHF	61.3 ± 0.3	124.7 ± 1.3	8.04 ± 0.42	153.7 ± 0.5	15.17 ± 1.10	20.24 ± 1.60
PLA/20TCHF	61.8 ± 0.4	125.9 ± 1.0	7.10 ± 0.55	152.6 ± 0.3	15.92 ± 0.75	21.24 ± 2.02
PLA/30TCHF	61.3 ± 0.4	122.1 ± 0.3	17.80 ± 0.76	150.7 ± 0.6	19.59 ± 0.14	29.87 ± 2.35
PLA/40TCHF	61.3 ± 0.8	121.8 ± 0.6	12.27 ± 0.94	149.9 ± 0.5	14.35 ± 0.59	25.52 ± 1.49
PLA/50TCHF	60.7 ± 0.5	123.2 ± 0.9	10.27 ± 0.35	149.6 ± 0.4	12.05 ± 0.99	25.72 ± 0.80

The TGA and first derivate thermogravimetric (DTG) curves of the injection-molded pieces of PLA and its green composite pieces with CHF and TCHF are shown in Figure 6, whereas Table 4 shows the relevant values extracted from the curves. In relation to the thermal degradation profiles of CHF and TCHF, in both cases they presented three stages; however, the torrefied fillers showed significantly higher stability. The first mass loss, which occurred below 100 °C, being in both cases less than 5% but slightly higher for CHF, is associated with the evaporation of moisture and volatiles present in the lignocellulosic fillers. The second one occurred between approximately 240 and 380 °C, was the most intense, and was associated with a mass loss of over 60% and 35% for CHF and TCHF, respectively. This degradation stage, which was also more intense for CHF, is mainly caused by the thermal decomposition of the molecular bonds of hemicellulose, cellulose, and small amounts of lignin, which overlapped with the thermal degradation of PLA. The third stage started at 385 °C and continued for the whole thermal range tested. This last mass loss is due to the degradation of the lignin fraction present in CHF and TCHF [11,69]. In this regard, de Brito et al. [70] showed that green coffee cake (GCC), a solid by-product of the coffee processing industry, resulting from the extraction of oil by the cold pressing technique, also presented three regions of mass loss, similar to other lignocellulosic biomasses. The first one was observed between 50 and 100 °C, the second one between 200 and 300 °C, and the last one between 275 and 400 °C. The authors ascribed the second and third mass losses to hemicellulose depolymerization in combination with temperature-induced glycosidic cleavage and to lignin components and cellulose degradation, respectively. Finally, after the degradation steps, a significant residue of approximately 25% and 40% was found for CHF and TCHF, respectively, remarking the higher stability of the carbon formed during the torrefaction step. The TGA results also indicated that TCHF was thermally stable up to more than 250 °C, which makes it useful as a reinforcing agent in most WPCs.

In the case of PLA, it can be observed that the biopolymer degraded in a quick step process at approximately 350 °C. The main products generated during the thermal degradation of PLA have been ascribed to lactic acid, oligomers of lactic acid (OLAs), acetaldehydes, carbon dioxide, carbon monoxide, and ketones [60]. The TGA curves of the PLA/TCHF pieces showed a slight increase in the thermal stability for the lowest filler content. In particular, the DTG curves showed an increase in the pyrolysis peak related to the degradation temperature (T_{deg}) from 347.4 °C, for neat PLA, up to 354.7 °C, for the PLA/20TCHF composite. The increase in the thermal stability of the green composite pieces can be related to the "barrier" effect of the decomposition of the biopolymer material by the lignocellulosic fillers, which promoted an increase of the degradation temperature [70]. However, a decrease in the temperature at the mass loss of 5% ($T_{5\%}$), which corresponds to the onset degradation temperature, was observed for TCHF contents above 30 wt%. The lower T_5% values attained at higher filler contents can be related to the "chimney" effect of the coffee waste lignocellulosic fillers, which prevails their "barrier" effect, and favored the movement of heat in the PLA matrix [57]. In particular, $T_{5\%}$ was reduced from 320.4 °C, for the unfilled PLA piece, down to approximately 300 °C for TCHF contents of 40 wt% and 50 wt%. Therefore, a slight improvement in the PLA's thermal stability was attained with the addition of 20 wt% of TCHF, whereas a negative effect was observed for the pieces filled with higher TCHF contents. It is noteworthy that a higher thermal stability was observed in comparison to the PLA composites prepared with CSS by Totaro et al. [71], who ascribed the decrease to both the metals present and the moisture released by the fillers, which can catalyze degradation and induce hydrolysis of the ester groups, respectively. Since lower thermal stability values were attained for the PLA/20CHF piece, the improvement attained herein is an indication that the less thermally stable compounds, such as hemicellulose and cellulose, were successfully removed during the torrefaction step [33], leading to a material with higher lignocellulosic content and, hence, improved thermal stability. Furthermore, as expected, one can observe that adding TCHF caused the char residue to increase. Such high char content is also related to the torrefaction step, which can generate a higher percentage of carbon compared with the non torrefied lignocellulosic filler [33].

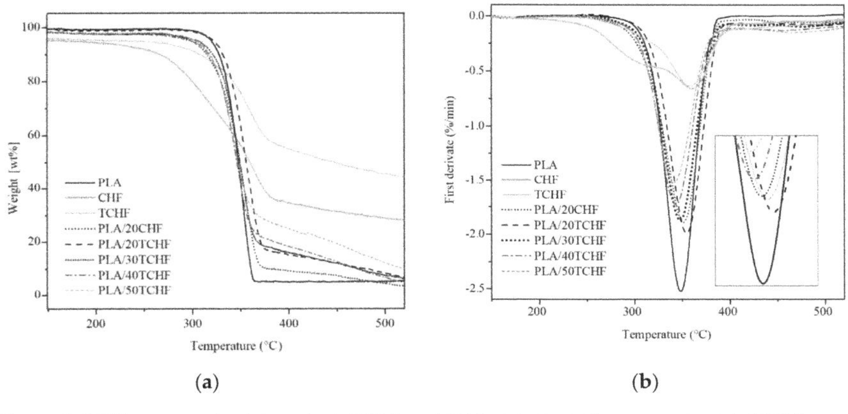

Figure 6. (**a**) Thermogravimetric analysis (TGA) and (**b**) first derivate thermogravimetric (DTG) curves of the injection-molded pieces of polylactide (PLA) with coffee husk flour (CHF) and torrefied coffee husk flour (TCHF).

Table 4. Main thermal degradation parameters of the injection-molded pieces of polylactide (PLA) with coffee husk flour (CHF) and torrefied coffee husk flour (TCHF) in terms of temperature at mass loss of 5 % ($T_{5\%}$), degradation temperature (T_{deg}), and residual mass at 700 °C.

Piece	$T_{5\%}$ (°C)	T_{deg} (°C)	Residual Mass (%)
CHF	237.8 ± 1.2	360.6 ± 2.2	25.2 ± 0.3
TCHF	252.1 ± 1.3	357.8 ± 1.3	39.6 ± 0.3
PLA	320.4 ± 2.3	347.4 ± 1.4	5.6 ± 0.3
PLA/20CHF	298.9 ± 1.6	351.1 ± 2.2	3.2 ± 0.6
PLA/20TCHF	322.4 ± 1.7	354.7 ± 2.2	5.6 ± 0.5
PLA/30TCHF	306.0 ± 1.3	347.7 ± 1.7	8.1 ± 0.4
PLA/40TCHF	299.8 ± 1.8	343.5 ± 1.8	9.7 ± 0.7
PLA/50TCHF	301.1 ± 1.3	341.4 ± 1.6	10.9 ± 0.8

3.4. Thermomechanical Performance of the Green Composite Pieces

Table 5 shows the VST and HDT values of the injection-molded pieces of neat PLA and PLA/TCHF, which are important thermomechanical parameters that can be used to estimate their upper useful temperature limit for uses in, for instance, packaging or disposables. VST, which reflects the material's softening by temperature rises, increased progressively from 56.6 °C, for neat PLA, to 59.9 °C, for PLA/50TCH, respectively, following an evolution similar to that found in hardness. This result is in agreement with the study of Huang et al. [52], who showed that the VST of HDPE increased with the loading of coffee husk fiber (CHF), which was attributed to a fiber reinforcement that prevented deformation of the composites. Concerning HDT, the values evolved in a similar manner to VST, showing a slight and progressive increase from 51 °C, for neat PLA, to 54.1 °C for PLA/50TCHF. This enhanced thermomechanical resistance can be attributed to the higher modulus and amount of crystallinity, since both parameters increased with increasing TCHF content. Similar results were obtained by Sarasini et al. [72], who showed that the softening point of PBAT/poly(3-hydroxybutyrate-*co*-3-hydroxyvalerate) (PHBV) blends increased up to 3 °C with the incorporation of CSS.

Table 5. Thermomechanical properties of the injection-molded pieces of polylactide (PLA) with torrefied coffee husk flour (TCHF) in terms of Vicat softening point (VST) and heat deflection temperature (HDT).

Piece	VST (°C)	HDT (°C)
PLA	56.6 ± 0.6	51.0 ± 0.8
PLA/20TCHF	57.8 ± 0.8	52.3 ± 0.8
PLA/30TCHF	58.1 ± 0.9	52.8 ± 0.9
PLA/40TCHF	59.4 ± 0.6	53.2 ± 0.7
PLA/50TCHF	59.9 ± 0.3	54.1 ± 0.8

Finally, Figure 7 shows the mechanical performance of the green composite pieces under dynamic load conditions as a function of temperature. Figure 7a displays the evolution of the storage modulus with temperature. In the temperature range between 30 and 50 °C, below the glass transition region, the incorporation of TCHF produced an increase in the storage modulus of PLA. This behavior indicates that the addition of the TCHF filler led to a stiffer material since the lignocellulosic particles restricted the biopolymer chain mobility and also promoted crystallization as described above during the thermal characterization. From approximately 55 °C, a prominent decrease in the storage modulus due the glass transition process can be seen [41]. In addition, from nearly 90 °C, an increment in the storage modulus was detected. This phenomenon is associated with the cold crystallization of PLA. During this process, the biopolymer chains reorganize into a more compact structure, leading to an increase in the storage modulus [73]. Furthermore, as more TCHF powder was added, in agreement with the previous DSC analysis, a slight decrease in the cold crystallization temperature was noticed.

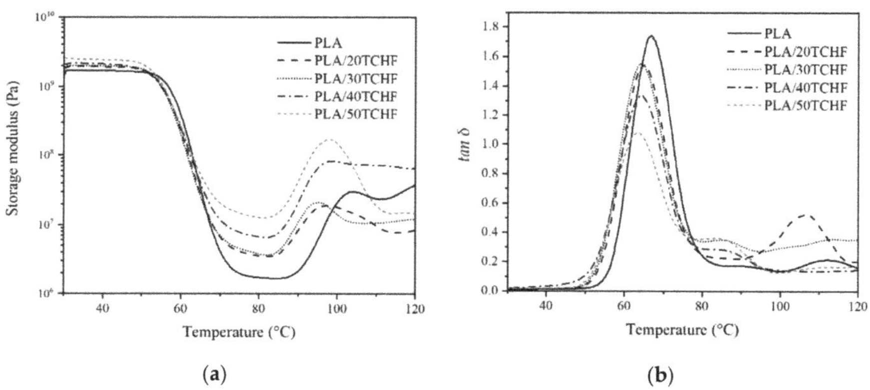

Figure 7. Evolution as a function of temperature of the (**a**) storage modulus and (**b**) dynamic damping factor (*tan δ*) of the injection-molded pieces of polylactide (PLA) with torrefied coffee husk flour (TCHF).

Figure 7b displays the evolution as a function of temperature of the damping factor (*tan δ*) for the unfilled PLA piece and the PLA/TCHF composite pieces. The peak for the neat PLA progressively shifted to lower values as more filler was added. It particularly decreased from 68.7 °C, for the neat PLA piece, down to values in the 62–64 °C range. This peak is ascribed to the alpha (α)-transition of

PLA, being in relation to the PLA's T_g. One can expect that with a greater interaction between filler and biopolymer matrix, which can be achieved by modifying the fiber surface during the torrefaction step, there would be a restriction on the movement of the macromolecular chains located on the surface of the fillers and, hence, an increase in T_g [74]. However, the reduction attained in T_g, also observed by DSC, suggests that the fillers allowed an increase in the free volume in the green composites, thus indicating low compatibility or poor interaction with PLA [70]. This behavior confirms that the movement of the amorphous PLA was promoted by the coffee waste derived lignocellulosic fillers and it suggests that shorter biopolymer chains were formed due to a chain scission. Furthermore, the nucleating effect of TCHF was further corroborated by the gradual reduction in the *tan* δ peak intensities of the green composite pieces. The latter indicates that the number of molecules experimenting glass transition was reduced by the fillers' presence [75].

4. Conclusions

The results achieved in this study demonstrate that coffee husk, a large food processing by-product, can be used in the form of flour as a reinforcing filler in green composites. Torrefaction, a mild form of pyrolysis at 250 °C in reduced atmosphere, yielded more thermally stable biomass that was also more hydrophobic and, hence, more compatible with PLA. Characterization of the injection-molded PLA/TCHF pieces showed a positive effect on the overall mechanical improvement and a thermal stability enhancement with the addition of 20 wt% of TCHF. However, the green composites filled with 30 wt%, 40 wt%, and 50 wt% showed lower performance, although they are still interesting due to the high amount of waste that can be valorized. Both elasticity and hardness increased, whereas ductility and toughness decreased, for all of the TCHF contents due to the lack of interaction as observed by FESEM. In contents of up to 30 wt%, the lignocellulosic fillers induced a nucleating effect that favored the formation of PLA crystals and showed sufficient thermal stability. Furthermore, the addition of TCHF led to slightly higher VST and HDT values and the thermomechanical resistance of PLA was improved, meaning the resultant pieces were able to withstand temperatures of up to nearly 60 °C. In a Circular Bioeconomy context, the resultant PLA/TCHF composite pieces prepared with 20 wt% of the torrefied coffee waste derived filler can be used for rigid packaging, cups and containers, and food contact disposables, such as cutlery and plates, where novel compostable and sustainable materials with high stiffness and low thermal resistance are required.

Author Contributions: Conceptualization, O.F. and J.A.A.-P.; methodology, D.L.O.-B., Y.J.R.-O. and J.A.A.-P.; validation, O.F. and S.T.-G.; formal analysis, O.F.; investigation, D.L.O.-B., Y.J.R.-O. and J.A.A.-P.; data curation, D.L.O.-B. and Y.J.R.-O.; writing-original draft preparation, J.A.A.-P. and O.F.; writing-review and editing, S.T.-G.; visualization, O.F. and J.A.A.-P.; funding acquisition, O.F., J.A.A.-P. and S.T.-G. All authors have read and agreed to the published version of the manuscript.

Funding: This research work was funded by the Spanish Ministry of Science and Innovation (MICI), project numbers RTI2018-097249-B-C21 and MAT2017-84909-C2-2-R, and by the Industrial University of Santander.

Acknowledgments: S.T.-G. acknowledges MICI for his Juan de la Cierva–Incorporación contract (IJCI-2016-29675). Microscopy services at UPV are also acknowledged for their help in collecting the FESEM images.

Conflicts of Interest: The authors declare no conflict of interest.

References

1. Guillard, V.; Gaucel, S.; Fornaciari, C.; Angellier-Coussy, H.; Buche, P.; Gontard, N. The next generation of sustainable food packaging to preserve our environment in a circular economy context. *Front. Nutr.* **2018**, *5*. [CrossRef] [PubMed]
2. Jurgilevich, A.; Birge, T.; Kentala-Lehtonen, J.; Korhonen-Kurki, K.; Pietikäinen, J.; Saikku, L.; Schösler, H. Transition towards circular economy in the food system. *Sustainability* **2016**, *8*, 69. [CrossRef]
3. Ramesh, M. Flax (Linum usitatissimum L.) fibre reinforced polymer composite materials: A review on preparation, properties and prospects. *Prog. Mater. Sci.* **2019**, *102*, 109–166. [CrossRef]

4. Agüero, A.; Lascano, D.; Garcia-Sanoguera, D.; Fenollar, O.; Torres-Giner, S. Valorization of linen processing by-products for the development of injection-molded green composite pieces of polylactide with improved performance. *Sustainability* **2020**, *12*, 652. [CrossRef]
5. Melendez-Rodriguez, B.; Torres-Giner, S.; Aldureid, A.; Cabedo, L.; Lagaron, J.M. Reactive melt mixing of poly(3-hydroxybutyrate)/rice husk flour composites with purified biosustainably produced poly(3-hydroxybutyrate-*co*-3-hydroxyvalerate). *Materials* **2019**, *12*, 2152. [CrossRef]
6. Jothibasu, S.; Mohanamurugan, S.; Vijay, R.; Lenin Singaravelu, D.; Vinod, A.; Sanjay, M.R. Investigation on the mechanical behavior of areca sheath fibers/jute fibers/glass fabrics reinforced hybrid composite for light weight applications. *J. Ind. Text.* **2018**, *49*, 1036–1060. [CrossRef]
7. Kumaran, P.; Mohanamurugan, S.; Madhu, S.; Vijay, R.; Lenin Singaravelu, D.; Vinod, A.; Sanjay, M.R.; Siengchin, S. Investigation on thermo-mechanical characteristics of treated/untreated Portunus sanguinolentus shell powder-based jute fabrics reinforced epoxy composites. *J. Ind. Text.* **2019**. [CrossRef]
8. Quiles-Carrillo, L.; Montanes, N.; Garcia-Garcia, D.; Carbonell-Verdu, A.; Balart, R.; Torres-Giner, S. Effect of different compatibilizers on injection-molded green composite pieces based on polylactide filled with almond shell flour. *Compos. Part B Eng.* **2018**, *147*, 76–85. [CrossRef]
9. Quiles-Carrillo, L.; Montanes, N.; Sammon, C.; Balart, R.; Torres-Giner, S. Compatibilization of highly sustainable polylactide/almond shell flour composites by reactive extrusion with maleinized linseed oil. *Ind. Crops Prod.* **2018**, *111*, 878–888. [CrossRef]
10. Montava-Jordà, S.; Quiles-Carrillo, L.; Richart, N.; Torres-Giner, S.; Montanes, N. Enhanced interfacial adhesion of polylactide/poly(ε-caprolactone)/walnut shell flour composites by reactive extrusion with maleinized linseed oil. *Polymers* **2019**, *11*, 758. [CrossRef]
11. Quiles-Carrillo, L.; Montanes, N.; Lagaron, J.M.; Balart, R.; Torres-Giner, S. On the use of acrylated epoxidized soybean oil as a reactive compatibilizer in injection-molded compostable pieces consisting of polylactide filled with orange peel flour. *Polym. Int.* **2018**, *67*, 1341–1351. [CrossRef]
12. Torres-Giner, S.; Hilliou, L.; Melendez-Rodriguez, B.; Figueroa-Lopez, K.J.; Madalena, D.; Cabedo, L.; Covas, J.A.; Vicente, A.A.; Lagaron, J.M. Melt processability, characterization, and antibacterial activity of compression-molded green composite sheets made of poly(3-hydroxybutyrate-*co*-3-hydroxyvalerate) reinforced with coconut fibers impregnated with oregano essential oil. *Food Packag. Shelf Life* **2018**, *17*, 39–49. [CrossRef]
13. Chaitanya, S.; Singh, I. Processing of PLA/sisal fiber biocomposites using direct- and extrusion-injection molding. *Mater. Manuf. Process.* **2017**, *32*, 468–474. [CrossRef]
14. Thyavihalli Girijappa, Y.G.; Mavinkere Rangappa, S.; Parameswaranpillai, J.; Siengchin, S. Natural fibers as sustainable and renewable resource for development of eco-friendly composites: A comprehensive review. *Front. Mater.* **2019**, *6*. [CrossRef]
15. Ramesh, M.; Deepa, C.; Kumar, L.R.; Sanjay, M.R.; Siengchin, S. Life-cycle and environmental impact assessments on processing of plant fibres and its bio-composites: A critical review. *J. Ind. Text.* **2020**. [CrossRef]
16. Adekomaya, O.; Jamiru, T.; Sadiku, R.; Huan, Z. A review on the sustainability of natural fiber in matrix reinforcement – A practical perspective. *J. Reinf. Plast. Compos.* **2015**, *35*, 3–7. [CrossRef]
17. Agüero, Á.; Garcia-Sanoguera, D.; Lascano, D.; Rojas-Lema, S.; Ivorra-Martinez, J.; Fenollar, O.; Torres-Giner, S. Evaluation of different compatibilization strategies to improve the performance of injection-molded green composite pieces made of polylactide reinforced with short flaxseed fibers. *Polymers* **2020**, *12*, 821. [CrossRef]
18. Lachenmeier, D.W.; Teipel, J.; Scharinger, A.; Kuballa, T.; Walch, S.G.; Grosch, F.; Bunzel, M.; Okaru, A.O.; Schwarz, S. Fully automated identification of coffee species and simultaneous quantification of furfuryl alcohol using NMR spectroscopy. *J. AOAC Int.* **2020**, *103*, 306–314. [CrossRef]
19. Murthy, P.S.; Madhava Naidu, M. Sustainable management of coffee industry by-products and value addition—A review. *Resour. Conserv. Recycl.* **2012**, *66*, 45–58. [CrossRef]
20. Janissen, B.; Huynh, T. Chemical composition and value-adding applications of coffee industry by-products: A review. *Resour. Conserv. Recycl.* **2018**, *128*, 110–117. [CrossRef]
21. Mussatto, S.I.; Machado, E.M.S.; Martins, S.; Teixeira, J.A. Production, composition, and application of coffee and its industrial residues. *Food Bioprocess Technol.* **2011**, *4*, 661. [CrossRef]

22. Klingel, T.; Kremer, J.I.; Gottstein, V.; Rajcic de Rezende, T.; Schwarz, S.; Lachenmeier, D.W. A review of coffee by-products including leaf, flower, cherry, husk, silver skin, and spent grounds as novel foods within the European Union. *Foods* **2020**, *9*, 665. [CrossRef] [PubMed]
23. Roussos, S.; de Los Angeles Aquiáhuatl, M.; Del Refugio Trejo-Hernández, M.; Gaime Perraud, I.; Favela, E.; Ramakrishna, M.; Raimbault, M.; Viniegra-González, G. Biotechnological management of coffee pulp—isolation, screening, characterization, selection of caffeine-degrading fungi and natural microflora present in coffee pulp and husk. *Appl. Microbiol. Biotechnol.* **1995**, *42*, 756–762. [CrossRef]
24. Narita, Y.; Inouye, K. Review on utilization and composition of coffee silverskin. *Food Res. Int.* **2014**, *61*, 16–22. [CrossRef]
25. Woldesenbet, A.G.; Woldeyes, B.; Chandravanshi, B.S. Bio-ethanol production from wet coffee processing waste in Ethiopia. *SpringerPlus* **2016**, *5*, 1903. [CrossRef]
26. Mendoza Martinez, C.L.; Alves Rocha, E.P.; Oliveira Carneiro, A.d.C.; Borges Gomes, F.J.; Ribas Batalha, L.A.; Vakkilainen, E.; Cardoso, M. Characterization of residual biomasses from the coffee production chain and assessment the potential for energy purposes. *Biomass Bioenergy* **2019**, *120*, 68–76. [CrossRef]
27. Hachicha, R.; Rekik, O.; Hachicha, S.; Ferchichi, M.; Woodward, S.; Moncef, N.; Cegarra, J.; Mechichi, T. Co-composting of spent coffee ground with olive mill wastewater sludge and poultry manure and effect of Trametes versicolor inoculation on the compost maturity. *Chemosphere* **2012**, *88*, 677–682. [CrossRef]
28. Ballesteros, L.F.; Teixeira, J.A.; Mussatto, S.I. Chemical, functional, and structural properties of spent coffee grounds and coffee silverskin. *Food Bioprocess Technol.* **2014**, *7*, 3493–3503. [CrossRef]
29. Murthy, P.S.; Naidu, M.M. Production and application of xylanase from penicillium sp. utilizing coffee by-products. *Food Bioprocess Technol.* **2012**, *5*, 657–664. [CrossRef]
30. Low, J.H.; Rahman, W.A.W.A.; Jamaluddin, J. Structural elucidation of tannins of spent coffee grounds by CP-MAS 13C NMR and MALDI-TOF MS. *Ind. Crops Prod.* **2015**, *69*, 456–461. [CrossRef]
31. Franklin, G.; Dias, A.C.P. Chlorogenic acid participates in the regulation of shoot, root and root hair development in Hypericum perforatum. *Plant Physiol. Biochem.* **2011**, *49*, 835–842. [CrossRef]
32. García, A.; Gandini, A.; Labidi, J.; Belgacem, N.; Bras, J. Industrial and crop wastes: A new source for nanocellulose biorefinery. *Ind. Crops Prod.* **2016**, *93*, 26–38. [CrossRef]
33. Moustafa, H.; Guizani, C.; Dupont, C.; Martin, V.; Jeguirim, M.; Dufresne, A. Utilization of torrefied coffee grounds as reinforcing agent to produce high-quality biodegradable pbat composites for food packaging applications. *ACS Sustain. Chem. Eng.* **2017**, *5*, 1906–1916. [CrossRef]
34. Wang, Z.; Dadi Bekele, L.; Qiu, Y.; Dai, Y.; Zhu, S.; Sarsaiya, S.; Chen, J. Preparation and characterization of coffee hull fiber for reinforcing application in thermoplastic composites. *Bioengineered* **2019**, *10*, 397–408. [CrossRef] [PubMed]
35. García-García, D.; Carbonell, A.; Samper, M.D.; García-Sanoguera, D.; Balart, R. Green composites based on polypropylene matrix and hydrophobized spend coffee ground (SCG) powder. *Compos. Part B Eng.* **2015**, *78*, 256–265. [CrossRef]
36. Franca, A.S.; Oliveira, L.S. Coffee processing solid wastes: Current uses and future perspectives. In *Agricultural Wastes*; Nova Science Publishers, Inc.: Hauppauge, NY, USA, 2009; pp. 155–190.
37. Gouvea, B.M.; Torres, C.; Franca, A.S.; Oliveira, L.S.; Oliveira, E.S. Feasibility of ethanol production from coffee husks. *Biotechnol. Lett.* **2009**, *31*, 1315–1319. [CrossRef]
38. Arias, B.; Pevida, C.; Fermoso, J.; Plaza, M.G.; Rubiera, F.; Pis, J.J. Influence of torrefaction on the grindability and reactivity of woody biomass. *Fuel Process. Technol.* **2008**, *89*, 169–175. [CrossRef]
39. Chen, W.-H.; Peng, J.; Bi, X.T. A state-of-the-art review of biomass torrefaction, densification and applications. *Renew. Sustain. Energy Rev.* **2015**, *44*, 847–866. [CrossRef]
40. Ribeiro, J.M.C.; Godina, R.; Matias, J.C.d.O.; Nunes, L.J.R. Future perspectives of biomass torrefaction: Review of the current state-of-the-art and research development. *Sustainability* **2018**, *10*, 2323. [CrossRef]
41. Hamad, K.; Kaseem, M.; Ayyoob, M.; Joo, J.; Deri, F. Polylactic acid blends: The future of green, light and tough. *Prog. Polym. Sci.* **2018**, *85*, 83–127. [CrossRef]
42. Quiles-Carrillo, L.; Montanes, N.; Pineiro, F.; Jorda-Vilaplana, A.; Torres-Giner, S. Ductility and toughness improvement of injection-molded compostable pieces of polylactide by melt blending with poly(ε-caprolactone) and thermoplastic starch. *Materials* **2018**, *11*, 2138. [CrossRef]
43. Chen, Y.; Geever, L.M.; Killion, J.A.; Lyons, J.G.; Higginbotham, C.L.; Devine, D.M. Review of multifarious applications of poly(lactic acid). *Polym. Plast. Technol. Eng.* **2016**, *55*, 1057–1075. [CrossRef]

44. Risyon, N.P.; Othman, S.H.; Basha, R.K.; Talib, R.A. Characterization of polylactic acid/halloysite nanotubes bionanocomposite films for food packaging. *Food Packag. Shelf Life* **2020**, *23*, 100450. [CrossRef]
45. Cardoso, R.M.; Silva, P.R.L.; Lima, A.P.; Rocha, D.P.; Oliveira, T.C.; do Prado, T.M.; Fava, E.L.; Fatibello-Filho, O.; Richter, E.M.; Muñoz, R.A.A. 3D-Printed graphene/polylactic acid electrode for bioanalysis: Biosensing of glucose and simultaneous determination of uric acid and nitrite in biological fluids. *Sens. Actuators B Chem.* **2020**, *307*, 127621. [CrossRef]
46. Harris, A.M.; Lee, E.C. Heat and humidity performance of injection molded PLA for durable applications. *J. Appl. Polym. Sci.* **2010**, *115*, 1380–1389. [CrossRef]
47. Kumar, S.; Bhatnagar, N.; Ghosh, A.K. Effect of enantiomeric monomeric unit ratio on thermal and mechanical properties of poly(lactide). *Polym. Bull.* **2016**, *73*, 2087–2104. [CrossRef]
48. Rocha, D.B.; de Souza, A.G.; Szostak, M.; Rosa, D.d.S. Polylactic acid/Lignocellulosic residue composites compatibilized through a starch coating. *Polym. Compos.* **2020**. [CrossRef]
49. Siakeng, R.; Jawaid, M.; Asim, M.; Siengchin, S. Accelerated weathering and soil burial effect on biodegradability, colour and textureof coir/pineapple leaf fibres/pla biocomposites. *Polymers* **2020**, *12*, 458. [CrossRef] [PubMed]
50. Rojas-Lema, S.; Quiles-Carrillo, L.; Garcia-Garcia, D.; Melendez-Rodriguez, B.; Balart, R.; Torres-Giner, S. Tailoring the properties of thermo-compressed polylactide films for food packaging applications by individual and combined additions of lactic acid oligomer and halloysite nanotubes. *Molecules* **2020**, *25*, 1976. [CrossRef] [PubMed]
51. Torres-Giner, S.; Martinez-Abad, A.; Gimeno-Alcañiz, J.V.; Ocio, M.J.; Lagaron, J.M. Controlled delivery of gentamicin antibiotic from bioactive electrospun polylactide-based ultrathin fibers. *Adv. Eng. Mater.* **2012**, *14*, B112–B122. [CrossRef]
52. Huang, L.; Mu, B.; Yi, X.; Li, S.; Wang, Q. Sustainable use of coffee husks for reinforcing polyethylene composites. *J. Polym. Environ.* **2018**, *26*, 48–58. [CrossRef]
53. Berthet, M.A.; Commandré, J.M.; Rouau, X.; Gontard, N.; Angellier-Coussy, H. Torrefaction treatment of lignocellulosic fibres for improving fibre/matrix adhesion in a biocomposite. *Mater. Des.* **2016**, *92*, 223–232. [CrossRef]
54. Lessa, E.F.; Nunes, M.L.; Fajardo, A.R. Chitosan/waste coffee-grounds composite: An efficient and eco-friendly adsorbent for removal of pharmaceutical contaminants from water. *Carbohydr. Polym.* **2018**, *189*, 257–266. [CrossRef]
55. Suaduang, N.; Ross, S.; Ross, G.M.; Pratumshat, S.; Mahasaranon, S. Effect of spent coffee grounds filler on the physical and mechanical properties of poly(lactic acid) bio-composite films. *Mater. Today Proc.* **2019**, *17*, 2104–2110. [CrossRef]
56. Wu, C.-S. Renewable resource-based green composites of surface-treated spent coffee grounds and polylactide: Characterisation and biodegradability. *Polym. Degrad. Stab.* **2015**, *121*, 51–59. [CrossRef]
57. Tsuji, H.; Kawashima, Y.; Takikawa, H.; Tanaka, S. Poly($_{\rm L}$-lactide)/nano-structured carbon composites: Conductivity, thermal properties, crystallization, and biodegradation. *Polymer* **2007**, *48*, 4213–4225. [CrossRef]
58. Dominici, F.; García García, D.; Fombuena, V.; Luzi, F.; Puglia, D.; Torre, L.; Balart, R. Bio-polyethylene-based composites reinforced with alkali and palmitoyl chloride-treated coffee silverskin. *Molecules* **2019**, *24*, 3113. [CrossRef]
59. Songtipya, L.; Limchu, T.; Phuttharak, S.; Songtipya, P.; Kalkornsurapranee, E. Poly(lactic acid)-based composites incorporated with spent coffee ground and tea leave for food packaging application: A waste to wealth. In Proceedings of the IOP Conference Series: Materials Science and Engineering, Bali, Indonesia, 30 October–2 November 2018.
60. Balart, J.F.; García-Sanoguera, D.; Balart, R.; Boronat, T.; Sánchez-Nacher, L. Manufacturing and properties of biobased thermoplastic composites from poly(lactid acid) and hazelnut shell wastes. *Polym. Compos.* **2018**, *39*, 848–857. [CrossRef]
61. Wang, L.; Qiu, J.; Sakai, E.; Wei, X. The relationship between microstructure and mechanical properties of carbon nanotubes/polylactic acid nanocomposites prepared by twin-screw extrusion. *Compos. Part A Appl. Sci. Manuf.* **2016**, *89*, 18–25. [CrossRef]
62. Arbelaiz, A.; Txueka, U.; Mezo, I.; Orue, A. Biocomposites based on poly(lactic acid) matrix and reinforced with lignocellulosic fibers: The effect of fiber type and matrix modification. *J. Nat. Fibers* **2020**, 1–14. [CrossRef]

63. Johari, A.P.; Mohanty, S.; Kurmvanshi, S.K.; Nayak, S.K. Influence of different treated cellulose fibers on the mechanical and thermal properties of poly(lactic acid). *ACS Sustain. Chem. Eng.* **2016**, *4*, 1619–1629. [CrossRef]
64. Cataldo, V.A.; Cavallaro, G.; Lazzara, G.; Milioto, S.; Parisi, F. Coffee grounds as filler for pectin: Green composites with competitive performances dependent on the UV irradiation. *Carbohydr. Polym.* **2017**, *170*, 198–205. [CrossRef]
65. Arrigo, R.; Jagdale, P.; Bartoli, M.; Tagliaferro, A.; Malucelli, G. Structure–property relationships in polyethylene-based composites filled with biochar derived from waste coffee grounds. *Polymers* **2019**, *11*, 1336. [CrossRef]
66. Cisneros-López, E.O.; Pérez-Fonseca, A.A.; González-García, Y.; Ramírez-Arreola, D.E.; González-Núñez, R.; Rodrigue, D.; Robledo-Ortíz, J.R. Polylactic acid–agave fiber biocomposites produced by rotational molding: A comparative study with compression molding. *Adv. Polym. Technol.* **2018**, *37*, 2528–2540. [CrossRef]
67. Sonseca, A.; Madani, S.; Rodríguez, G.; Hevilla, V.; Echeverría, C.; Fernández-García, M.; Muñoz-Bonilla, A.; Charef, N.; López, D. Multifunctional PLA blends containing chitosan mediated silver nanoparticles: Thermal, mechanical, antibacterial, and degradation properties. *Nanomaterials* **2020**, *10*, 22. [CrossRef]
68. Yaacab, N.D.; Ismail, H.; Ting, S.S. Potential use of paddy straw as filler in poly lactic acid/paddy straw powder biocomposite: Thermal and thermal properties. *Procedia Chem.* **2016**, *19*, 757–762. [CrossRef]
69. Cai, H.; Ba, Z.; Yang, K.; Zhang, Q.; Zhao, K.; Gu, S. Pyrolysis characteristics of typical biomass thermoplastic composites. *Results Phys.* **2017**, *7*, 3230–3235. [CrossRef]
70. De Brito, E.B.; Tienne, L.G.P.; Cordeiro, S.B.; Marques, M.d.F.V. Development of polypropylene composites with green coffee cake fibers subjected to water vapor explosion. *Waste Biomass Valorization* **2020**. [CrossRef]
71. Totaro, G.; Sisti, L.; Fiorini, M.; Lancellotti, I.; Andreola, F.N.; Saccani, A. Formulation of green particulate composites from pla and pbs matrix and wastes deriving from the coffee production. *J. Polym. Environ.* **2019**, *27*, 1488–1496. [CrossRef]
72. Sarasini, F.; Tirillò, J.; Zuorro, A.; Maffei, G.; Lavecchia, R.; Puglia, D.; Dominici, F.; Luzi, F.; Valente, T.; Torre, L. Recycling coffee silverskin in sustainable composites based on a poly(butylene adipate-co-terephthalate)/ poly(3-hydroxybutyrate-*co*-3-hydroxyvalerate) matrix. *Ind. Crops Prod.* **2018**, *118*, 311–320. [CrossRef]
73. Bouzidi, F.; Guessoum, M.; Fois, M.; Haddaoui, N. Viscoelastic, thermo-mechanical and environmental properties of composites based on polypropylene/poly(lactic acid) blend and copper modified nanoclay. *J. Adhes. Sci. Technol.* **2018**, *32*, 496–515. [CrossRef]
74. Kodama, M.; Karino, I. Effects of polar groups of polymer matrix on reinforcement-matrix interaction in kevlar fiber-reinforced composites. *J. Appl. Polym. Sci.* **1986**, *32*, 5057–5069. [CrossRef]
75. Torres-Giner, S.; Montanes, N.; Fenollar, O.; García-Sanoguera, D.; Balart, R. Development and optimization of renewable vinyl plastisol/wood flour composites exposed to ultraviolet radiation. *Mater. Des.* **2016**, *108*, 648–658. [CrossRef]

applied sciences

MDPI

Article

Low Temperature Decomposition of Polystyrene

Hideki Kimukai [1], Yoichi Kodera [2], Koushirou Koizumi [3], Masaki Okada [4], Kazunori Yamada [4], Toshihiko Hiaki [4] and Katsuhiko Saido [1,4,*]

[1] Albatross Alliance, 2234-1 Minamiboso, Shirahama, Chiba 2950102, Japan; hkimukai@hotmail.com
[2] National Institute of Advanced Industrial Science and Technology (AIST), Tsukuba, Ibaraki 3058569, Japan; y-kodera@aist.go.jp
[3] College of Science & Technology, Nihon University, Funabashi, Chiba 2748501, Japan; Koizumi.koushirou@nihon-u.ac.jp
[4] College of Industrial Technology, Nihon University, Narashino, Chiba 2758575, Japan; okada.masaki@nihon-u.ac.jp (M.O.); yamada.kazunori@nihon-u.ac.jp (K.Y.); hiaki.toshihiko@nihon-u.ac.jp (T.H.)
* Correspondence: katsu.saido@gmail.com

Received: 20 June 2020; Accepted: 23 July 2020; Published: 24 July 2020

Abstract: Styrene oligomers (SOs), of styrene (styrene monomer, SM), 1,3-diphenylpropane (styrene dimer, SD_1), 2,4-diphenyl-1-butene (styrene dimer, SD_2) and 2,4,6-triphenyl-1-hexene (styrene trimer, ST), had been detected in the natural environments far from industrial area. To confirm SOs formation through thermal decomposition of polystyrene (PS) wastes in the nature, purified polystyrene (SO-free PS) has been shown to decompose at 30 to 150 °C. The SO ratio of SM:SD:ST was about 1:1:5 with ST as the main product. Mass spectrometry with selected ion monitoring was used for the quantitative analysis of the trace amounts of SOs. The rate of PS decomposition was obtained as $k\left(\text{year}^{-1}\right) = 5.177\ exp(-5029/T(\text{K}))$ based on the amount of ST. Decomposition kinetics indicated that not only does drifting lump PS break up into micro/nano pieces in the ocean, but that it also subsequently undergoes degradation into basic structure units SO. According to the simulation at 30 °C, the amounts of SOs in the ocean will be over 400 MT in 2050.

Keywords: low-temperature decomposition; polystyrene; styrene oligomer; plastic debris; chemical contamination

---

1. Introduction

Total plastic production in the 1950s has been shown to be around several million metric tons (million MT), but at present, cumulative production has been estimated as 6.98×10^9-MT during the period 1950 to 2015 [1,2]. Waste plastic from land sources is continually flowing into world oceans via rivers due to accidents or carelessness. In 1972, Carpenter and Smith pointed out marine contamination from plastics drifting on the surfaces of the Sargasso Sea [3]. In 2001, Moore et al. [4] reported debris plastics to form garbage patches in the Pacific Ocean and to increase by 17-fold (by weight) or 95 times (by pieces) as much the amounts of contaminants in the past 30 years. Thompson et al. [5], Takada [6], Lavender [7] and Isobe et al. [8] have shown this drifting plastic to break up into a great many small pieces by the action of waves and effects of light to form micro/nanoplastics which in turn lead to massive plastic contamination in bulk. All drifting plastic quite likely break up into small pieces in this manner [9]. By 2050, debris plastics will have attained a weight exceeding that of all fish throughout the oceans of the world [10].

The authors have been engaged in the collection of coastal waters and beach sands from around the world for the past 20 years so as to determine what chemicals are generated from plastics in oceans worldwide [11–15]. However, the manner in which various chemicals are derived from plastics has yet to be fully clarified. Plastics decomposition has long been a topic of intensive research [16–22] at temperatures of 250 °C or higher.

In order to estimate the amounts of SOs from drifting PS in the ocean, the rate of PS decomposition needed to be determined at a living temperature range in the nature. However, to date, no kinetic research on PS decomposition has been conducted at the lower temperature range in the natural environment. Plastic debris undergoes degradation called as weathering in the ocean and on a beach, being exposed to salt water or to the sun light in the presence of the air. Temperature is an essential factor among various potential factors to govern plastic degradation. PS decomposition at low temperature was thus examined in the present study using polyethylene glycol (PEG1540) as a heating medium.

In this study, kinetic parameters of PS decomposition at a lower temperature range, 30–150 °C, was determined to obtain the rate of SOs formation from PS. The amounts of SOs and PS in the ocean with time were simulated based on the kinetic parameters obtained although there are various factors other than reaction temperature.

Using purified PS without contamination of SOs, low-temperature decomposition was carried at 30 to 150 °C. The products obtained were styrene oligomer (SOs), of styrene (styrene monomer, SM), 1,3-diphenylpropane (styrene dimer, SD_1), 2,4-diphenyl-1-butene (styrene dimer, SD_2) and 2,4,6-triphenyl-1-hexene (styrene trimer, ST). The composition ratio of SM:SD:ST was 1:1:5 and the main product was ST. The rate of ST formation by PS decomposition was measured at 30 to 150 °C. Kinetic parameters of the PS decomposition were determined and the activation energy of the conversion of PS into ST was given as 45.0 kJ mol^{-1}. One MT of PS was found to decompose at a rate of 0.3 g per year at 30 °C.

Simulation results indicated that the total amount of degraded PS in the ocean has been as much as 430-MT during the period, 1950 to 2050. The results indicated PS to have little stability toward heat, kinetically. Drifting macro PS not only breaks up into micro/nanosized pieces, but subsequently degrades into basic structure units of SOs in the ocean.

2. Experimental

2.1. Preparation of Reaction Samples

Commercial pellets of PS (Number–average molecular weight: 500,000, Teijin Chemicals, Ltd., Tokyo, Japan) were found to contain 90 mg kg^{-1} of unreacted SOs along with various additives. For the removal of these chemicals, 50 g PS were dissolved in 2000 mL benzene and reprecipitated with 4000 mL methanol at room temperature. This operation was conducted three times and SO remained at less than 0.1 mg kg^{-1}. Purified PS was used in this operation. Figure 1a shows the procedure for purifying PS and Figure 1b shows the subsequent treatment.

2.2. Reagent

Polyethylene glycol (PEG1540, Average molecular weight 1350–1650, Wako Pure Chemical Co., Osaka, Japan) was used as the heat medium due to high solubility for PS, thermal stability and low volatility. Benzene and methanol for dissolving PS or reprecipitation and tetrahydrofuran (THF) as the GPC eluent all these were of reagent grades and manufactured by Wako Pure Chemical Co. SD_2 and ST were prepared by the decomposition of PS and purified by distillation under reduced pressure by boiling point fractionation [23]. The purity of SD_2, ST was determined by gas chromatography (GC) with a flame ionization detector (FID) as 99.8% or more before use. The internal standard, phenanthrene, and diphenyl were used after being purified using a special grade of Kanto Chemical Co., Inc. (Tokyo, Japan) following sublimation treatment. SM and SD_1 (Wako Pure Chemical Co.Osaka, Japan) were used after the distillation of reagents prior to use.

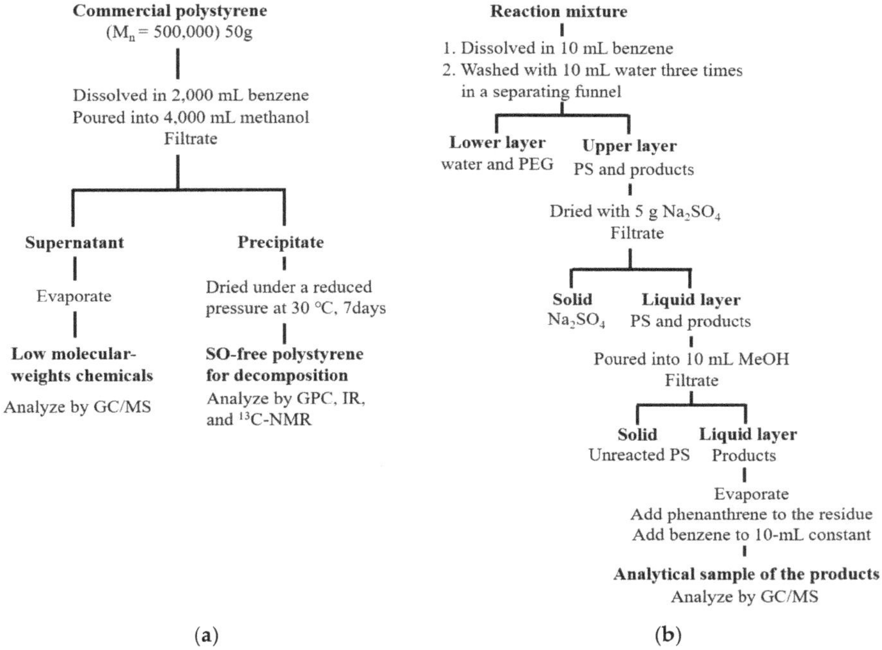

Figure 1. Procedure for purifying (**a**) styrene oligomers (SO)-free polystyrene (PS) and (**b**) subsequent treatment.

2.3. Decomposition Method

Figure 2 shows the schematic diagram of the experimental equipment for PS decomposition higher than 50 °C. Silicone oil bath was used for heating. A 20-mL glass flask was used for SO-free PS decomposition. Chromyl/alumel thermocouples were used to monitor the temperature of a reaction mixture in the flask and that of an oil bath. The flask has a nitrogen gas inlet and a gas seal for nitrogen gas flowing out. A heat medium, 4.9 g of PEG a stirrer were placed in the flask and nitrogen gas was introduced at a rate of 50 mL min^{-1}.-When the flask had reached a predetermined temperature, 0.1 g SO-free PS (cut off small pieces under 5 mm) and diphenyl as a surrogate (1.0×10^{-6} g in benzene) were charged into the flask. The reaction solution was stirred at 500 rpm. Temperature was adjusted to a predetermined temperature ± 1 °C using a digital thermometer.

PS decomposition at 30 °C was carried out without solvent because the melting point of PEG1540 is about 45 °C. PS samples were stored in a flask placed in a thermostatic chamber (Yamato Scientific Co., Ltd.,Tokyo, Japan) control accuracy ±1 °C) under nitrogen gas. The reaction at 30 °C was continued for three years.

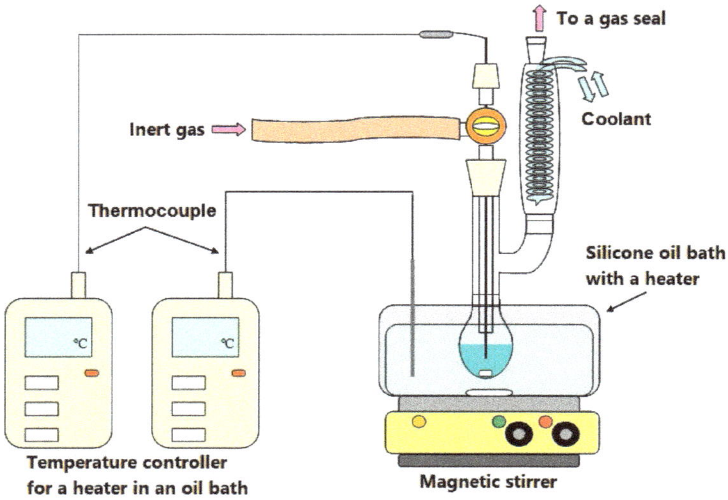

Figure 2. Schematic diagram of a decomposition equipment.

2.4. Apparatus and Operating Conditions

SO-free PS was heated at a fixed temperature for a given period in a flask. The reaction mixture was recovered with 10 mL benzene and was transferred to a 50 mL separating funnel and washed three times with 10 mL purified water to remove PEG. After drying with 5 g anhydrous sodium sulfate overnight, the solution was filtered and transferred to 10 mL methanol to precipitate unreacted PS. The solvent was completely removed from the solution by evaporation at 25 °C and the residue was collected with benzene in a 10-mL volumetric flask. Following the addition of phenanthrene, 1 μL of the solution was injected into GC/MS with a micro-syringe and analyzed. The details are shown in Figure 1b. The GC used was HP6890 (Agilent Technologies, Inc., Santa Clara, CA, USA). The mass spectrometer (MS) was JMS-AMII manufactured by JEOL, Ltd. (Tokyo, Japan) and the separation column was DB-1 manufactured by Agilent J & W. The operating conditions are shown in Table 1.

Table 1. Apparatus and operation conditions of GC/MS analysis.

Apparatus	JEOL JMS-AM II with GC/MS (HP6890)
Selected Ion (SI, *m/z*)	78, 104, 105, 152, 178, 193, 196, 207, 208, 312
Injection	1 μL
Column	DB-1, L 30 m, ID 0.32 mm, Thick 0.25 μm
Injection method	Spitless
Injection temperature	250 °C
Column temperature	Initial temp. 40 °C (holding 5 min), max temp. 290 °C (holding 5 min), program rate 15 °C/min
Interface temperature	250 °C
Ion source temperature	200 °C
Ion acceleration current	70 eV
Current	300 μA
PM voltage	600 V
Carrier gas	He, 1.4 mL/min

3. Results and Discussion

3.1. Accuracy

The mass spectrum obtained by GC/MS, TIM was used to determine fragment ion (*m/z*) of the target chemicals (SM, $SD_{1,2}$, ST). The quantitative analysis was conducted by using the monitoring ion: Q (quantitative ion), *m/z*: 104, 154, 178, 196, 208, 312 and qualitative ion: *m/z*: 78, 105, 117, 152, 193 ions. Calibration curves were prepared by the internal standard method using the detected ion peak area ratio (I/Q). For quantitative analysis of the target chemicals by mass spectrometry, Selected Ion Monitoring (SIM) was used because target detection performance of SIM was 100 times higher than that in Total Ion Monitoring (TIM). It was thus possible to obtain very low concentrations (10^{-9} level) of SOs by the SIM.

A standard solution of each compound (SM, $SD_{1,2}$, ST) was prepared to obtain a target solution using a pipette and volumetric flask, and the calibration curve of each compound was prepared. Curve linearity ranged from 0.2 μg to 10 mg kg^{-1} with a correlation coefficient, r = 0.9996 to 0.9999. The detection limit was found at S/N = 2 to be 10 μg kg^{-1}.

3.2. Effects of Temperature and Time

Polymer decomposition such as that for PS has long been studied [16–22]. Since the polymer has low thermal conductivity, a preliminary heating time is necessary to achieve the target temperature for decomposition [21]. The generated SOs has an extremely low value of less than μg kg^{-1} thus making analysis difficult in this study, this difficulty was overcome by a new decomposition using PEG as the heat medium. Among the various factors governing plastic degradation in the nature, we focused thermal effect on PS decomposition in the experiments. We assume first-order reaction of PS to form SOs. The rate of ST formation was calculated at each reaction temperature. ln *k* and the reciprocal of the absolute temperature (T^{-1}) are shown in Figure 3, which shows a linear relationship in the entire temperature range of 30 to 150 °C. PS activation energy of 45 kJ mol^{-1} was obtained from the slope of the line by using the Arrhenius equation. A low activation energy 45.0 kJ mol^{-1} indicated a secondary reaction, a backbiting reaction [24] to be a dominant factor in low temperature decomposition.

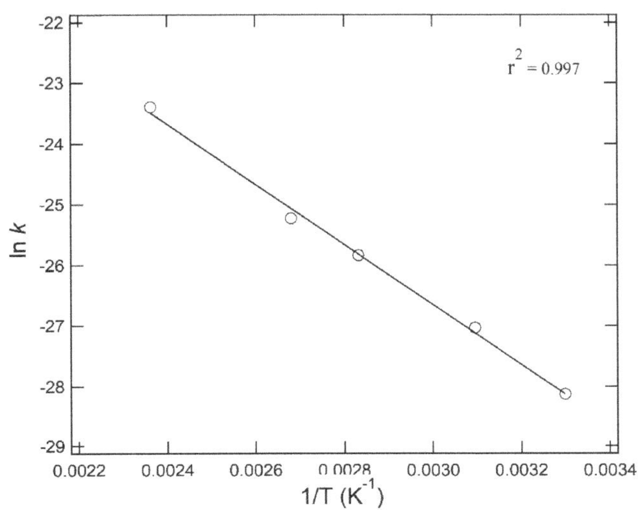

Figure 3. Arrhenius plot of styrene trimer (ST) formation through PS-decomposition at 30 to 150 °C.

SO composition did not change in the range 30 to 150 °C. The largest component was ST followed by SD and SM. The SO composition ratio of SM:SD:ST was about 1:1:5. Trace amounts of ethylbenzene, propyl benzene and benzaldehyde were observed along with secondary reactions when temperature reached 200 °C and higher. After the reaction, the PS was recovered and prepared into a film and analyzed by IR (infrared absorption spectrum). The IR analysis showed no change in the PS main chain (2924, 2850 cm^{-1} aliphatic stretching vibration, 1450 cm^{-1} aliphatic vibration). Alternatively Gel Permeation Chromatography (GPC, Column: Asahipak GF-7MHQ, Detector: RI, eluent: THF 0.6 mL/min, Sample: 2 mg in 100 mL THF, 20 μL injection, 30 °C) detected number–average molecular weight decreased by 20% when reaction conditions were between 200–280 °C for 30 min or more [21].

3.3. Degree of PS Degradation in Ocean, Determined from Rate Constant of ST Formation

Various chemical species possibly derived from artificial sources have been detected in the nature. Figures 4 and 5 show typical TIM-GC/MS chromatograms of analytical samples by extraction of sand and pebble in each sampling site far from industrial area.

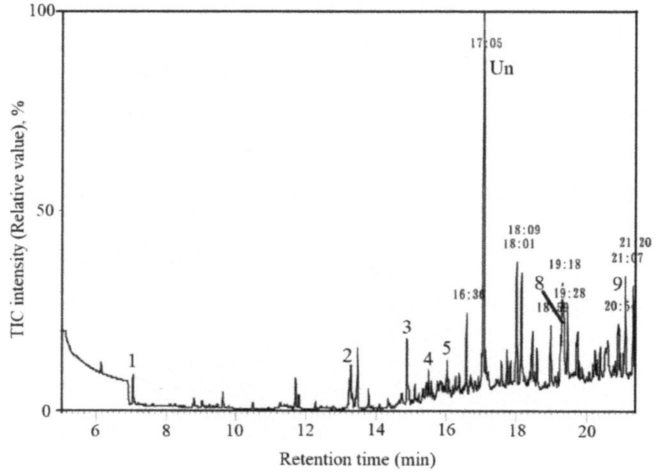

Figure 4. TIM-GC/MS chromatograms of the sample extracted from sand in Waikiki beach, Hawaii, 21°16′ N, 158°8′ E, Date: 25 March 1975.

Figure 4 shows the chromatogram of an extraction sample of sand from Waikiki beach at Oahu, Hawaii in 1975 and Figure 5 shows the chromatogram of an extraction sample of pebbles by Showa base, Antarctic Continent in 1984. Compared with the mass fragments of each standard sample, Peak 1 was identified as SM, 3 as diethyl phthalate, 4 as SD_1, 5 as SD_2, 8 as bisphenol A, 9 as ST. 2 as diphenyl added as a surrogate and 6 as phenanthrene added as an internal standard and Un means unknown chemicals. SM has been shown a breakdown product formed by cinnamon mold flora and possibly may be present in oceans as a single contaminant [25]. However, the other styrene oligomers are not naturally present in ocean. Considering the constant ratio of SM:SD:ST, the authors concluded that SOs have been shown to be the degradation products of PS from land-based sources. The chemical contamination generated from PS had already been present in the nature over 45 years ago.

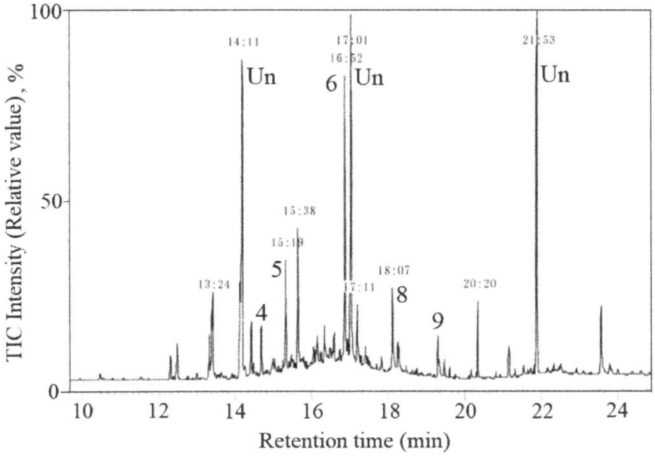

Figure 5. TIM-GC/MS chromatograms of the sample extracted from pebble by Showa base, Antarctic continent, 60°00′ S, 39°35′ E, 1984.

Jambeck et al. [26] estimated total waste plastics inflow in 2010 into world oceans to range from 4.8 to 12.7 × 10^6 MT. However, in all these computations [3–6,26], there has always been considerable margin for error. There is really virtually no completely relivable information on the quantities of plastics that have undergone degradation and sedimented to the ocean floor.

The rate of ST formation was thought same as the rate of PS decomposition at ambient temperatures in oceans and on beaches since PS decomposition at low temperature, 30–150 °C, give ST as the dominant product [22]. The decomposition rate, k, of PS (basically the same as the formation rate of ST) was found to be $S_t = P_0[1 - exp(-kt)] \approx P_0kt$, where $k\left(\text{min}^{-1}\right) = 9.850 \times 10^{-6}\ exp(-5029/T)$ where P_0 is the initial amount of purified PS in a flask, t the reaction time and T the temperature in Kelvin. Rate of ST formation was 6.15 × 10^{-13} min^{-1}. Converting the unit of rate into annual rate, the annual decomposition rate, k, of PS was given as $k\left(\text{year}^{-1}\right) = 5.177\ exp(-5029/T)$ where P_0 is the amount of PS in the ocean, t time in years since 1950 and T the temperature in Kelvin. The ST rate was 3.233 × 10^{-7} year^{-1} at 30 °C. One MT PS decomposed at a rate of 0.3 g per year at 30 °C. PS production of seven percent in the total plastics production [2]. Jambeck et al. assumed three percent of the total plastic consumption as plastic inflow to the ocean [26]. The cumulative amounts of PS and SOs in the ocean were simulated as shown in Figure 6.

The simulation was conducted by using three differential equations (Equations (1) through (3)). The production share of PS at seven percent and inflow ratio at three percent were assumed to be constant during 1950 to 2050.

$$w(t) = 1.50 \times 10^6 + 4.845 \times 10^4 t^{2.107} \tag{1}$$

$$\frac{dP(t)}{dt} = 3.150 \times 10^3 + 101.7t^{2.107} - 3.232 \times 10^{-7}P(t) \tag{2}$$

$$\frac{dS(t)}{dt} = 3.232 \times 10^{-7}P(t) \tag{3}$$

where $w(t)$ (MT) is the annual plastic production at the eclipsed year t since 1950, $P(t)$ (MT) is the cumulative amount of PS accumulated in the ocean at eclipsed year since 1950, t, $S(t)$ (MT) is the total amount of SOs in the ocean at the year t.

Equation (1) is a fitting result to the statistical data of global plastic production [1,2]. In Equation (3), 3.240 × 10^{-7} (year^{-1}) is the rate of ST formation at 30 °C that equals to the rate of SOs formation.

Equation (2) was given by multiplying Equation (1) by the share of PS 7% and inflow ratio 3%. Equation (3) is based on the kinetic parameters of ST formation at 30 °C.

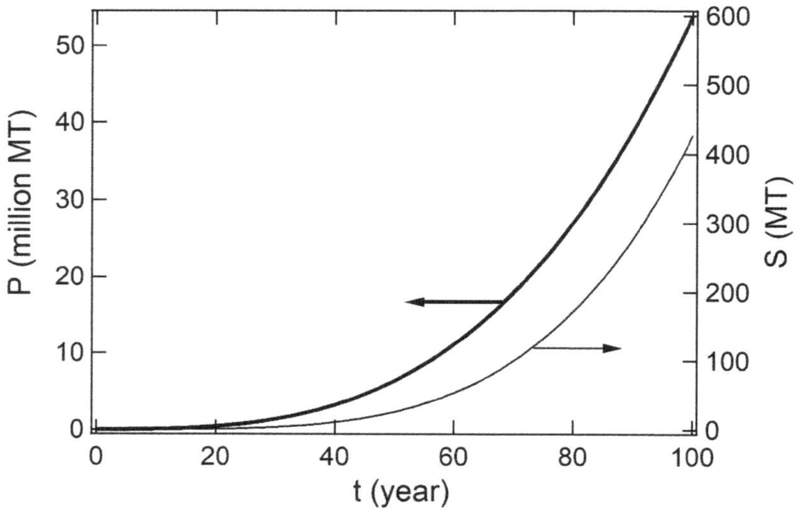

Figure 6. Simulated amounts of (P, left) PS debris and generated (S, right) SOs that have been accumulated in oceans at time *t* (year) since 1950 to 2050.

ST gradually decomposed to generate SM and SD and SD generated SM [22]. PS had been considered non-degradable in the nature, but kinetically PS, which contains no additives such as stabilizers, has been shown to degrade under the environmental temperature range to produce SOs. SOs formation would be enhanced under the severer conditions in the nature such as UV irradiation and exposure to oxygen in the air.

There has been the common misconception that plastic is stable and does not decompose at ambient temperatures. Accordingly, plastics have been considered to remain permanently intact in oceans. However, this study has clearly indicated this not to be the case, based on careful examination of PS degradation. In this study, even purified PS decomposition was shown to start even at 30 °C. Not only dose drifting lumps of PS break into micro/nano pieces, but also subsequently degrades into basic structure units.

It is widely known that the reaction rate doubles with a temperature increase of 10 °C. Tropical and subtropical ocean temperatures reach 30 °C and coastal sand at these latitudes reach 60 °C. The amounts of SO in the world oceans should thus be considered to be significantly higher than this PS value, suggesting ocean contamination to intensify as a result of plastics. Increasing amounts of various polymers go into the ocean and coastal area. Intensive studies are required to evaluate the biologic impacts of SOs from PS and the other chemical species possibly derived from various plastics.

4. Conclusions

SOs were often detected in sea water and sand of coastal area. PS decomposition at a low temperature range, 30 to 150 °C, was conducted to confirm SOs formation from PS wastes in the nature. In this temperature range, ST was a dominant product. ST would be an intermediate yielding SD_1, SD_2 and SM because SM became dominant in PS decomposition at the higher temperature range. The Arrhenius plot of ln *k* with the reciprocal of a reaction temperature in Kelvin showed a linear relationship within the low temperature range. The rate coefficient, *k* ($year^{-1}$), was obtained as 5.177

$exp(-5029/T)$ based on the rate of ST formation. SOs in the ocean was estimated over 400 MT in 2050 based on the simulation including thermal decomposition at 30 °C.

Author Contributions: Conceptualization, H.K., Y.K. and K.S.; data curation, H.K.; M.O. and K.S.; investigation, H.K., Y.K., K.K. and K.S.; writing of the original draft preparation, H.K., Y.K. and K.S.; writing of review and editing, H.K., Y.K., T.H. and K.Y. All authors have read and agreed to the published version of the manuscript.

Funding: This research received no external funding.

Conflicts of Interest: The authors declare no conflicts of interest.

References

1. Plastics Europe. The Compelling Facts about Plastics 2006 (Plastics Europe, 2008). Available online: https://www.plasticseurope.org/application/files/2815/1689/9283/2006compelling_fact_PubJan2008.pdf (accessed on 20 June 2020).
2. Plastics Europe. Plastics-the Facts 2016 (Plastics Europe, 2016). Available online: https://www.plasticseurope.org/application/files/4315/1310/4805/plastic-the-fact-2016.pdf (accessed on 20 June 2020).
3. Carpenter, E.D.; Smith, K.L., Jr. Plastics on the Sargasso Sea surface. *Science* **1972**, *175*, 1240–1241. [CrossRef] [PubMed]
4. Moore, C.J.; Moore, S.L.; Leecaster, M.K.; Weisberg, S.B. A comparison of plastic and plankton in the north pacific central gyre. *Mar. Pollut. Bull.* **2001**, *42*, 1297–1300. [CrossRef]
5. Thompson, R.C.; Olsen, Y.; Mitchel, R.P.; Davis, A.; Rowland, S.J.; John, A.W.G.; McGonigle, D.; Russell, A.E. Lost at sea: Where is all the plastic? *Science* **2004**, *304*, 838. [CrossRef] [PubMed]
6. Takada, H. Call for pellets! International pellet watch global monitoring of POPs using beached plastic resin pellets. *Mar. Pollut. Bull.* **2006**, *52*, 1547–1548. [CrossRef]
7. Law, K.L. Plastics in the marine environment. *Annu. Rev. Mar. Sci.* **2017**, *9*, 205–229. [CrossRef]
8. Isobe, A.; Iwasaki, S.; Uchida, K.; Tokai, T. Abundance of non-conservative microplastics in the upper ocean from 1957 to 2066. *Nat. Commun.* **2019**, *10*, 417. [CrossRef] [PubMed]
9. Law, K.L.; Thompson, R.C. Microplastics in the sea. *Science* **2014**, *345*, 144–145. [CrossRef]
10. Jennings, S.; Melin, F.; Blanchard, J.L.; Forster, R.M.; Dulvy, N.K.; Wilson, R.W. Global-scale predictions of community and ecosystem properties from simple ecological theory. *Proc. R. Soc. B* **2008**, *275*, 1375–1383. [CrossRef]
11. Saido, K.; Amamiya, K.; Sato, H.; Okabe, A.; Ogawa, N.; Kamaya, Y.; Kogure, K.; Nishimura, M.; Okukawa, K.; Kusui, T. Analysis of styrene oligomer contaminations generated from marine debris polystyrene on the coast of Okinawa. *Bunseki Kagaku* **2012**, *61*, 629–636. [CrossRef]
12. Kwon, B.G.; Kiozumi, K.; Chung, S.-Y.; Kodera, Y.; Kim, J.-O.; Saido, K. Global styrene oligomers monitoring as new chemical contamination from polystyrene plastic marine pollution. *J. Hazard. Mater.* **2015**, *300*, 359–367. [CrossRef]
13. Kwon, B.G.; Chung, S.-Y.; Park, S.-S.; Saido, K. Qualitative assessment to determine internal and external factors influencing the origin of styrene oligomers pollution by polystyrene plastic in coastal marine environments. *Environ. Pollut.* **2018**, *234*, 167–173. [CrossRef] [PubMed]
14. Amamiya, K.; Saido, K.; Chung, S.-Y.; Hiaki, T.; Lee, D.S.; Kwon, B.-G. Evidence of transport of styrene oligomers originated from polystyrene plastic to oceans by runoff. *Sci. Total Environ.* **2019**, *667*, 57–63. [CrossRef] [PubMed]
15. Amamiya, K.; Koizumi, K.; Yamada, K.; Hiaki, T.; Kusui, T.; Saido, K. Analysis of Drifting Polystyrene Degradation Surround Japan. *Austin J. Environ. Txicol.* **2020**, *6*, id1030.
16. Staudinger, H.; Steinhofer, A. Über hochpolymere Verbindungen. 107. Beiträge zur Kenntnis der Polystyrole. *Justus Liebigs Ann. der Chem.* **1935**, *517*, 35–53. [CrossRef]
17. Madorsky, S.L. Rate of thermal degradation of polystyrene and polyethylene in a vacuum. *J. Polym. Sci.* **1952**, *9*, 133–156. [CrossRef]
18. John, J.R.M.; Eddy, J.; Callaerts, A.; Buekens, A. Kinetic study of thermal decomposition of polystyrene by means of pyrolysis gas chromatography. *Macro. Chem. Rpid Commun.* **1982**, *3*, 349–356. [CrossRef]
19. Faravelli, F.; Pinciroli, M.; Pisano, F.; Bozzano, G.; Dente, M.; Ranizi, E.G. Themal degradation of polystyrene. *J. Anal. Appl. Pyrolysis* **2000**, *60*, 103–121. [CrossRef]

20. Dewangga, P.B.; Rochmadi; Purnomo, C.W. Pyrolysis of polystyrene plastic waste using bentonite catalyst. *Earth Environ. Sci.* **2019**, *399*, 0120110. [CrossRef]
21. Saido, K.; Taguchi, H.; Kodera, Y.; Ishihara, Y.; Ryu, I.-J.; Chung, S.-Y. Novel method for polystyrene reactions at low temperature. *Macromol. Res.* **2003**, *11*, 87–91. [CrossRef]
22. Saido, K.; Kodera, Y.; Taguchi, H.; Tomono, K.; Ishihara, Y.; Kuroki, Y. New trimer depolymerization pathway in thermal degradation of polystyrene. *230th ACS Polym. Prep* **2005**, *46*, 685.
23. Saido, K.; Motohashi, S.; Kuroki, T.; Ikemura, T.; Kirisawa, M. Preparation and properties of heat-resistant phthalic ester plasticizers derived from the thermal decomposition products of waste polystyrene. *Ind. Eng. Chem. Prod. Res. Dev.* **1985**, *24*, 102–106. [CrossRef]
24. Shibasaki, Y.; Yang, M.; Sun, L.; Shi, L. Role of back-biting in the thermal degradation of vinyl polymers. *Polym. Recycl.* **2000**, *5*, 189–204.
25. Lafeuille, J.-L.; Buniak, M.-L.; Vioujas, M.-C.; Lefevre, S. Natural formation of styrene by cinnamon mold flora. *J. Food Sci.* **2009**, *74*, 276–283. [CrossRef] [PubMed]
26. Jambeck, J.R.; Geyer, R.; Wilcox, C.; Siegler, T.R.; Perryman, M.; Andrady, A.; Narayan, R.; Law, K.L. Plastic waste inputs from land into the ocean. *Science* **2015**, *347*, 768–771. [CrossRef] [PubMed]

applied sciences

MDPI

Article

Atomization of Microfibrillated Cellulose and Its Incorporation into Poly(3-hydroxybutyrate-*co*-3-hydroxyvalerate) by Reactive Extrusion

Pedro A. V. Freitas [1], Hector Barrrasa [1], Fátima Vargas [2], Daniel Rivera [3], Maria Vargas [1] and Sergio Torres-Giner [1,*]

[1] Research Institute of Food Engineering for Development (IIAD), Universitat Politècnica de València (UPV), 46022 Valencia, Spain; pedvidef@doctor.upv.es (P.A.V.F.); hecbarhe@etsiamn.upv.es (H.B.); mavarco@tal.upv.es (M.V.)
[2] Biorefineries Department, AINIA, 46980 Paterna, Spain; fvargas@ainia.es
[3] Microencapsulation Technologies Department, AINIA, 46980 Paterna, Spain; jdrivera@ainia.es
* Correspondence: storresginer@upv.es

Featured Application: The sensitivity of our society has increased by the effect of petrochemical plastics on the environment, which has fostered the development of biodegradable materials derived from natural resources. This study puts forth the potential use of the atomization process to microdisperse cellulose structures that can be later incorporated into biopolyester films through a process of reactive extrusion, without altering the optical properties, and improving their performance.

Abstract: The present study focuses on the preparation and characterization of poly(3-hydroxybutyrate-*co*-3-hydroxyvalerate) (PHBV) films that were reinforced with cellulose microstructures to obtain new green composite materials for sustainable food packaging applications. The atomization of suspensions of microfibrillated cellulose (MFC) successfully allowed the formation of ultrathin cellulose structures of nearly 3 µm that were, thereafter, melt-mixed at 2.5, 5, and 10 wt % with PHBV and subsequently processed into films by thermo-compression. The most optimal results were attained for the intermediate MFC content of 5 wt %, however, the cellulose microstructures showed a low interfacial adhesion with the biopolyester matrix. Thus, two reactive compatibilizers were explored in order to improve the properties of the green composites, namely the multi-functional epoxy-based styrene-acrylic oligomer (ESAO) and the combination of triglycidyl isocyanurate (TGIC) with dicumyl peroxide (DCP). The chemical, optical, morphological, thermal, mechanical, and barrier properties against water and aroma vapors and oxygen were analyzed in order to determine the potential application of these green composite films in food packaging. The results showed that the incorporation of MFC yielded contact transparent films, whereas the reactive extrusion with TGIC and DCP led to green composites with enhanced thermal stability, mechanical strength and ductility, and barrier performance to aroma vapor and oxygen. In particular, this compatibilized green composite film was thermally stable up to ~280 °C, whereas it showed an elastic modulus (E) of above 3 GPa and a deformation at break (ε_b) of 1.4%. Moreover, compared with neat PHBV, its barrier performance to limonene vapor and oxygen was nearly improved by nine and two times, respectively.

Keywords: PHBV; cellulose; green composites; atomization; reactive extrusion; food packaging

Citation: Freitas, P.A.V.; Barrrasa, H.; Vargas, F.; Rivera, D.; Vargas, M.; Torres-Giner, S. Atomization of Microfibrillated Cellulose and Its Incorporation into Poly(3-hydroxybutyrate-*co*-3-hydroxyvalerate) by Reactive Extrusion. *Appl. Sci.* **2022**, *12*, 2111. https://doi.org/10.3390/app12042111

Academic Editor: Alessandro Pegoretti

Received: 19 January 2022
Accepted: 14 February 2022
Published: 17 February 2022

1. Introduction

In recent years, there has been a growing interest in the development of sustainable alternative materials for food packaging due to the environmental impacts related to the high disposal of petrochemical polymers. Recycling of single-use plastics currently represents the best option to reduce plastic waste, however, after reprocessing, recycled plastics show a significant performance reduction and associate food safety risks due to chemical migration issues [1]. A possible solution, which is also compatible with the recycling technologies, is the replacement of polymers that are derived from petroleum with biopolymers that are biodegradable [2]. These materials are macromolecules that are obtained from natural sources, are biodegradable, or show both features, which can be composted in industrial facilities and, in some cases, in domestic composting conditions and natural environments [3]. In addition, biopolymers present an improved sustainability profile since they can be produced with lower energy consumption and a reduced carbon footprint [4]. Among the currently available biopolymers, carbohydrates, such as starch [5], pectin [6], chitosan [7,8], and cellulose derivatives [9–11], or proteins, such as casein derivatives [12] and wheat gluten [13,14], have been widely explored as sustainable candidates in biodegradable food packaging applications.

Polyhydroxyalkanoates (PHAs) are semi-crystalline aliphatic polymers that are sustainable candidates to replace classical polyolefins of fossil origin for food packaging applications. In particular, PHAs are renewable polymers since these can be synthesized by bacteria, and other microorganisms, using sugars as the production source [15]. Moreover, PHAs are biodegradable in composting facilities and also in the environment avoiding the "white pollution" that is manifested, for example, as marine debris and microplastics [16]. Poly(3-hydroxybutyrate) (PHB) is the most common type of PHA and this homopolyester has been explored to be used as a biodegradable thermoplastic material since it has properties similar to those of polypropylene (PP) [17]. However, the use of PHB results in rigid and brittle materials due to its high crystallinity, whereas it is also difficult to process due to its relatively low thermal stability [18]. In order to overcome these limitations, its copolymer with 3-hydroxyvalerate (3HV), that is, poly(3-hydroxybutyrate-*co*-3-hydroxyvalerate) (PHBV), offers higher ductility and reduced crystallinity [19,20]. Thus, by increasing the percentage of 3HV units, materials with greater flexibility and lower melting point can be obtained, improving the performance and processing window [21].

Furthermore, different strategies have been developed for the incorporation of micro- and nanostructured cellulose reinforcements to give rise to new green composites with improved properties. The term 'green composite' refers to a material that is composed of a biopolymer and a natural fiber or particle and, in this latter case, cellulose nanomaterials are excellent candidates as reinforcing elements since they are cost effective, renewable, and biodegradable [22]. There are different types of ultrathin celluloses, that is, cellulose nanofibers (CNFs), cellulose microfibers (CMFs), also called microfibrillated cellulose (MFC), and nanocrystals of hydrolytically extracted cellulose (cellulose nanocrystals, CNCs) depending on the size and the presence of amorphous and crystalline parts [23,24]. Another type of nanocellulose is bacterial cellulose (BC), whose morphology can be modified by controlling its biosynthesis process [25]. Among nanocellulose materials, MFC has recently been used as a reinforcing filler in petrochemical thermoplastics [26].

The use of MFC in green composites can be particularly advantageous in order to achieve significant improvements in the mechanical and oxygen barrier performance, which are the most limiting factors of biopolymers. However, when ultrafine cellulose structures are used as a mixing element to reinforce polymer matrices, there are two main technological challenges to consider. On the one hand, ultrafine cellulose structures are mostly prepared in the form of aqueous suspensions that are usually not practical in terms of processability and industrial use. During the drying or dewatering process, these tend to form aggregates due to the high hydrophilic nature of cellulose, which has a surface that is very rich in hydroxyl groups (–OH) that are prone to form strong intermolecular hydrogen bonds. This effect is amplified in the case of nanostructured materials due to its large surface-to-volume aspect ratio and high surface energy [27]. For example, it has been observed that during the spray-drying process, cellulose nanoparticles agglomerate by a capillarity effect and the formation of hydrogen bonds and van der Waals forces [28]. Alternatively, lyophilization or freeze drying led to ice crystal growth, which tends to favor the agglomeration of nanocelluloses [29]. On the other hand, the presence of a large number of -OH groups on the cellulose surface results in a polar fiber surface with a relatively low compatibility with the polymer matrices. This habitually results in a non-uniform dispersion and distribution of microfibers during the melt-transformation processes by which nanocelluloses are incorporated into polymers, which leads to a significant loss of performance in the composites [30]. Different strategies have been carried out in order to improve the interfacial adhesion of green composites based on MFC or CNC, mainly by means of chemical pre-treatments and compatibilizing agents or coupling agents [31]. For example, some studies have focused on modifications on the surface of the cellulose through chemical or physical pre-treatments to reduce its hydrophilicity [32]. However, although these methods can positively result in a decrease in moisture absorption and an increase in the mechanical properties, they are usually expensive and involve the use of additional processing steps or toxic chemicals that could be an impediment to the subsequent use of the resulting composites [27]. Other recent studies have reported the use of compatibilizers such as polymers grafted with maleic anhydride (MAH) [33] or multi-functional additives based on epoxy groups [34] and isocyanates [35].

This research reports a novel and potentially scalable technique to prepare green composites of PHBV and MFC based on two steps. Initially, an aqueous MFC suspension was atomized to yield micron-sized droplets containing the dry cellulose microstructures that evaporated upon deposition, leaving no water residues. Secondly, the resultant spray-dried cellulose ultrathin structures were incorporated at different contents, that is, 2.5, 5, and 10 wt %, into PHBV by melt-mixing and their processing and properties were analyzed. Finally, the green composite with the optimal MFC content was compatibilized with two different reactive additives in order to enhance the performance of the films for potential uses in food packaging applications.

2. Materials and Methods

2.1. Materials

Commercial PHBV with food-grade status was supplied as ENMAT Y1000P in the form of pellets by Tianan Biologic Materials (Ningbo, China). The biopolymer presents a density of 1.23 g/cm^3 and a melt flow index (MFI) of 5–10 g/10 min (190 °C, 2.16 kg). The 3HV fraction in the copolyester is ~2 mol% and the molecular weight (M_W) is ~2.8 × 10^5 g/mol. The aqueous suspension, Exilva F01-V grade MFC, was provided by Borregaard ChemCell AS (Sarpsborg, Norway) in a paste form with a solid content of 10 wt %.

Joncryl® ADR 4368C was provided by BASF S.A. (Barcelona, Spain) in the form of solid flakes. It is a proprietary epoxy-based styrene-acrylic oligomer (ESAO) with an average functionality of (*f*) > 4, where the styrene and acrylate blocks are each present between 1 and 20. M_W is 6800 g/mol, the glass transition temperature (T_g) is 54 °C, and the epoxide equivalent weight (EEW) is 285 g/mol. Recommended dosage by the manufacturer is 0.1–1 wt % for proper processability without impairing compostability and food contact status. Triglycidyl isocyanurate (TGIC, reference 379506), with a M_W of 297.26 g/mol, and dicumyl peroxide (DCP, reference 329541), with a M_W of 270.37 g/mol and 98% purity, were both purchased from Sigma-Aldrich S.A. (Madrid, Spain). Magnesium nitrate ($Mg(NO_3)_2$) and D-limonene were all obtained from Panreac Quimica S.L.U. (Castellar del Vallés, Barcelona, Spain). Phosphorus pentoxide (P_2O_5), 99% purity, reference 214701, was obtained from Sigma-Aldrich S.A.

2.2. Atomization of Cellulose Suspension

Prior to the atomization process, the as-received cellulose suspension was diluted in water to a concentration of 1.8 wt % to reduce viscosity and facilitate product handling. To carry out the dehydration process, a laboratory spray-drying equipment (Minispray Dryer Büchi B-290 Advanced, BUCHI Ibérica S.L.U., Barcelona, Spain) was used. This equipment has an air inlet of up to 220 °C, an evaporation flow-rate of water of up to 1 L/h, a nozzle of 0.7 mm, and a drying flow-rate of air of up to 38 m^3/h. The calorific power is 2300 w. The gas used for atomization was compressed air with a variable flow-rate between

200 and 1000 L/h and operating at a variable working pressure between 3 and 7 bar. The drying conditions of the MFC suspension were set as follows: inlet temperature of 110 °C and outlet temperature ranging between 55 and 60 °C. Figure 1 shows an image of the equipment (Figure 1a) and a diagram of the process carried out to atomize the cellulose suspension (Figure 1b).

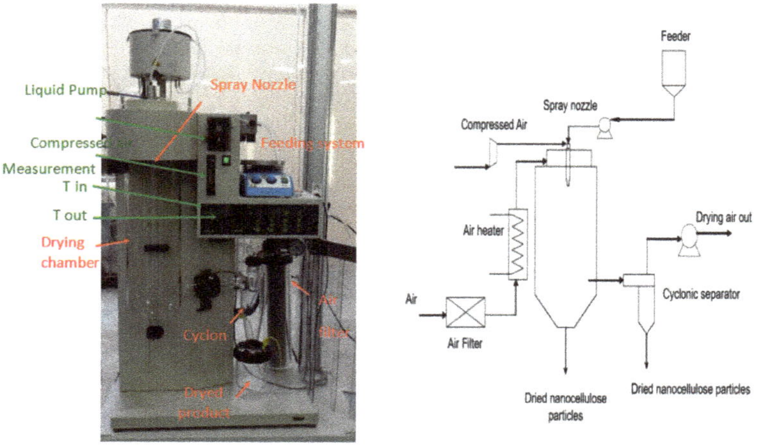

(**a**) (**b**)

Figure 1. Atomization unit with indications of its components (**a**) and process scheme (**b**).

2.3. Preparation of Green Composite Films

2.3.1. Melt Mixing

The PHBV pellets were dried in a vacuum oven (vacuum TEM-TJP Selecta, S.A., Barcelona, Spain) at 60 °C for 4 h. Thereafter, the dried pellets and the atomized cellulose powder were placed in a desiccator containing P_2O_5 at 25 °C for one week to remove the remaining water.

The PHBV pellets were melt-mixed with MFC at contents of 2.5, 5, and 10 wt %. The melt-blending process was carried out in an internal mixer (HAAKE™ PolyLab™ QC, Thermo Fisher Scientific, Herzogenaurach, Germany) at a temperature of 170 °C with a rotor speed of 50 rpm. The mixing time was set at 5 min after analyzing the stability of neat PHBV during melt mixing. The processed amount of each composition was approximately 50 g. PHBV samples with different MFC contents and compatibilizers were used to produce the green composites. Samples without MFC and compatibilizers were also prepared as control materials. Table 1 shows the set of compositions prepared.

Subsequently, the samples obtained from the mixer were cold-milled in order to have the product in a more easy-to-handle format for subsequent processing into films. For this, each dough was ground at 2 pulses of 30 s in a milling machine (Model M20, IKA, Staufen, Germany). The powder-like material resulting from the milling process was stored in a desiccator at 25 °C and 0% relative humidity (RH) with P_2O_5 for one week.

2.3.2. Thermo-Compression

Thermo-compression was carried out using a hydraulic press (Model LP20, Labtech Engineering, Bangpoo, Thailand). An amount of 2.5 g of the obtained doughs was thermo-compressed as follows: preheated for 3 min at 200 °C, compressed at ~100 bar and 200 °C for 4 min and, finally, cooled for 3 min until reaching a temperature of ~80 °C. The resultant films, sizing 10 cm × 10 cm and with a thickness of 120−140 μm, were stored in a desiccator with P_2O_5 (0% RH) at 25 ± 2 °C for a minimum period of 15 days to eliminate the remaining humidity and also reduce the effect of physical aging on PHBV.

Table 1. Summary of compositions according to the weight content (wt %) of poly(3-hydroxybutyrate-*co*-3-hydroxyvalerate) (PHBV) and microfibrillated cellulose (MFC) in which epoxy-based styrene-acrylic oligomer (ESAO), triglycidyl isocyanurate (TGIC), and dicumyl peroxide (DCP) were added as parts per hundred resins (phr) of green composite.

Sample	PHBV (wt %)	MFC (wt %)	MAH (phr)	ESAO (phr)	TGIC (phr)	DCP (phr)
PHBV	100	0	0	0	0	0
PHBV_2.5%MFC	97.5	2.5	0	0	0	0
PHBV_5%MFC	95	5	0	0	0	0
PHBV_10%MFC	90	10	0	0	0	0
PHBV_ESAO	100	0	0	1	0	0
PHBV_TGIC+DCP	100	0	0	0	1	0.25
PHBV_5%MFC_ESAO	95	5	0	1	0	0
PHBV_5%MFC_TGIC+DCP	95	5	0	0	1	0.25

2.4. Material Characterization

2.4.1. Microscopy

Film thickness was determined at 10 random positions with a digital electronic micrometer, having ±0.001 mm accuracy (Palmer model COMECTA, Barcelona, Spain).

Atomic force microscopy (AFM) was carried out in a Bruker Multimode 8 microscope (ELECMI, Zaragoza, Spain) to determine the topography of MFC at the nanoscale. To this end, a drop of the cellulose suspension was placed on the scanning probe and scanned with an image area of 10 μm × 10 μm and 5 μm × 5 μm in "tapping" mode using a cantilever of silicon nitride at a nominal resonance frequency of 0.6 Hz.

The morphology of the atomized cellulose microstructures and the fracture surfaces of the films were studied by means of field emission scanning electron microscopy (FESEM) (JEOL, model JSM-5410, Tokyo, Japan) equipped with focused ion gun (AURIGA Compact, Zeiss, Oxford Instruments, Abingdon, UK). To obtain the cross-sections of the composites, the films were cryo-fractured by immersion in liquid nitrogen. The samples were mounted on the sample holder using double-sided carbon tape, coated with a platinum layer (EM MED020 sputter coater, Leica Biosystems, Barcelona, Spain), and observed using an accelerated voltage of 2 kV. The SmartTiff program (version 2, Zeiss, Oxford Instruments) was used to measure the cellulose fiber diameters. To this end, at least 20 micrographs of each formulation were analyzed.

2.4.2. Infrared Spectroscopy

Attenuated total reflection-Fourier transform infrared (ATR-FTIR) spectroscopy chemical analysis was performed on the film surfaces. Spectra were recorded with a Vector 22 from Bruker S.A. (Madrid, Spain) coupling a PIKE MIRacle™ ATR accessory from PIKE Technologies (Madison, WI, USA). Ten scans were averaged from 4000 to 400 cm^{-1} at a resolution of 4 cm^{-1}.

2.4.3. Optical Evaluation

Optical properties of the films were obtained by determining the film reflection spectra (R) using the black (R_0) and white (R_g) backgrounds according the Kubelka–Munk theory of multiple scattering. The internal transmittance (T_i) of the films, ranging from 400 to 700 nm, was determined using Equation (1). The film lightness (L^*) and the color coordinates a^* (redness-greenness) and b^* (yellowness-blueness) were determined using Equations (2) and (3). The colorimetric and transparency properties of the films were also evaluated in terms of the hue angle (h_{ab}^*) and chroma (C_{ab}^*) following Equations (4) and (5).

$$T_i = \sqrt{(a + R_0)^2 - b^2} \quad (1)$$

$$a = \frac{1}{2}\left[R + \left(\frac{R_0 - R + R_g}{R_0 \times R_g}\right)\right] \tag{2}$$

$$b = \sqrt{a^2 - 1} \tag{3}$$

$$h_{ab}{}^* = arctg\left(\frac{b^*}{a^*}\right) \tag{4}$$

$$C_{ab}{}^* = \sqrt{a^{*2} + b^{*2}} \tag{5}$$

Color difference (ΔE^*) between the neat PHBV film and the films of PHBV containing MFC, or those processed with the compatibilizers, was evaluated using Equation (6).

$$\Delta E^* = \sqrt{(\Delta L^*)^2 + (\Delta a^*)^2 + (\Delta b^*)^2} \tag{6}$$

where $\Delta L^* = (L^* - L_0^*)$; $\Delta a^* = (a^* - a_0^*)$; $\Delta b^* = (b^* - b_0^*)$; and L_0^*, a_0^*, and b_0^* are the color coordinates of the neat PHBV film. The color differences were evaluated according to the following criteria [36]: $\Delta E^* < 1$ indicates unnoticeable color change; $1 \leq \Delta E^* < 2$ suggests that only an experienced observer can notice the difference; $2 \leq \Delta E^* < 3.5$ means that an inexperienced observer notices the difference; $3.5 \leq \Delta E^* < 5$ indicates a clear noticeable difference; and $\Delta E^* \geq 5$ suggests that the observer notices different colors. All measurements were performed in triplicate.

2.4.4. Thermal Analysis

Differential scanning calorimeter (DSC) was used to determine the phase transitions of the PHBV samples. Measurements were performed in a DSC 1 Star^e System analyzer (Mettler-Toledo GmbH, Greifensee, Switzerland) operating under a nitrogen atmosphere (20 mL/min). To this end, 5–7 mg of each sample was weighted in aluminum pans and heated from −40 to 200 °C to remove the thermal history, cooled to −40 °C to determine the crystallization temperature (T_c), and then heated (second heating step) to 200 °C for determining the glass transition temperature (T_g) and melting temperature (T_m). The heating and cooling rates were set at 10 and 50 °C/min, respectively. The degree of crystallinity (X_C) was measured using the second heating scans and following Equation (7):

$$X_C = \left[\frac{\Delta H_m - \Delta H_{cc}}{\Delta H_m^0 \cdot (1 - w)}\right] \cdot 100 \tag{7}$$

where ΔH_m (J/g) is the melting enthalpy, ΔH_{cc} (J/g) is the enthalpy of cold crystallization, ΔH_m^0 (J/g) represents the theoretical enthalpy of a fully crystalline sample of PHBV with a value of 146.6 J/g [37], and the term $1 - w$ represents the weight fraction of PHBV.

Thermal degradation of the PHBV films was assessed by thermogravimetric analysis (TGA) in a thermogravimetric analyzer (TGA 1 Star^e System analyzer, Mettler-Toledo GmbH). Film samples, with a weight of 3–5 mg, were subjected to a heating program from 25 to 700 °C at a heating rate of 10 °C/min under a nitrogen atmosphere (10 mL/min). The initial or onset temperature (T_{onset}), corresponding to the temperature yielding a mass loss of 5%, the temperature at the maximum degradation rate (T_{deg}), and the residual mass at 700 °C were determined by analyzing the TGA curves and their first derivative thermogravimetry (DTG) curves. All thermal tests were performed in triplicate.

2.4.5. Mechanical Characterization

The tensile properties of the films were measured in a universal testing machine (Stable Micro Systems, TA.XT plus, Stable Micro Systems, Godalming, UK) following the guidelines of the ASTM D882. Rectangular film samples sizing 25 mm × 10 mm were grabbed by two grips initially separated by 50 mm and stretched at a cross-head speed of 12.5 mm/min to determine the values of elastic modulus (E), tensile stress at yield (σ_y), and elongation at break (ε_b). Eight film samples were evaluated for each formulation.

2.4.6. Barrier Measurements

Water vapor permeability (WVP) of the films was determined using the ASTM 2011 gravimetric method. To this end, 5 mL of distilled water was placed inside Payne permeability cups (Ø = 3.5 cm, Elcometer Sprl (Hermalle-sous-Argenteau, Belgium) and the films were placed in the cups so that on one side they were exposed to water (100% RH), avoiding direct contact with water. Thus, the cups containing the films were secured with silicon rings and stored in desiccators containing $Mg(NO_3)_2$ over-saturated solution (53% RH) in a chamber with temperature control at 25 °C. Identical cups with aluminum films were used as control samples to estimate the vapor loss through the sealing. The cups were weighed periodically using an analytical balance of ±0.0001 g accuracy (ME36S, Sartorius, Fisher Scientific, Hampton, NH, USA). The water vapor permeance was calculated considering the water vapor transmission rate (WVTR), which was determined from the steady-state permeation slope obtained from the regression analysis of weight loss data per unit area versus time, where the weight loss was calculated as the total cell loss minus the loss through the sealing. The water vapor permeance was corrected for permeant partial pressure and the film thickness to obtain the WVP values.

Limonene permeability (LP) was measured following the same procedure described above for water vapor, replacing water with 5 mL of D-limonene inside the Payne permeability cups. The permeability cups containing the films were placed at the controlled conditions of 25 °C and 53% RH and cups with aluminum films were used as control samples to estimate solvent loss through the sealing. Limonene permeation rate (LPR) was obtained from the steady-state permeation slopes and weight loss was calculated as the total cell loss minus the loss through the sealing, which was thereafter corrected for permeant partial pressure and the film thickness to yield LP. Both vapor permeability measurements were performed in triplicate.

Oxygen permeability (OP) of the films was determined using an oxygen permeation analyzer (OxySense® Model 8101e, Systech Illinois, Thame, UK) at 25 °C and 53% of RH, according to ASTM D3985-05. The exposed area of the films was 50 cm^2 and the oxygen transmission rate (OTR) was obtained every 15 min until equilibrium was reached. The gas permeability measurements were carried out in duplicate.

2.5. Statistical Analysis

The experimental data were submitted to analysis of variance (ANOVA) at a confidence level of 95% using Minitab statistical program (version 17). Fisher's least significant difference (LSD) was used to determine whether there were significant differences among the formulations, using the least significant difference (α) of 5% ($p < 0.05$).

3. Results and Discussion

3.1. Atomization of Cellulose Suspension

The morphology of the suspended cellulose before atomization was observed using by AFM. This technique consists of a mechano-optical instrument that is capable of resolving the topography of a material by means of a nanometer-scale probe at high magnification. Figure 2 shows two AFM images that were obtained from the dilute resuspension of the as-received sample in water for the scanned areas of 10 μm × 10 μm (Figure 2a,b) and 5 μm × 5 μm (Figure 2c,d). It can be observed that samples are formed by a complex entangled network of micro- and nanofibers with mean diameter dimensions ranging from 10 nm, for individual nanofibrils, to 146.2 nm for bundles of these nanofibrils. In Figure 2e,f, corresponding to the 3D phase representation of each scanned area, one can see that the highest topography of these figures corresponds to the bundles of nanofibrils, whereas the lowest valleys are related to the smaller nanofibril bundles or individual ones.

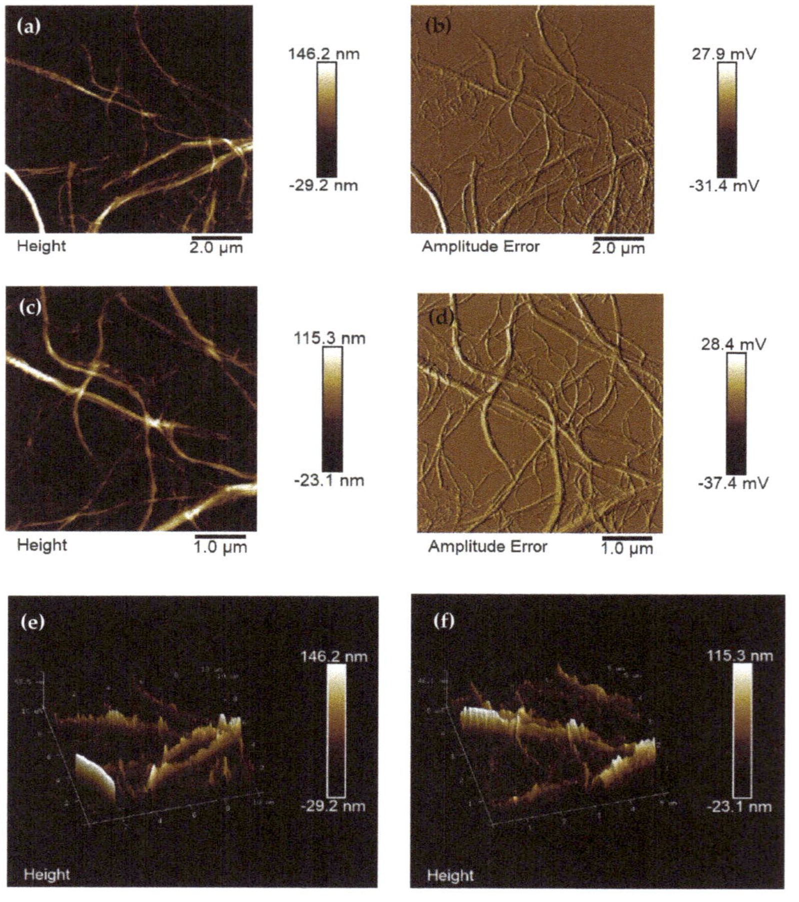

Figure 2. Atomic force microscopy (AFM) image of microfibrillated cellulose (MFC) in tapping mode for a scanned 10 μm × 10 μm area in height (**a**) and amplitude (**b**) and for a scanned 5 μm × 5 μm area at height (**c**) and amplitude (**d**). Three-dimensional phase representation of the scanned 10 μm × 10 μm area (**e**) and 5 μm × 5 μm area (**f**).

The morphology of the resultant atomized cellulose powder obtained from the diluted cellulose suspension in water was observed by FESEM and its micrographs, at different magnifications, are shown in Figure 3. As it can be seen in Figure 3a, atomization gave rise to micrometer aggregates of cellulose sizing approximately 3 ± 1 μm. The FESEM image that was taken at higher magnification, shown in Figure 3b, evidences the formation of microparticles with a rod-like morphology and an aspect ratio of nearly four. Figure 3c, showing the magnification of the microaggregate surface, reveals that these structures were formed by the so-called MFC, originally present in the suspended cellulose with fibers with thicknesses in the nano-range. Therefore, atomization was able to successfully yield microparticles of cellulose, in the powder form and with a low aspect ratio, being constituted by MFC. Similar results have been recently obtained by Shariatnia et al. [38] during the atomization of 0.5–1 wt % CNC suspensions by spraying techniques at high pressure (244.8 kPa). The authors reported that the morphology of the nanocrystals was maintained, although they also tended to form micrometric agglomerates due to the high attractive forces and their high aspect ratio.

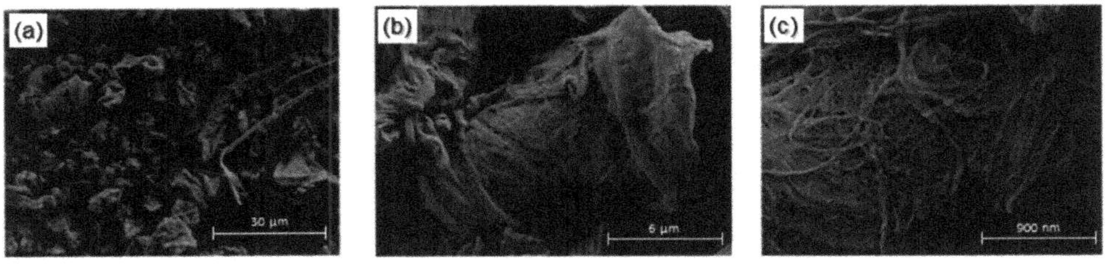

Figure 3. Field emission scanning electron microscopy (FESEM) micrographs of the atomized microfibrillated cellulose (MFC) observed at different magnifications, as follows: (**a**) taken at 1000× with a scale marker of 30 µm; (**b**) 5000× and 6 µm; (**c**) 20,000× and 900 nm.

3.2. Melt Processing of Green Composites

Figure 4 shows the torque vs. time curves of the different PHBV formulations with MFC and/or the different compatibilizing agents during melt-mixing at 50 rpm and 170 °C. Figure 4a shows the torque evolution for a processing time of 25 min for the neat PHBV sample, which was conducted to ascertain the stability of the biopolyester in terms of melt shear-thinning strength at high temperatures. One can observe that, during processing, the torque sharply increased for about 30 s until the biopolymer pellets were plasticized and then it decreased, showing a peak centered at approximately 1.2 min. The latter peak can be related to the complete melting of the PHBV crystals and spanned up to 2 min. Thereafter, the torque of the liquid PHBV mass progressively reduced, this effect was more noticeable from the processing times that were longer than 5 min. Finally, a plateau was nearly reached after approximately 20 min. The lower torque values indicate that a progressive decrease in viscosity was achieved in the samples, which can be ascribed to thermo-mechanical degradation of PHBV due to chain scission and resultant M_W reduction, confirming the low processing window of this biopolyester. Indeed, the inherently poor thermal stability of PHB, and also of PHBV with low 3HV contents, is a major concern since thermal decomposition by chain scission starts at temperatures close to the melting point, making its processing in the melt state a considerable challenge [39]. However, a minimum residence time would be needed to both facilitate a correct mixing and dispersion of the microparticles in the PHBV matrix during melt compounding [40] as well as to achieve or promote the reactivity of the compatibilizing agents [41]. In particular, the dispersion of the cellulose microfibrils depends on time and shear force that will open the physical entanglements. According to this result, a residence time of 5 min was selected to process all of the formulations as a trade-off solution between stability and mixing.

Thus, Figure 4b–d show the curves of torque vs. mixing time for all of the formulations, which were processed at 50 rpm and 170 °C for 5 min. As can be seen in the curves for the uncompatibilized green composite samples, shown in Figure 4b, the incorporation of the microstructures of cellulose yielded very similar curves to the neat PHBV sample, with a progressive increase in the torque during the first ~2 min. This effect can be ascribed to the mechanical friction increase due to the MFC presence in the mixer prior to melting since the dispersion and disaggregation of particles in the biopolymer matrix mainly occurred during this period. In the case of the unfilled compatibilized PHBV samples, included in Figure 4c, one can observe that these samples presented a second increase in torque, approximately after 1–1.5 min, which corresponds to the increase in viscosity associated with the chemical reaction of the compatibilizing agents. In the case of the PHBV sample that was processed with TGIC, this reaction took place at a slightly shorter time than for ESAO, which may be associated with the presence of the highly reactive DCP. Finally, Figure 4d shows the composite samples that were prepared with the compatibilizing agents, where a much higher viscosity increase can be observed than in the rest of the samples, especially noticeable in the PHBV/MFC composite samples that were processed with TGIC. This viscosity increase can be ascribed to the formation of new bonds in the PHBV chains or between PHBV and MFC. In particular, the generation of these molecular entanglements could result in an increase in torque with time due to the formation of macromolecular structures with higher melt strength.

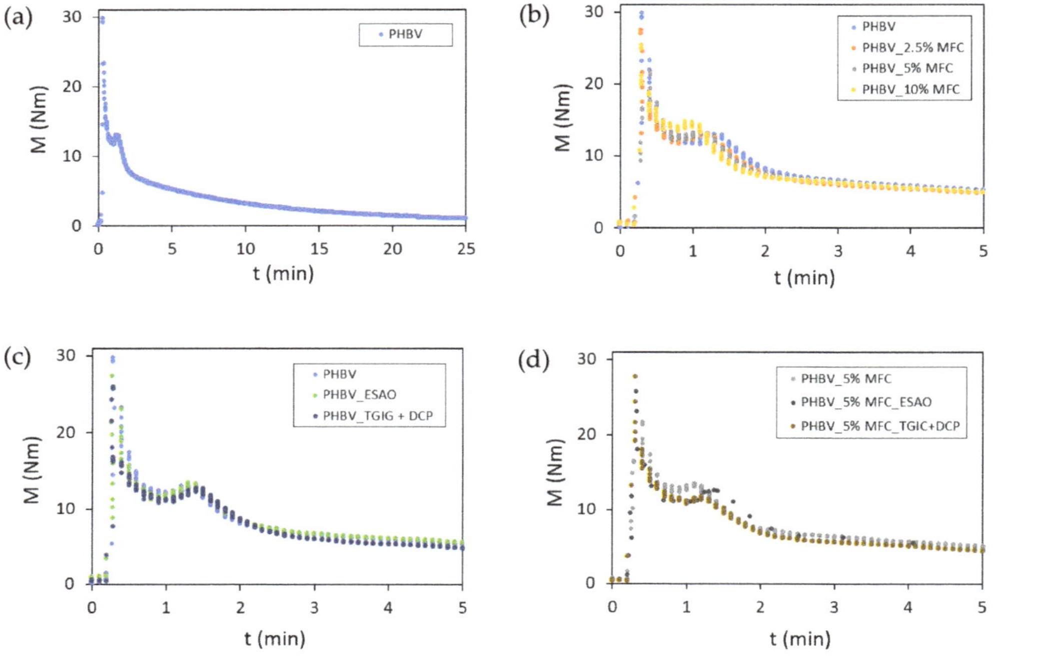

Figure 4. Representation of torque (M) as a function of time during melt mixing of the following: (**a**) poly(3-hydroxybutyrate-*co*-3-hydroxyvalerate) (PHBV) for 25 min; (**b**) PHBV without and with microfibrillated cellulose (MFC) at 2.5, 5, and 10 wt % for 5 min; (**c**) PHBV without and with epoxy-based styrene-acrylic oligomer (ESAO) and with triglycidyl isocyanurate (TGIC) and dicumyl peroxide (DCP) for 5 min; (**d**) PHBV/MFC at 5 wt % without and with ESAO and with TGIC and DCP for 5 min.

In this regard, one should consider that these compatibilizing agents are chemical substances that are able to establish covalent bonds with different groups of the biopolyester and, in the case of composites, give rise to grafting processes between the functional groups on the surface of the lignocellulose with the terminal groups of the biopolymer. This process is called "reactive extrusion" and it is observed in the processing curves by an increase in torque with time, that is, the molten biopolymer shows a greater resistance to flow after the melting of the biopolymer. Thus, the use of reactive compatibilizing agents produced an increase in viscosity, that is, a melt with a higher torque during mixing, either by an effect of extension or linear growth of the chains ("chain extension"), branching, or cross-linking [42]. Finally, one can also observe that, in terms of comparison performance, similar increases in torque were attained for the unfilled PHBV and PHBV/MFC samples with TGIC and DCP and the composite with ESAO. In the case of the isocyanurate, it exhibited higher reactivity with PHBV, while both exerted a large effect on the grafting of MFC onto PHBV. Also, it is noteworthy that the torque increase was seen to take place 20 s faster for TGIC and DCP than ESAO, which can be ascribed to the presence of the peroxide, offering the benefit to facilitate its processability at a larger extrusion equipment.

3.3. Chemical Analysis of Green Composites

ATR-FTIR was performed in order to evaluate the chemical interactions between PHBV, MFC, and the reactive compatibilizers, that is, ESAO and TGIC, and spectra of the films are gathered in Figure 5. One can observe in Figure 5a that all of the film samples showed the peaks of the characteristic FTIR spectrum of PHBV without noticeable changes [41], which can be due to the relatively low MFC contents. The strongest band was observed at 1718 cm^{-1}, which has been assigned to the stretching vibration of the carbonyl group (C=O) in PHA copolyesters and it corresponds to the intramolecular bonding of their crystalline state [43]. The complex and multiple peaks that were observed in the 1000–1200 cm^{-1} region arise mainly from the C–O stretching vibration in the ester groups of polyesters [44]. Finally, the peaks centered at ~2975 and ~2935 cm^{-1} have been ascribed to asymmetric stretching vibrations of methyl (–CH_3) and methylene (–CH_2–) groups, respectively, which also produced smaller symmetric stretching intensities at ~2875 and ~2850 cm^{-1}. Figure 5b compared the FTIR spectra of the uncompatibilized PHBV film with 5 wt % MFC and the green composite films compatibilized with ESAO and TGIC. It can be observed that the crystalline C–O stretching vibration was kept at 1718 cm^{-1}. However, the carbonyl peak was slightly broadened in the spectra corresponding to the compatibilized films and also generated a shoulder at approximately 1685 cm^{-1}. This band change, which was more noticeable in the green composite sample that was processed with ESAO, has been ascribed to an alteration of the hydrogen bonding in the molecular arrangement of PHAs [34]. This suggests that a stronger intramolecular disruption of the biopolymer chains by the presence of cellulose was achieved in the green composites that were processed with the compatibilizers, indicating a higher interaction of MFC with PHBV. Other authors have reported shifts in this peak due to the reaction between the epoxy groups of chain extenders and the carboxyl groups (–COO) in polyesters [45]. For both of the compatibilizers, and particularly more noticeable in the case of ESAO, all of the peak intensities increased in the 1500–800 cm^{-1} region. These intensity increases were seen to be more intense for the –O–C–O– stretching modes of the carboxyl group, particularly for the band arising at ~1184 cm^{-1} and other ester-related bands that were observed at ~1055 cm^{-1} and ~1043 cm^{-1} [46]. Therefore, these band signal increases suggest the formation of new ester groups in the compatibilized composite films of PHBV. However, one should further consider that the bands in this region are also due to the C–H bending modes [47] and, thus, the band assignments in this region are not straightforward. Additionally, the intensity of the latter bands can also be increased due to their close proximity to the bands that were ascribed to the newly formed OH-related bonds, which correspond to the –OH side groups that were produced during esterification reaction.

In relation to the chemical changes observed above, it has been described that in the case of ESAO, this compatibilizing agent presents multiple epoxide groups in the styrene chains of the oligomer that are capable of forming new ester groups with the terminal nucleophilic groups of the biopolyester, that is, with its alcohol groups (R–OH) and, fundamentally, acid groups (R–COOH) present at the termination of the polyester chains [48]. On the other hand, it has been shown that part of the other epoxide groups in the ESAO structure, which do not react with the biopolyester, can also react with the –OH groups that are present on the surface of the cellulose, or even with silanol groups (Si–OH) of certain minerals, giving rise to covalent chemical bonding or "grafting" that subsequently achieves a better compatibility of the biopolymer composite [34]. Similarly, reactive extrusion by means of isocyanurates, as is the case for TGIC, which is usually initiated by organic radicals from peroxides, such as DCP, have managed to achieve the referred grafting compatibilization [35]. This reaction is accomplished by means of the reaction of the –OH groups that are present both in the terminal groups of the biopolyester molecular chains and on the cellulose surface with the multiple epoxide groups present in TGIC [49]. Therefore, ester bonds are formed with the PHA chains by an esterification reaction with the glycidyl terminal groups of the carboxylic acid, which precedes the etherification of the –OH terminal group [34]. The second reaction generates C–O–C type bonds with subsequent formation of –OH side groups on the cellulose surface [50]. For DCP, Wei et al. [51] described the mechanism of cellulose coupling or grafting onto PHB and PHBV. Briefly, when the peroxide is exposed to heat during the extrusion process, it decomposes into free radicals, with high reactivity, which tend to abstract hydrogens (H′) from the molecular chains of the biopolymer and cellulose to subsequently initiate the referred grafting process between the two phases of the composites. The authors postulated that, specifically, grafted copolymers are formed at the interfaces of the cellulose and the biopolyester matrix.

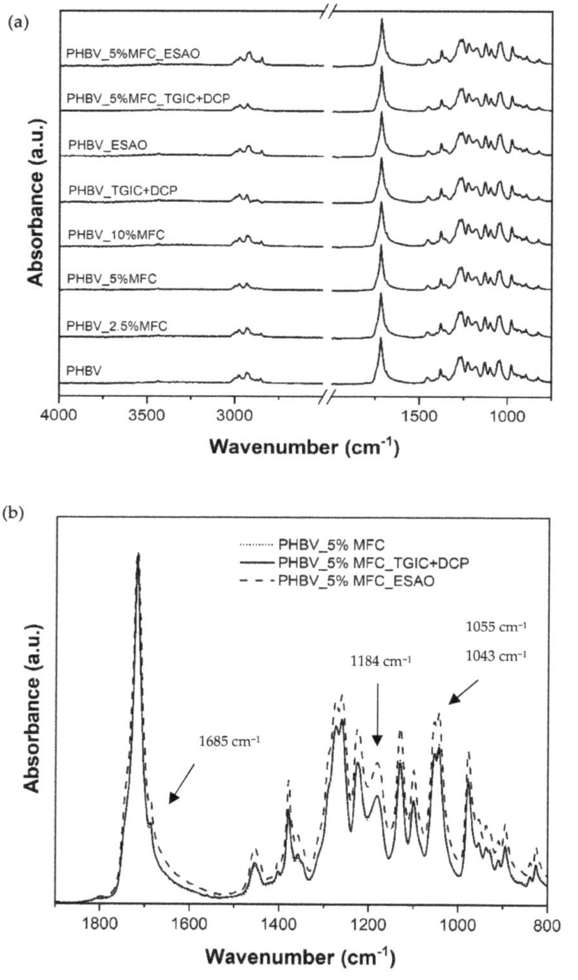

Figure 5. (**a**) Fourier transform infrared (FTIR) spectra, from bottom to top, of the films of the following: poly(3-hydroxybutyrate-*co*-3-hydroxyvalerate) (PHBV), PHBV with microfibrillated cellulose (MFC) at 2.5, 5, and 10 wt %, PHBV with triglycidyl isocyanurate (TGIC) and dicumyl peroxide (DCP) and with epoxy-based styrene-acrylic oligomer (ESAO), and PHBV/MFC at 5 wt % with TGIC and DCP and with ESAO. (**b**) Detail of the FTIR spectra for the films of PHBV/MFC at 5 wt % uncompatibilized and compatibilized with TGIC and DCP and with ESAO. Arrows indicate the wavenumbers of the chemical bonds described in the text.

3.4. Optical Properties of Green Composites

Figure 6 gathers the visual images of the PHBV and its composite films with different contents of MFC and/or the compatibilizers ESAO and TGIC with DCP. The films were obtained by thermo-compression at 200 °C for 4 min using the doughs resulting from the melt-mixing process. One can observe from the images that the optical properties that the neat PHBV sample, shown as a control in Figure 6a, presented very similar characteristics in terms of transparency or color to the PHBV film samples with MFC without compatibilizers (Figure 6b–d), with compatibilizers without MFC (Figure 6e,f), or with MFC and compatibilizers (Figure 6g,h). In all cases, thermo-compression yielded homogeneous PHBV films, of about 130 μm thickness, with high opacity due to the inherent crystallinity of PHBV [18]. It is also worth mentioning that, by simple visual observation, it was not possible to determine the presence of MFC due to their micrometer-scale size, as demonstrated previously in Figure 2.

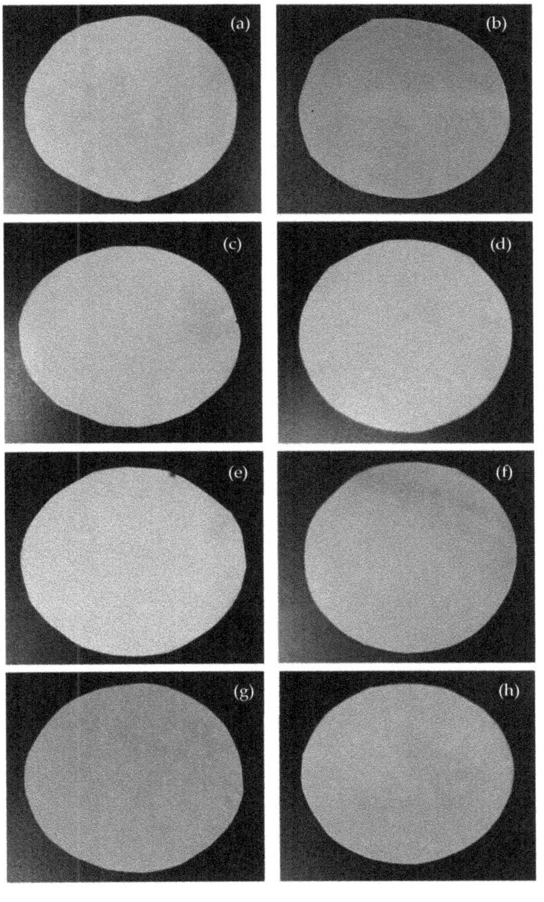

Figure 6. Films obtained by thermo-compression of poly(3-hydroxybutyrate-*co*-3-hydroxyvalerate) (PHBV) with microfibrillated cellulose (MFC) and compatibilized with epoxy-based styrene-acrylic oligomer (ESAO) and with triglycidyl isocyanurate (TGIC) and dicumyl peroxide (DCP) as follows: (**a**) PHBV; (**b**) PHBV_2.5%MFC; (**c**) PHBV_5%MFC; (**d**) PHBV_10%MFC; (**e**) PHBV_ESAO; (**f**) PHBV_TGIC+DCP; (**g**) PHBV_5%MFC_ESAO; (**h**) PHBV_5%MFC_TGIC+DCP.

In order to quantify the visual appearance of the PHBV/MFC composite films, Figure 7 represents their spectral distribution curves in terms of the percentage of Ti as a function of wavelength (λ). High Ti values are related to a higher ability of the sample to transmit light and, therefore, correspond to more transparent film samples, while low Ti values indicates higher opacity and a decrease in light passing through the film [52]. In this case, it can be observed that the films presented very similar Ti values throughout the spectral range, showing low-to-intermediate values. It is also worth noting that a slight reduction was observed for the samples with MFC contents of 5 wt % and, more noticeably, 10 wt %, due to the presence of the cellulose microparticles. In this regard, one of the desired characteristics of packaging materials is that they should protect food from the effects of light, especially ultraviolet radiation [53], for which the presence of MFC can be advantageous.

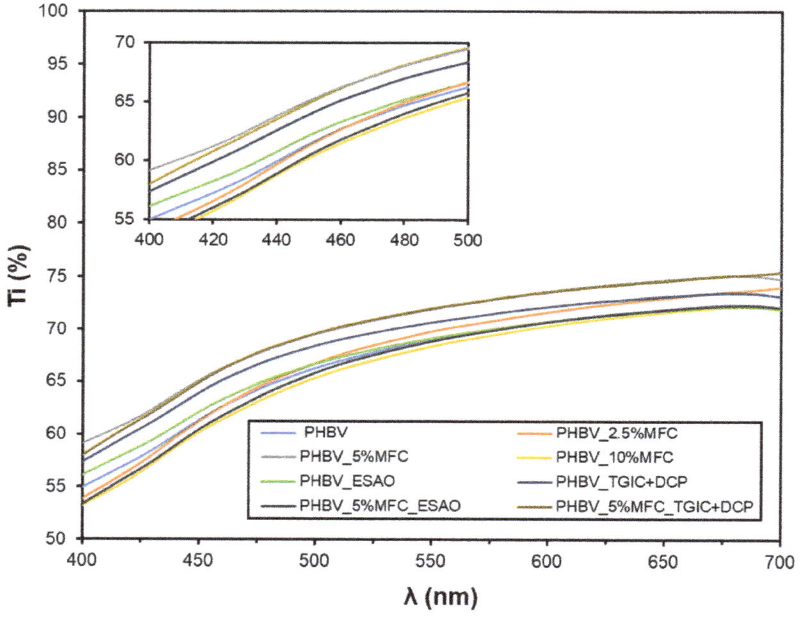

Figure 7. Spectral distribution curves of the percentage of internal transmittance (Ti) of the films of poly(3-hydroxybutyrate-*co*-3-hydroxyvalerate) (PHBV) with microfibrillated cellulose (MFC) and compatibilized with epoxy-based styrene-acrylic oligomer (ESAO) and with triglycidyl isocyanurate (TGIC) and dicumyl peroxide (DCP).

As can be observed in Table 2, the green composite films showed very similar values of brightness as PHBV, measured as L^*, in the range of 87–89, showing no significant differences ($p > 0.05$). Regarding the color coordinates a^* and b^*, the values that were obtained from all of the samples in the CIE-L$^*a^*b^*$ space indicated the color of the films as yellow ($b^* \simeq 8$–12) with a slight green hue ($a^* \simeq -1$). This corresponds to a $h_{ab}{}^*$ angle $\simeq$ 95–99, that is, of a yellow color with a very slightly green hue. On the other hand, the chroma or color saturation was very similar in all of the samples, with $C_{ab}{}^*$ values ranging between 8 and 12. In relation to the color difference between PHBV and its composites with MFC, these can be only appreciable by an experienced observer ($\Delta E_{ab}{}^* \geq 1$ and <2). The color differences were very similar in all cases, showing no significant differences ($p > 0.05$) and being slightly lower for the compabilized films. These values confirmed that the presence of the cellulose microstructures did not successfully alter the optical properties of PHBV.

Table 2. Color parameters (L^*, a^*, b^*) hue angle (h_{ab}^*), color saturation or chroma (C_{ab}^*), and color difference (ΔE_{ab}^*) of the films of poly(3-hydroxybutyrate-*co*-3-hydroxyvalerate) (PHBV) with microfibrillated cellulose (MFC) and compatibilized with epoxy-based styrene-acrylic oligomer (ESAO) and with triglycidyl isocyanurate (TGIC) and dicumyl peroxide (DCP).

Film	L^*	a^*	b^*	C_{ab}^*	h_{ab}^*	ΔE^*
PHBV	88.41 ± 0.11 [a]	−0.94 ± 0.01 [a]	10.20 ± 0.15 [a]	10.24 ± 0.15 [a]	95.25 ± 0.11 [a]	-
PHBV_2.5%MFC	87.75 ± 0.38 [a]	−1.12 ± 0.03 [b]	11.52 ± 0.87 [b]	11.57 ± 0.86 [b]	95.59± 0.47 [a]	1.53 ± 0.87 [a]
PHBV_5%MFC	88.40 ± 0.27 [a]	−0.91 ± 0.04 [a]	10.22 ± 0.98 [a]	10.25 ± 0.99 [a]	95.10 ± 0.32 [a]	1.27 ± 0.41 [a]
PHBV_10%MFC	87.89 ± 0.50 [a]	−0.86 ± 0.05 [a]	11.42 ± 0.64 [b]	11.45 ± 0.64 [b]	94.34 ± 0.41 [b]	1.35 ± 0.75 [a]
PHBV_ESAO	89.11 ± 0.24 [a]	−1.10 ± 0.05 [b]	8.75 ± 0.23 [c]	8.82 ± 0.23 [c]	97.15 ± 0.17 [c]	1.66 ± 0.34 [a]
PHBV_TGIC+DCP	89.39 ± 0.15 [a]	−1.22 ± 0.04 [c]	8.80 ± 0.44 [c]	8.88 ± 0.44 [c]	97.91 ± 0.38 [c]	1.75 ± 0.52 [a]
PHBV_5%MFC_ESAO	88.73 ± 0.07 [a]	−1.14 ± 0.02 [b]	11.24± 0.24 [b]	11.30 ± 0.24 [a]	95.76 ± 0.12 [a]	1.12 ± 0.17 [a]
PHBV_5%MFC_TGIC+DCP	88.86 ± 0.09 [a]	−1.38 ± 0.05 [d]	9.30 ± 0.22 [d]	9.40 ± 0.22 [c]	98.47 ± 0.23 [c]	1.12 ± 0.20 [a]

L^* lightness (L^* = 0 black, L^* = 100 white), a^* coordinate between red (+) and green (−), b^* between yellow (+) and blue (−), hue or tone (h_{ab}^*), color saturation or chroma (C_{ab}^*). Mean values and standard deviation. [a–d] Different superscripts in the same column indicate significant differences between formulations ($p < 0.05$).

3.5. Morphological Properties of Green Composites

Figure 8 shows the surface fractures of the PHBV films observed by FESEM. As can be seen in Figure 8a, the neat PHBV film showed a fracture surface corresponding to that of a brittle material, although it should be noted that the sample was cryo-fractured with liquid nitrogen, below the T_g of PHBV, that is, around −5 °C [41]. In this FESEM micrograph, some inorganic microparticles can be observed, which can correspond to boron nitride that is used by the manufacturer as a nucleating agent during the fabrication of commercial PHA [54].

In general, all of the composite films presented a very similar fracture surface, mainly determined by the high intrinsic brittleness of PHBV. In Figure 8b–d, corresponding to the cross-sections of the PHBV/MFC composite films without compatibilizers, the presence of the cellulose microstructures can be observed as isolated particles, especially in the samples with the highest contents. The formation of voids ("gaps") between the microparticles and the PHBV matrix was also noticeable as well as the presence of pores or holes corresponding to the separation of the microparticles during the fracture process, being both indicated by the arrows. The presence of a large interface, or pores, is representative of a poor filler-polymer adhesion and it is associated with a consequent loss in the mechanical properties of the composite [18]. Due to the lack of an adequate stress transmission between the filler and the matrix, MFC can act as mere filler elements, without offering increases in mechanical strength, or even as stress points that facilitate breakage. On the other hand, in Figure 8e,f, one can observe that the presence of the reactive compatibilizing agents improved the ductility of the PHBV matrix since some visible plastic deformation was evidenced by the presence of PHBV fibrils or threads generated by stretching during fracture. Finally, Figure 8g,h showed that a decrease, or complete elimination, of the gap between the microparticles and the biopolymer was attained, annotated by the circles, which was evident in the two compatibilized green composite samples.

3.6. Thermal Properties of Green Composites

Figure 9 shows the DSC curves of the analyzed PHBV films and their green composites that were compatibilized with ESAO and with TGIC and DCP. Table 3 gathers the thermal properties that were obtained from the curves. As it can be seen in Figure 9a, PHBV crystallized during cooling, showing a T_c of approximately 79 °C. Both the incorporation of MFC, as well as the two reactive compatibilizing agents, slightly affected the crystallization of the biopolymer chains. In particular, an increase in the crystallization process was observed in the composite films of PHBV with 2.5 wt % of MFC and with 5 wt % of MFC compatibilized with TGIC and DCP, reaching values close to 82 °C and 80 °C, respectively. A similar effect was reported in the study of Jun et al. [55], indicating that the presence of ultrathin structures of cellulose may favor the crystallization of PHBV by reducing the energy barrier during crystal formation. Thus, for these two film samples, it seems that the cellulose microstructures could serve as nucleating agents of the biopolymer crystals, favoring heterogeneous crystallization. In the rest of the samples, the T_c values were found to be very similar to that of PHBV, although somewhat in a lower range, that is, between 73 and 78 °C. This effect was significantly noticeable ($p < 0.05$) for the green composite sample filled with 10 wt % MFC and in the unfilled PHBV film samples that were compatibilized with TGIC and DCP and, more remarkably, ESAO. In particular, the latter sample reached the lowest T_c value, that is, ~73 °C. For the green composite films, this result suggests that the cellulose microparticles tended to agglomerate at the highest content, impairing the biopolymer chain arrangement during crystallization. In the case of the compatibilizing agents, these could facilitate the formation of longer biopolyester chains by a chain-extension effect or, even, a process of branching and cross-linking [42], which resulted in a macromolecular structure with chains of impaired movement. However, in the case of the compatibilized composite film samples, as a consequence of the aforementioned grafting process of the MFCs onto the macromolecular structure of the PHBV described during the FTIR analysis and further supported by the FESEM micrographs, the dispersion of the cellulose microstructures was enlarged and it could favor the formation of crystals. This effect was more noticeable in the case of the green composite film that was processed with TGIC and DCP, suggesting a higher compatibilizing effect.

Table 3. Thermal properties obtained by differential scanning calorimetry (DSC) of the films of poly(3-hydroxybutyrate-*co*-3-hydroxyvalerate) (PHBV) with microfibrillated cellulose (MFC) and compatibilized with epoxy-based styrene-acrylic oligomer (ESAO) and with triglycidyl isocyanurate (TGIC) and dicumyl peroxide (DCP).

Film	T_g (°C)	T_c (°C)	ΔH_c (J/g)	T_m (°C)	ΔH_m (J/g)	χ_c (%)
PHBV	-1.73 ± 0.11^a	78.65 ± 0.36^a	79.07 ± 0.70^a	170.53 ± 0.16^a	95.81 ± 8.64^a	65.35 ± 1.89^a
PHBV_2.5%MFC	-1.62 ± 0.07^a	82.26 ± 0.57^b	78.63 ± 0.64^a	167.80 ± 0.11^b	91.61 ± 8.42^b	65.78 ± 6.04^a
PHBV_5%MFC	-1.73 ± 0.04^a	76.40 ± 0.66^c	71.17 ± 0.23^b	168.81 ± 0.04^b	95.09 ± 1.22^a	68.28 ± 0.87^b
PHBV_10%MFC	-1.71 ± 0.21^a	75.59 ± 0.56^c	63.59 ± 0.71^c	169.06 ± 0.79^b	84.45 ± 1.60^c	64.01 ± 1.21^c
PHBV_ESAO	-1.72 ± 0.06^a	73.07 ± 0.54^d	73.18 ± 0.73^b	168.16 ± 0.49^b	91.92 ± 0.24^b	63.34 ± 0.16^d
PHBV_TGIC+DCP	-1.73 ± 0.14^a	77.07 ± 4.24^c	75.79 ± 0.19^d	166.31 ± 0.25^c	90.95 ± 2.18^b	62.85 ± 1.51^d
PHBV_5%MFC_ESAO	-1.77 ± 0.07^a	76.90 ± 1.66^c	78.87 ± 1.40^a	167.40 ± 0.56^c	93.66 ± 0.13^d	67.25 ± 0.10^b
PHBV_5%MFC_TGIC+DCP	-1.76 ± 0.30^a	79.93 ± 0.85^e	68.49 ± 0.34^e	165.76 ± 0.45^d	89.59 ± 1.23^b	65.15 ± 0.89^c

T_g = glass transition temperature; T_c = crystallization temperature; ΔH_c = enthalpy of crystallization, T_m = melting temperature, ΔH_m = enthalpy of fusion; χ_c = percent crystallinity. Mean values and standard deviation. [a–e] Different superscripts in the same column indicate significant differences between formulations ($p < 0.05$).

On the other hand, during the second heating of the samples, represented in Figure 9b, it can be observed that the biopolymer presented a second order change in the −5 to 5 °C range, which corresponds to the glass transition of PHBV. In the case of the neat PHBV, the T_g value was observed at approximately −1.7 °C, having no significant differences ($p < 0.05$) among the samples. Glass transition is a region where the biopolymer passes from a glassy state, without molecular mobility, towards a rubbery state, with greater molecular mobility and, therefore, greater toughness. Finally, it can be observed that the PHBV crystals formed during cooling melted at a temperature close to 171 °C. In general, all of the PHBV samples with MFC and/or compatibilizers presented values that were very similar to those of PHBV, although somewhat lower, in the range of 166–170 °C. This reduction in the melting point values can be related to certain crystal imperfections or a lower lamellar thickness growth of the PHBV crystals [56]. This observation suggests that both the presence of MFC and the effect of the compatibilizers on the molecular structure of the biopolymer affected the growth of the PHBV crystals. In terms of crystallinity, small differences were observed in most cases, but still significant ($p < 0.05$), being more noticeable in the green composite filled with 5 wt % MFC and the unfilled PHBV film samples that were processed with the compatibilizers. Thus, the percentage of crystallinity, that is χ_c, reached a value of approximately 65% in the neat PHBV sample. The presence of the cellulose microstructures slightly increased crystallinity due to their above-described nucleating agent, reaching values of up to ~68%. In the case of compatibilizers, the aforementioned increase in the molecular PHBV chains could slightly reduce the crystallinity, showing values in the range of 62–64%.

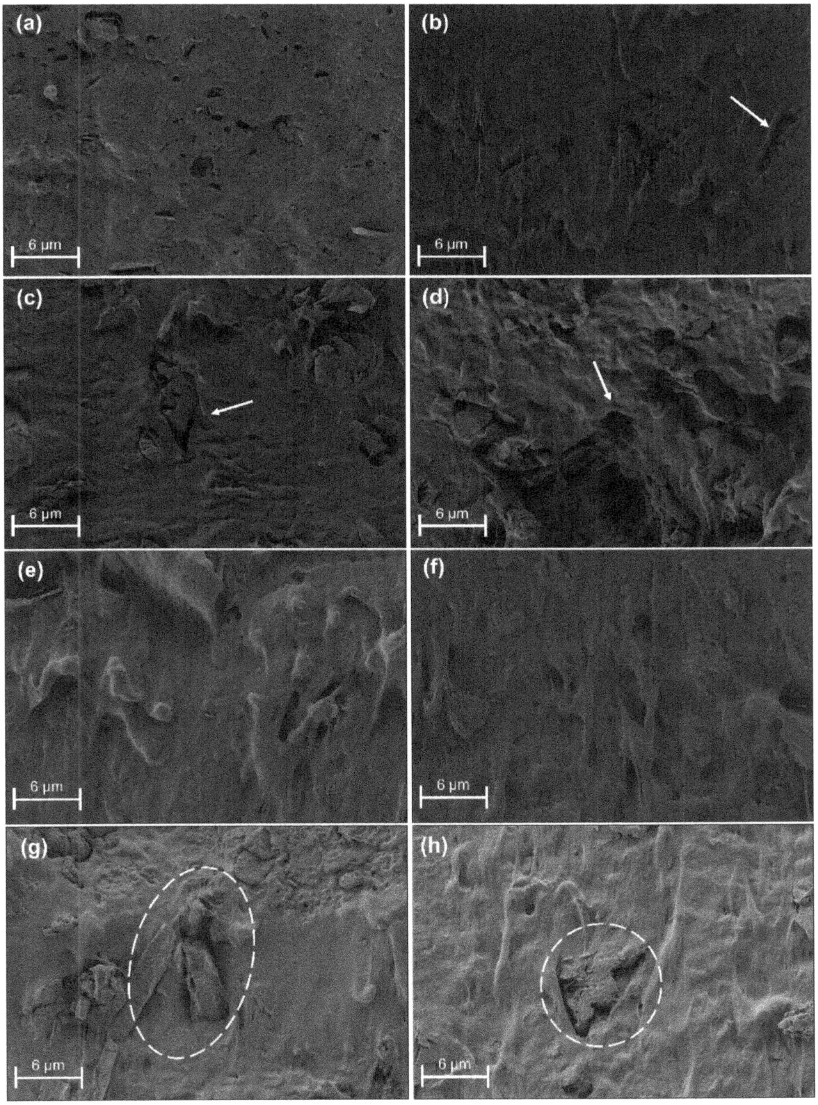

Figure 8. Field emission scanning electron microscopy (FESEM) images of the films of poly(3-hydroxybutyrate-*co*-3-hydroxyvalerate) (PHBV) with microfibrillated cellulose (MFC) and compatibilized with epoxy-based styrene-acrylic oligomer (ESAO) and with triglycidyl isocyanurate (TGIC) and dicumyl peroxide (DCP) as follows: (**a**) PHBV; (**b**) PHBV_2.5%MFC; (**c**) PHBV_5%MFC; (**d**) PHBV_10%MFC; (**e**) PHBV_ESAO; (**f**) PHBV_TGIC+DCP; (**g**) PHBV_5%MFC_ESAO; (**h**) PHBV_5%MFC_TGIC+DCP. Arrows and circles show the morphologies described in the text.

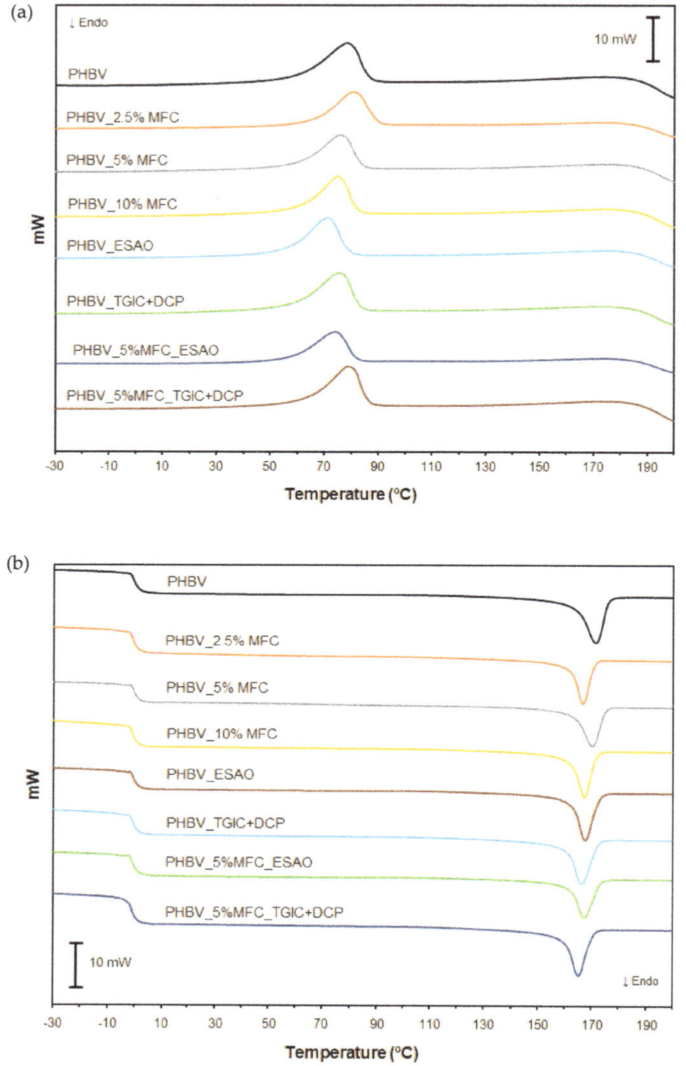

Figure 9. Differential scanning calorimetry (DSC) thermograms of the films of poly(3-hydroxybutyrate-*co*-3-hydroxyvalerate) (PHBV) with microfibrillated cellulose (MFC) and compatibilized with epoxy-based styrene-acrylic oligomer (ESAO) and with triglycidyl isocyanurate (TGIC) and dicumyl peroxide (DCP) as follows: (**a**) Cooling; (**b**) Second heating.

Furthermore, the TGA curves of the different PHBV films with MFC and/or the compatibilizing agents are shown in Figure 10. In these curves one can see that the biopolymer was stable up to temperatures close to 280–290 °C. In particular, thermal degradation occurred through a single and rapid degradation step that took place in the thermal range from 290 °C to 310 °C. As shown in Table 4, where the thermal stability data that was obtained from the TGA curves gathers, the PHBV sample showed T_{onset} and T_{deg} values of 277.5 °C and 296.3 °C, respectively. It has been described in the literature that the thermal degradation of PHAs generally follows a random chain scission model of the ester bond, involving a cis-type elimination reaction of –CH and a six-membered ring transition to form crotonic acid and its oligomers [57]. It can be observed from the TGA curves of the respective PHBV films with the compatibilizing agents that they exhibited a significant drop in thermal resistance, with T_{onset} values of 227.1 °C and 256.3 °C for the PHBV film samples that were compatibilized with ESAO and with TGIC and DCP, respectively. In the case of T_{deg}, these samples showed values in the range of 255–285 °C. These results thus indicate that the presence of the reactive compatibilizers resulted in the formation of macromolecular structures of lower thermal resistance. In this regard, it should also be noted that traces of these chemical agents, which could remain unreacted in excess, can catalyze the thermal degradation of PHBV. Regarding the PHBV/MFC composites and their respective composites compatibilized with ESAO and TGIC with DCP, it can be seen that the thermal stability of these films was similar to, or even in some cases higher than, that of the neat copolyester. The highest thermal properties were obtained for the PHBV composite with 10 wt % of MFC, reaching T_{onset} and T_{deg} values of 288.8 °C and 305.9 °C, respectively. This result is quite positive when comparing the present green composites to others based on lignocellulosic materials since these tend to show significantly lower thermal stability than that of unfilled biopolyesters and usually induce thermal decomposition [31]. This difference can be ascribed to the fact that MFC is based on pure alpha (α)-cellulose, without hemicellulose, and then with a higher thermal stability. Furthermore, it is also noteworthy that the presence of the compatibilizers in the PHBV/MFC composite samples did not contribute to a significant improvement of the thermal stability properties, presenting very similar values to that of the uncompatibilized PHBV composite with a 5 wt % MFC, but still significantly higher ($p < 0.05$) than that of the unfilled PHBV processed with the reactive agents. Therefore, this suggests that during the melt-mixing process of the green composites, by which the chemical anchoring of the cellulose microstructures on the PHBV structure was achieved, the reactive agents were successfully consumed, whereas these did not react completely in the case of the biopolymer samples without MFC. Finally, all of the PHBV films showed residual mass values of around 2%, with no significant differences ($p > 0.05$) among the film samples.

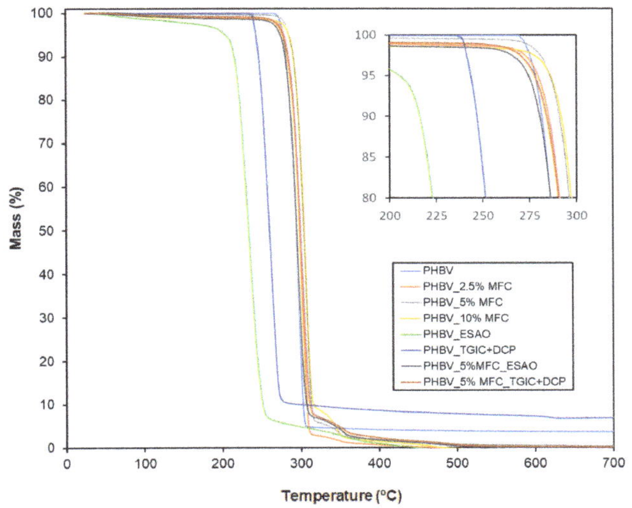

Figure 10. Thermogravimetric analysis (TGA) curves of the films of poly(3-hydroxybutyrate-*co*-3-hydroxyvalerate) (PHBV) with microfibrillated cellulose (MFC) and compatibilized with epoxy-based styrene-acrylic oligomer (ESAO) and with triglycidyl isocyanurate (TGIC) and dicumyl peroxide (DCP).

Table 4. Thermal properties obtained by thermogravimetric analysis (TGA) of the films of poly(3-hydroxybutyrate-*co*-3-hydroxyvalerate) (PHBV) with microfibrillated cellulose (MFC) and compatibilized with epoxy-based styrene-acrylic oligomer (ESAO) and with triglycidyl isocyanurate (TGIC) and dicumyl peroxide (DCP).

Film	T_{onset} (°C)	T_{deg} (°C)	Residual Mass (%)
PHBV	277.50 ± 1.54 ^{a}	296.25 ± 0.59 ^{a}	2.26 ± 0.04 ^{a}
PHBV_2.5%MFC	280.42 ± 1.77 ^{b}	301.67 ± 1.18 ^{b}	2.76 ± 0.12 ^{a}
PHBV_5%MFC	282.92 ± 1.12 ^{b}	301.67 ± 1.71 ^{b}	2.07 ± 0.57 ^{a}
PHBV_10%MFC	288.75 ± 2.95 ^{c}	305.83 ± 2.47 ^{c}	1.91 ± 0.84 ^{a}
PHBV_ESAO	227.08 ± 3.23 ^{d}	255.00 ± 2.46 ^{d}	2.02 ± 0.08 ^{a}
PHBV_TGIC+DCP	256.25 ± 1.27 ^{e}	276.67 ± 2.39 ^{e}	2.11 ± 0.15 ^{a}
PHBV_5%MFC_ESAO	274.17 ± 2.35 ^{a}	296.67 ± 1.18 ^{d}	2.16 ± 0.69 ^{a}
PHBV_5%MFC_TGIC+DCP	277.08 ± 0.59 ^{a}	300.83 ± 1.17 ^{d}	2.01 ± 0.11 ^{a}

T_{onset} = onset degradation temperature (measured at 5% mass loss); T_{deg} = degradation temperature. Mean values and standard deviation. $^{a–e}$ Different superscripts in the same column indicate significant differences between formulations ($p < 0.05$).

3.7. Mechanical Properties of Green Composites

Figure 11 shows the mechanical tensile stress (σ) versus strain (ε) curves, obtained at room temperature, of the PHBV films with MFCs and/or the compatibilizing agents. Table 5 summarizes the values of E, σ_y, and ε_b. As can be seen from the mechanical curve of the PHBV sample, the neat biopolymer film showed a mechanical behavior typical of a brittle material, with an E value of approximately 2.6 GPa and σ_y and ε_b values of 27.4 MPa and 1.4%. This mechanical curve showed nearly no plastic deformation since, after elastic deformation, the film broke with little or no deformation upon exceeding the yield point. The high brittleness of the PHB-based materials, as well as those of PHBV copolymers with low 3HV contents, that is, below 20 mol%, is due to their high crystallinity, which takes place mainly due to a secondary crystallization, and/or the physical aging of the amorphous region. In the first case, during the crystallization of PHB, the formation of large spherulites occurs due to the low nucleation density, together with the high degree of crystallinity and the slow crystallization rate [58]. Inside the spherulites, due to the difference between the radial and circumferential coefficient of thermal expansion, circular and radial cracks can form [59], which can act as stress concentration points and promote brittleness of the PHBV [60]. In terms of the physical aging process of PHB, produced during storage at room temperature, it implies a progressive restriction in the movement of interlamellar amorphous chains [61]. As a result of this, an increase of the rigid amorphous fraction is habitually observed during two weeks of storage at room temperature, that is, 25 °C [62]. In this regard, it should be noted that all of the PHBV films were allowed to age for at least 15 days, so this embrittlement effect was already reflected in the mechanical results of the biopolymer films.

As can be observed in Table 5, the incorporation of MFC resulted in a progressive increase in the values of E. In particular, the film samples of PHBV filled with 2.5, 5, and 10 wt % of MFC presented values of 2685 MPa, 2747 MPa, and 2873 MPa, respectively. However, because of their low compatibility with the PHBV matrix, as evidenced earlier during the morphological analysis of the film composites, both σ_y and ε_b were reduced to values within the range of 20–23 MPa and 0.8–1.1%. In the case of PHBV films without MFC, but processed with the compatibilizers, their tensile mechanical properties were similar to those of PHBV, with a slight but significant increase ($p < 0.05$) in E, having values of approximately 2.7 GPa. In the case of the σ_y and ε_b parameters, the values were similar to those of the PHBV film, in the range of 26–28 MPa and 1.2–1.4%, with no significant differences between them ($p > 0.05$). However, a significant improvement ($p < 0.05$) in terms of both mechanical strength and ductility was achieved in the film samples of PHBV/MFC that were compatibilized with ESAO and, more remarkable, with TGIC and DCP. In particular, for these composite films, the E values increased to 2863 MPa and 3071 MPa, respectively. A slight increase in the mechanical strength was also observed, reaching σ_y values of 27.8 MPa and 29.3 MPa. Furthermore, in the case of ductility, the composite films that were processed with ESAO and with TGIC and DCP yielded values of 0.9% and 1.4%, being the latter significantly higher ($p < 0.05$) than that of the uncompatibilized PHBV/MFC composite. Moreover, for the PHBV composite samples that were processed with TGIC and DCP, the samples were slightly more deformable than the neat PHBV film, but still not significant ($p > 0.05$). The results attained herein confirm the improvement that was achieved in the PHA composite samples with MFC compatibilized by means of these two multi-functional agents, that is, with multiple reactive groups in their chemical structure that are capable of interacting through covalent bonds favoring the grafting of reinforcing fillers onto the biopolymer matrix [34,35]. The positive effect on the mechanical strength provided by the cellulose microstructures obtained by atomization is also noteworthy, even without the use of compatibilizers. Thus, the incorporation of MFC provided higher mechanical strength to PHBV, with a relatively low impairment in ductility, due to their high degree of dispersion and micrometric size.

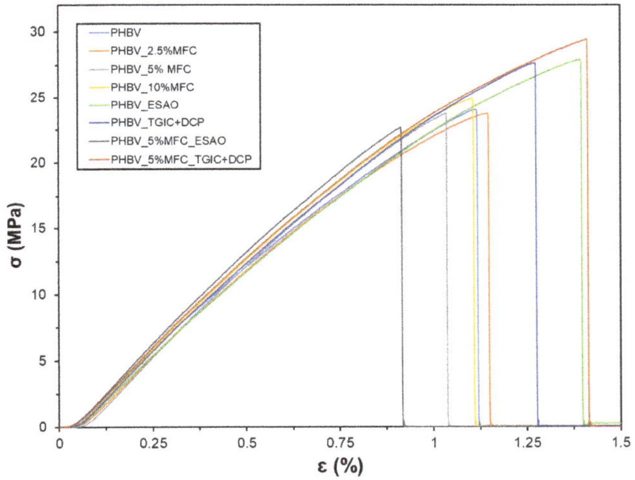

Figure 11. Stress (σ) versus strain (ε) curves of the films of poly(3-hydroxybutyrate-*co*-3-hydroxyvalerate) (PHBV) with microfibrillated cellulose (MFC) and compatibilized with epoxy-based styrene-acrylic oligomer (ESAO) and with triglycidyl isocyanurate (TGIC) and dicumyl peroxide (DCP).

Table 5. Mechanical properties of the films of poly(3-hydroxybutyrate-*co*-3-hydroxyvalerate) (PHBV) with microfibrillated cellulose (MFC) and compatibilized with epoxy-based styrene-acrylic oligomer (ESAO) and with triglycidyl isocyanurate (TGIC) and dicumyl peroxide (DCP).

Film	E (MPa)	σ_y (MPa)	ε_b (%)
PHBV	2613.5 ± 92.7 ^{a}	27.4 ± 2.1 ^{a}	1.37 ± 0.21 ^{a}
PHBV_2.5%MFC	2684.9 ± 77.1 ^{b}	22.2 ± 1.6 ^{b}	1.09 ± 0.13 ^{b}
PHBV_5%MFC	2746.5 ± 69.7 ^{c}	21.6 ± 0.9 ^{c}	0.92 ± 0.20 ^{c}
PHBV_10%MFC	2873.1 ± 73.5 ^{d}	20.1 ± 0.7 ^{d}	0.78 ± 0.14 ^{d}
PHBV_ESAO	2670.2 ± 21.7 ^{b}	27.3 ± 1.2 ^{a}	1.27 ± 0.11 ^{a}
PHBV_TGIC+DCP	2710.9 ± 76.4 ^{ab}	27.6 ± 1.7 ^{a}	1.29 ± 0.15 ^{a}
PHBV_5%MFC_ESAO	2863.1 ± 55.6 ^{d}	27.8 ± 1.1 ^{a}	0.91 ± 0.09 ^{c}
PHBV_5%MFC_TGIC+DCP	3070.7 ± 41.7 ^{e}	29.3 ± 2.3 ^{e}	1.43 ± 0.08 ^{a}

E = elastic modulus; σ_y = tensile stress at yield; ε_b = deformation at break. Mean values and standard deviation. $^{a–e}$ Different superscripts in the same column indicate significant differences between formulations ($p < 0.05$).

3.8. Barrier Properties of Green Composites

Table 6 shows the barrier properties against water and limonene vapors and oxygen of the PHBV films with MFC and compatibilizers. In the food packaging field, among other applications, the barrier performance is one the main parameters to consider, especially for the shelf-life extension of foodstuff [63]. Both water vapor and oxygen barrier properties are important in order to avoid physical and chemical deterioration, whereas limonene transport properties are measured to ascertain the aroma barrier. As can be seen in the table, all of the film samples showed significant differences ($p < 0.05$) and the neat PHBV films presented values for WVP, LP, and OP of 1.84×10^{-15} kg·m·m^{-2}·Pa^{-1}·s^{-1}, 1.41×10^{-14} kg·m·m^{-2}·Pa^{-1}·s^{-1}, and 2.09×10^{-19} m^3·m·m^{-2}·Pa^{-1}·s^{-1}, respectively. These permeability values are very similar to those that have been reported previously for films of PHBV with 2–3 mol% 3HV, also obtained by thermo-compression [41]. In terms of water vapor, one can observe that the incorporation of both MFC and the reactive compatibilizers increased the permeability of PHBV. In this regard, one should consider that PHBV is a medium-to-high barrier material against water vapor, having a slightly better performance than petrochemical PET, that is, 3.01×10^{-15} kg·m·m^{-2}·Pa^{-1}·s^{-1} [64] and relatively close to those of polyolefins, such as low-density polyethylene (LDPE) (1.20×10^{-15} kg·m·m^{-2}·Pa^{-1}·s^{-1}) and polypropylene (PP) (0.73×10^{-15} kg·m·m^{-2}·Pa^{-1}·s^{-1}) [65]. However, cellulose is a hydrophilic material due to its large number of free –OH groups and the presence of nanocelluloses in PHB-based films has been reported to cause a wettability increase [66]. In the case of the compatibilizers, the lower water-barrier performance can be ascribed to the lower crystallinity attained in these samples, as shown above during DSC analysis, due to development of a macromolecular structure that is composed of longer, branched, and/or cross-linked chains [34,42]. However, it is also worth noting that the green composite film that was compatibilized with TGIC and DCP showed higher barrier performance to water vapor than the uncompatibilized green composite films, but it was still two times more permeable than neat PHBV. This result can be ascribed to the higher dispersion of MFC in this compatibilized sample, by which this presented lower nanofiber aggregates and displays a reduced number of water diffusion paths.

Table 6. Barrier properties of the films of poly(3-hydroxybutyrate-*co*-3-hydroxyvalerate) (PHBV) with microfibrillated cellulose (MFC) and compatibilized with epoxy-based styrene-acrylic oligomer (ESAO) and with triglycidyl isocyanurate (TGIC) and dicumyl peroxide (DCP).

Film	Thickness (μm)	WVP × 10^{15} (kg·m·m^{-2}·Pa^{-1}·s^{-1})	LP × 10^{14} (kg·m·m^{-2}·Pa^{-1}·s^{-1})	OP × 10^{19} (m^3·m·m^{-2}·Pa^{-1}·s^{-1})
PHBV	126 ± 5	1.84 ± 0.39^{a}	1.41 ± 0.12^{a}	2.09 ± 0.48^{a}
PHBV_2.5%MFC	133 ± 4	4.19 ± 0.40^{b}	1.34 ± 0.06^{a}	1.60 ± 0.19^{b}
PHBV_5%MFC	128 ± 7	5.27 ± 0.53^{c}	1.07 ± 0.09^{b}	1.24 ± 0.36^{c}
PHBV_10%MFC	129 ± 6	5.89 ± 0.47^{c}	1.66 ± 0.15^{c}	1.08 ± 0.12^{d}
PHBV_ESAO	137 ± 2	4.46 ± 0.19^{b}	2.04 ± 0.17^{d}	2.27 ± 0.15^{a}
PHBV_TGIC+DCP	135 ± 8	5.69 ± 0.29^{c}	3.80 ± 0.31^{e}	2.70 ± 0.26^{a}
PHBV_5%MFC_ESAO	136 ± 7	5.48 ± 0.58^{c}	0.99 ± 0.39^{b}	1.36 ± 0.12^{c}
PHBV_5%MFC_TGIC+DCP	128 ± 3	3.29 ± 0.30^{d}	0.16 ± 0.01^{f}	0.98 ± 0.17^{f}

WVP = water vapor permeability; LP = limonene permeability; OP = oxygen permeability. Mean values and standard deviation. $^{a–f}$ Different superscripts in the same column indicate significant differences between formulations ($p < 0.05$).

In the case of limonene, as opposed to moisture, this is a strong plasticizing component for PHAs and, additionally, solubility plays a key role in permeability [67]. Thus, the presence of MFC contributed to improve the LP values of PHBV, particularly at a filler content of 5 wt %. However, similar to WVP, the use of the reactive compatibilizers in the unfilled PHBV sample films significantly increased ($p < 0.05$) their permeability to aroma. Interestingly, the compatibilized green composite films, particularly those that were processed with TGIC and DCP, showed a significant improvement ($p < 0.05$) in the aroma barrier. The latter film sample showed an LP value of 1.59×10^{-15} kg·m·m^{-2}·Pa^{-1}·s^{-1}, which is nearly one order of magnitude lower than the permeability of neat PHBV. In this regard, one should consider that the aroma barrier performance of nanocellulose is remarkably higher than that of PHB and PHBV [68] so that the presence of highly dispersed MFC in the compatibilized film sample could successfully reduce limonene absorption in the biopolyester matrix. Finally, in terms of oxygen barrier, one can see that permeability values followed a similar trend to the one observed in the case of aroma. Thus, the incorporation of MFC significantly reduced ($p < 0.05$) the OP values, showing the highest improvement at fillers of 10 wt % with a value of 1.08×10^{-19} m^3·m·m^{-2}·Pa^{-1}·s^{-1}. This result agrees with previous studies reporting on the oxygen-barrier enhancements of biopolyesters by the incorporation of nanocellulose materials. For instance, Arrieta et al. [66] incorporated 5 wt % CNC in PLA, attaining a reduction in OTR of about 43%. Similar improvements were achieved by Cao et al. [69], showing 47–64% reductions in the OP values of PLA by the addition of nanocellulose with different aspect ratios and at contents of 1, 3, 5, and 10 wt %. Higher reductions were achieved by Jiang et al. [70], who showed that the addition of 0.7 wt % CNC led to a decrease of 88.9% in PHB-based films. The improvement and contribution mechanism of ultrathin cellulose materials to the gas barrier property of polymers is based on the creation of a tortuous path for the diffusing gas molecules, since these nanomaterials are impervious to gases. Furthermore, the unfilled PHBV films that were processed with the compatibilizers showed similar OP values but were still significantly higher ($p < 0.05$) than the neat PHBV film, explained in terms of their lower crystallinity. Similarly, the compatibilized green composite sample with TGIC and DCP attained the highest oxygen barrier, having a value of 9.82×10^{-20} m^3·m·m^{-2}·Pa^{-1}·s^{-1}, which represents a barrier improvement of approximately two times compared with neat PHBV. This permeability value corresponds to a polymer with a medium-oxygen-barrier performance, being around three times lower than PET (3.27×10^{-19} m^3·m·m^{-2}·Pa^{-1}·s^{-1}), but still two orders of magnitude higher than poly(ethylene-*co*-vinyl alcohol) containing 32 mol% of ethylene (EVOH32, 0.77 and 91×10^{-21} m^3·m·m^{-2}·Pa^{-1}·s^{-1} for 0 and 75% RH, respectively) [65].

4. Conclusions

As demonstrated in the present work, atomization allows us to obtain micrometric structures based on ultrathin cellulose fibers. These cellulose microparticles showed a high capacity to improve the physical properties of biopolyesters, mainly in terms of mechanical strength and barrier against aroma and oxygen, without significantly altering their optical properties or reducing their thermal stability. Therefore, the atomization and subsequent incorporation of MFC into biopolyester can successfully represent a novel and sustainable solution to increase the performance and applications of green materials in the development of compostable and/or biodegradable packaging within the framework of the Circular Economy. However, it was also evidenced that the direct incorporation of these cellulose microstructures into biopolyester matrices still present problems of chemical incompatibility, which is mainly reflected in a reduction of the mechanical and barrier performance. In this regard, reactive compatibilizing agents that are based on multiple epoxide groups, such as ESAO and TGIC, have demonstrated their ability to improve the adhesion of cellulose microstructures to PHBV matrices through a reactive grafting mechanism. In particular, the combined use of TGIC with DCP, which facilitates the initiation of the reaction through the generation of free radicals, yielded the most notable improvements, resulting in PHBV/MFC composite films with improved processability and increased mechanical and barrier performances. Future studies will focus on the application of the resultant compatibilized green composite films as rigid packaging articles for food preservation applications, in particular for determining the shelf life of foodstuff whose rapid deterioration represents significant economic losses and whose market price allows the cost of innovation to be absorbed. Moreover, migration studies should also be performed in the resultant green composite films with food simulants in order to corroborate their safety and potential use in food packaging.

Author Contributions: P.A.V.F.: methodology, formal analysis, writing—original draft preparation; H.B.: methodology, writing—original draft preparation; F.V.: methodology, formal analysis; D.R.: methodology, formal analysis; M.V.: supervision, funding acquisition; S.T.-G.: conceptualization, formal analysis, investigation, writing—original draft preparation, writing—review and editing, supervision. All authors have read and agreed to the published version of the manuscript.

Funding: This research was funded by Spanish Ministry of Science and Innovation (MICI), grant numbers PID2019-105207RB-I00 and RTC2019-007268-2.

Institutional Review Board Statement: Not applicable.

Informed Consent Statement: Not applicable.

Data Availability Statement: Data are contained within the article and are also available on request.

Acknowledgments: P.A.V.F. is grateful to Generalitat Valenciana (GVA) for the Grisolia P/2019/115 grant. S.T.-G. acknowledges MICI for his Ramón y Cajal contract (RYC2019-027784-I). The Integrated Employment Service (SIE) of the Universtitat Politècnica de València (UPV) is also acknowledged for the collaboration through its Educational Cooperation Program.

Conflicts of Interest: The authors declare no conflict of interest.

References

1. Geueke, B.; Groh, K.; Muncke, J. Food packaging in the circular economy: Overview of chemical safety aspects for commonly used materials. *J. Clean. Prod.* **2018**, *193*, 491–505. [CrossRef]
2. Hottle, T.A.; Bilec, M.M.; Landis, A.E. Biopolymer production and end of life comparisons using life cycle assessment. *Resour. Conserv. Recycl.* **2017**, *122*, 295–306. [CrossRef]
3. Torres-Giner, S.; Figueroa-Lopez, K.J.; Melendez-Rodriguez, B.; Prieto, C.; Pardo-Figuerez, M.; Lagaron, J.M. Emerging Trends in Biopolymers for Food Packaging. In *Sustainable Food Packaging Technology*; Wiley: Hoboken, NJ, USA, 2021; pp. 1–33.
4. Durkin, A.; Taptygin, I.; Kong, Q.; Gunam Resul, M.F.M.; Rehman, A.; Fernández, A.M.L.; Harvey, A.P.; Shah, N.; Guo, M. Scale-up and Sustainability Evaluation of Biopolymer Production from Citrus Waste Offering Carbon Capture and Utilisation Pathway. *ChemistryOpen* **2019**, *8*, 668–688. [CrossRef] [PubMed]
5. Hosseini, S.N.; Pirsa, S.; Farzi, J. Biodegradable nano composite film based on modified starch-albumin/MgO; antibacterial, antioxidant and structural properties. *Polym. Test.* **2021**, *97*, 107182. [CrossRef]
6. Sharifi, K.A.; Pirsa, S. Biodegradable film of black mulberry pulp pectin/chlorophyll of black mulberry leaf encapsulated with carboxymethylcellulose/silica nanoparticles: Investigation of physicochemical and antimicrobial properties. *Mater. Chem. Phys.* **2021**, *267*, 124580. [CrossRef]
7. Mohammadi, B.; Pirsa, S.; Alizadeh, M. Preparing chitosan–polyaniline nanocomposite film and examining its mechanical, electrical, and antimicrobial properties. *Polym. Polym. Compos.* **2019**, *27*, 507–517. [CrossRef]
8. Pirsa, S.; Mohammadi, B. Conducting/biodegradable chitosan-polyaniline film; Antioxidant, color, solubility and water vapor permeability properties. *Main Group Chem.* **2021**, *20*, 133–147. [CrossRef]
9. Pirsa, S.; Farshchi, E.; Roufegarinejad, L. Antioxidant/Antimicrobial Film Based on Carboxymethyl Cellulose/Gelatin/TiO_2–Ag Nano-Composite. *J. Polym. Environ.* **2020**, *28*, 3154–3163. [CrossRef]
10. Pirsa, S.; Chavoshizadeh, S. Design of an optical sensor for ethylene based on nanofiber bacterial cellulose film and its application for determination of banana storage time. *Polym. Adv. Technol.* **2018**, *29*, 1385–1393. [CrossRef]
11. Pirsa, S. Nanocomposite base on carboxymethylcellulose hydrogel: Simultaneous absorbent of ethylene and humidity to increase the shelf life of banana fruit. *Int. J. Biol. Macromol.* **2021**, *193*, 300–310. [CrossRef]
12. Yorghanlu, R.A.; Hemmati, H.; Pirsa, S.; Makhani, A. Production of biodegradable sodium caseinate film containing titanium oxide nanoparticles and grape seed essence and investigation of physicochemical properties. *Polym. Bull.* **2021**. [CrossRef]
13. Jabraili, A.; Pirsa, S.; Pirouzifard, M.K.; Amiri, S. Biodegradable Nanocomposite Film Based on Gluten/Silica/Calcium Chloride: Physicochemical Properties and Bioactive Compounds Extraction Capacity. *J. Polym. Environ.* **2021**, *29*, 2557–2571. [CrossRef]
14. Rezaei, M.; Pirsa, S.; Chavoshizadeh, S. Photocatalytic/Antimicrobial Active Film Based on Wheat Gluten/ZnO Nanoparticles. *J. Inorg. Organomet. Polym. Mater.* **2019**, *30*, 2654–2665. [CrossRef]
15. Arcos-Hernández, M.V.; Laycock, B.; Donose, B.C.; Pratt, S.; Halley, P.; Al-Luaibi, S.; Werker, A.; Lant, P.A. Physicochemical and mechanical properties of mixed culture polyhydroxyalkanoate (PHBV). *Eur. Polym. J.* **2013**, *49*, 904–913. [CrossRef]
16. Thushari, G.G.N.; Senevirathna, J.D.M. Plastic pollution in the marine environment. *Heliyon* **2020**, *6*, e04709. [CrossRef]
17. Bucci, D.Z.; Tavares, L.B.B.; Sell, I. Biodegradation and physical evaluation of PHB packaging. *Polym. Test.* **2007**, *26*, 908–915. [CrossRef]
18. Torres-Giner, S.; Montanes, N.; Fombuena, V.; Boronat, T.; Sanchez-Nacher, L. Preparation and characterization of compression-molded green composite sheets made of poly(3-hydroxybutyrate) reinforced with long pita fibers. *Adv. Polym. Technol.* **2018**, *37*, 1305–1315. [CrossRef]

19. Ray, S.S.; Bousmina, M. Biodegradable polymers and their layered silicate nanocomposites: In greening the 21st century materials world. *Prog. Mater. Sci.* **2005**, *50*, 962–1079.
20. Keshavarz, T.; Roy, I. Polyhydroxyalkanoates: Bioplastics with a green agenda. *Curr. Opin. Microbiol.* **2010**, *13*, 321–326. [CrossRef]
21. Laycock, B.; Arcos-Hernandez, M.V.; Langford, A.; Buchanan, J.; Halley, P.J.; Werker, A.; Lant, P.A.; Pratt, S. Thermal properties and crystallization behavior of fractionated blocky and random polyhydroxyalkanoate copolymers from mixed microbial cultures. *J. Appl. Polym. Sci.* **2014**, *131*, 40836. [CrossRef]
22. Tajeddin, B. Cellulose-Based Polymers for Packaging Applications. In *Lignocellulosic Polymer Composites*; Wiley: Hoboken, NJ, USA, 2014; pp. 477–498.
23. Henriksson, M.; Henriksson, G.; Berglund, L.A.; Lindström, T. An environmentally friendly method for enzyme-assisted preparation of microfibrillated cellulose (MFC) nanofibers. *Eur. Polym. J.* **2007**, *43*, 3434–3441. [CrossRef]
24. Fortunati, E.; Armentano, I.; Zhou, Q.; Iannoni, A.; Saino, E.; Visai, L.; Berglund, L.A.; Kenny, J.M. Multifunctional bionanocomposite films of poly(lactic acid), cellulose nanocrystals and silver nanoparticles. *Carbohydr. Polym.* **2012**, *87*, 1596–1605. [CrossRef]
25. Gatenholm, P.; Klemm, D. Bacterial Nanocellulose as a Renewable Material for Biomedical Applications. *MRS Bull.* **2010**, *35*, 208–213. [CrossRef]
26. Suzuki, K.; Sato, A.; Okumura, H.; Hashimoto, T.; Nakagaito, A.N.; Yano, H. Novel high-strength, micro fibrillated cellulose-reinforced polypropylene composites using a cationic polymer as compatibilizer. *Cellulose* **2014**, *21*, 507–518. [CrossRef]
27. Miao, C.; Hamad, W.Y. Cellulose reinforced polymer composites and nanocomposites: A critical review. *Cellulose* **2013**, *20*, 2221–2262. [CrossRef]
28. Venkatraman, P.; Gohn, A.M.; Rhoades, A.M.; Foster, E.J. Developing high performance PA11/cellulose nanocomposites for industrial-scale melt processing. *Compos. Part B Eng.* **2019**, *174*, 106988. [CrossRef]
29. Oksman, K.; Aitomäki, Y.; Mathew, A.P.; Siqueira, G.; Zhou, Q.; Butylina, S.; Tanpichai, S.; Zhou, X.; Hooshmand, S. Review of the recent developments in cellulose nanocomposite processing. *Compos. Part A Appl. Sci. Manuf.* **2016**, *83*, 2–18. [CrossRef]
30. Fonseca-Valero, C.; Ochoa-Mendoza, A.; Arranz-Andrés, J.; González-Sánchez, C. Mechanical recycling and composition effects on the properties and structure of hardwood cellulose-reinforced high density polyethylene eco-composites. *Compos. Part A Appl. Sci. Manuf.* **2015**, *69*, 94–104. [CrossRef]
31. Agüero, Á.; Garcia-Sanoguera, D.; Lascano, D.; Rojas-Lema, S.; Ivorra-Martinez, J.; Fenollar, O.; Torres-Giner, S. Evaluation of Different Compatibilization Strategies to Improve the Performance of Injection-Molded Green Composite Pieces Made of Polylactide Reinforced with Short Flaxseed Fibers. *Polymers* **2020**, *12*, 821. [CrossRef]
32. Pothan, L.A.; Thomas, S. Polarity parameters and dynamic mechanical behaviour of chemically modified banana fiber reinforced polyester composites. *Compos. Sci. Technol.* **2003**, *63*, 1231–1240. [CrossRef]
33. Bengtsson, M.; Gatenholm, P.; Oksman, K. The effect of crosslinking on the properties of polyethylene/wood flour composites. *Compos. Sci. Technol.* **2005**, *65*, 1468–1479. [CrossRef]
34. Torres-Giner, S.; Montanes, N.; Boronat, T.; Quiles-Carrillo, L.; Balart, R. Melt grafting of sepiolite nanoclay onto poly(3-hydroxybutyrate-*co*-4-hydroxybutyrate) by reactive extrusion with multi-functional epoxy-based styrene-acrylic oligomer. *Eur. Polym. J.* **2016**, *84*, 693–707. [CrossRef]
35. Melendez-Rodriguez, B.; Torres-Giner, S.; Aldureid, A.; Cabedo, L.; Lagaron, J.M. Reactive Melt Mixing of Poly(3-Hydroxybutyrate)/Rice Husk Flour Composites with Purified Biosustainably Produced Poly(3-Hydroxybutyrate-*co*-3-Hydroxyvalerate). *Materials* **2019**, *12*, 2152. [CrossRef] [PubMed]
36. Agüero, A.; Morcillo, M.d.C.; Quiles-Carrillo, L.; Balart, R.; Boronat, T.; Lascano, D.; Torres-Giner, S.; Fenollar, O. Study of the Influence of the Reprocessing Cycles on the Final Properties of Polylactide Pieces Obtained by Injection Molding. *Polymers* **2019**, *11*, 1908. [CrossRef] [PubMed]
37. Torres-Giner, S.; Hilliou, L.; Melendez-Rodriguez, B.; Figueroa-Lopez, K.J.; Madalena, D.; Cabedo, L.; Covas, J.A.; Vicente, A.A.; Lagaron, J.M. Melt processability, characterization, and antibacterial activity of compression-molded green composite sheets made of poly(3-hydroxybutyrate-*co*-3-hydroxyvalerate) reinforced with coconut fibers impregnated with oregano essential oil. *Food Packag. Shelf Life* **2018**, *17*, 39–49. [CrossRef]
38. Shariatnia, S.; Veldanda, A.; Obeidat, S.; Jarrahbashi, D.; Asadi, A. Atomization of cellulose nanocrystals aqueous suspensions in fused deposition modeling: A scalable technique to improve the strength of 3D printed polymers. *Compos. Part B Eng.* **2019**, *177*, 107291. [CrossRef]
39. Larsson, M.; Markbo, O.; Jannasch, P. Melt processability and thermomechanical properties of blends based on polyhydroxyalkanoates and poly(butylene adipate-*co*-terephthalate). *RSC Adv.* **2016**, *6*, 44354–44363. [CrossRef]
40. Torres-Giner, S. Preparation of conductive carbon black-filled polymer nanocomposites via melt compounding. In *Conductive Materials and Composites*; Nova Science Publishers, Inc.: Hauppauge, NY, USA, 2016; pp. 117–164.
41. Quiles-Carrillo, L.; Montanes, N.; Lagaron, J.M.; Balart, R.; Torres-Giner, S. In Situ Compatibilization of Biopolymer Ternary Blends by Reactive Extrusion with Low-Functionality Epoxy-Based Styrene–Acrylic Oligomer. *J. Polym. Environ.* **2019**, *27*, 84–96. [CrossRef]
42. Quiles-Carrillo, L.; Duart, S.; Montanes, N.; Torres-Giner, S.; Balart, R. Enhancement of the mechanical and thermal properties of injection-molded polylactide parts by the addition of acrylated epoxidized soybean oil. *Mater. Des.* **2018**, *140*, 54–63. [CrossRef]

43. Melendez-Rodriguez, B.; Castro-Mayorga, J.L.; Reis, M.A.M.; Sammon, C.; Cabedo, L.; Torres-Giner, S.; Lagaron, J.M. Preparation and Characterization of Electrospun Food Biopackaging Films of Poly(3-hydroxybutyrate-*co*-3-hydroxyvalerate) Derived From Fruit Pulp Biowaste. *Front. Sustain. Food Syst.* **2018**, *2*, 38. [CrossRef]
44. Rojas-Lema, S.; Arevalo, J.; Gomez-Caturla, J.; Garcia-Garcia, D.; Torres-Giner, S. Peroxide-Induced Synthesis of Maleic Anhydride-Grafted Poly(butylene succinate) and Its Compatibilizing Effect on Poly(butylene succinate)/Pistachio Shell Flour Composites. *Molecules* **2021**, *26*, 5927. [CrossRef] [PubMed]
45. Sun, Q.; Mekonnen, T.; Misra, M.; Mohanty, A.K. Novel Biodegradable Cast Film from Carbon Dioxide Based Copolymer and Poly(Lactic Acid). *J. Polym. Environ.* **2016**, *24*, 23–36. [CrossRef]
46. Invernizzi, C.; Rovetta, T.; Licchelli, M.; Malagodi, M. Mid and Near-Infrared Reflection Spectral Database of Natural Organic Materials in the Cultural Heritage Field. *Int. J. Anal. Chem.* **2018**, *2018*, 7823248. [CrossRef]
47. Colthup, N. *Introduction to Infrared and Raman Spectroscopy*; Elsevier: Amsterdam, The Netherlands, 2012.
48. Villalobos, M.; Awojulu, A.; Greeley, T.; Turco, G.; Deeter, G. Oligomeric chain extenders for economic reprocessing and recycling of condensation plastics. *Energy* **2006**, *31*, 3227–3234. [CrossRef]
49. Samaniego, K.; Matos, A.; Sánchez-Safont, E.; Candal, M.V.; Lagaron, J.M.; Cabedo, L.; Famez-Perez, J. Role of Plasticizers on PHB/bio-TPE Blends Compatibilized by Reactive Extrusion. *Materials* **2022**, *15*, 1226. [CrossRef]
50. Quiles-Carrillo, L.; Montanes, N.; Lagaron, J.M.; Balart, R.; Torres-Giner, S. On the use of acrylated epoxidized soybean oil as a reactive compatibilizer in injection-molded compostable pieces consisting of polylactide filled with orange peel flour. *Polym. Int.* **2018**, *67*, 1341–1351. [CrossRef]
51. Wei, L.; McDonald, A.G.; Stark, N.M. Grafting of Bacterial Polyhydroxybutyrate (PHB) onto Cellulose via In Situ Reactive Extrusion with Dicumyl Peroxide. *Biomacromolecules* **2015**, *16*, 1040–1049. [CrossRef]
52. Ortega-Toro, R.; Jiménez, A.; Talens, P.; Chiralt, A. Properties of starch–hydroxypropyl methylcellulose based films obtained by compression molding. *Carbohydr. Polym.* **2014**, *109*, 155–165. [CrossRef]
53. Villalobos, R.; Chanona, J.; Hernández, P.; Gutiérrez, G.; Chiralt, A. Gloss and transparency of hydroxypropyl methylcellulose films containing surfactants as affected by their microstructure. *Food Hydrocoll.* **2005**, *19*, 53–61. [CrossRef]
54. Ivorra-Martinez, J.; Quiles-Carrillo, L.; Boronat, T.; Torres-Giner, S.; Covas, J.A. Assessment of the Mechanical and Thermal Properties of Injection-Molded Poly(3-hydroxybutyrate-*co*-3-hydroxyhexanoate)/Hydroxyapatite Nanoparticles Parts for Use in Bone Tissue Engineering. *Polymers* **2020**, *12*, 1389. [CrossRef]
55. Jun, D.; Guomin, Z.; Mingzhu, P.; Leilei, Z.; Dagang, L.; Rui, Z. Crystallization and mechanical properties of reinforced PHBV composites using melt compounding: Effect of CNCs and CNFs. *Carbohydr. Polym.* **2017**, *168*, 255–262. [CrossRef] [PubMed]
56. Melendez-Rodriguez, B.; Figueroa-Lopez, K.J.; Bernardos, A.; Martínez-Máñez, R.; Cabedo, L.; Torres-Giner, S.; Lagaron, J.M. Electrospun Antimicrobial Films of Poly(3-hydroxybutyrate-*co*-3-hydroxyvalerate) Containing Eugenol Essential Oil Encapsulated in Mesoporous Silica Nanoparticles. *Nanomaterials* **2019**, *9*, 227. [CrossRef] [PubMed]
57. Bugnicourt, E.; Cinelli, P.; Lazzeri, A.; Alvarez, V.A. Polyhydroxyalkanoate (PHA): Review of synthesis, characteristics, processing and potential applications in packaging; Budapest University of Technology and Economics. *Express Polym. Lett.* **2014**, *8*, 791–808. [CrossRef]
58. De Koning, G.J.M.; Lemstra, P.J. Crystallization phenomena in bacterial poly[(R)-3-hydroxybutyrate]: 2. Embrittlement and rejuvenation. *Polymer* **1993**, *34*, 4089–4094. [CrossRef]
59. Martinez-Salazar, J.; Sanchez-Cuesta, M.; Barham, P.J.; Keller, A. Thermal expansion and spherulite cracking in 3-hydroxybutyrate/3-hydroxyvalerate copolymers. *J. Mater. Sci. Lett.* **1989**, *8*, 490–492. [CrossRef]
60. Barham, P.J.; Keller, A. The relationship between microstructure and mode of fracture in polyhydroxybutyrate. *J. Polym. Sci. B Polym. Phys.* **1986**, *24*, 69–77. [CrossRef]
61. Biddlestone, F.; Harris, A.; Hay, J.N.; Hammond, T. The physical ageing of amorphous poly(hydroxybutyrate). *Polym. Int.* **1996**, *39*, 221–229. [CrossRef]
62. Di Lorenzo, M.L.; Righetti, M.C. Effect of thermal history on the evolution of crystal and amorphous fractions of poly[(R)-3-hydroxybutyrate] upon storage at ambient temperature. *Eur. Polym. J.* **2013**, *49*, 510–517. [CrossRef]
63. Torres-Giner, S.; Gil, L.; Pascual-Ramírez, L.; Garde-Belza, J.A. Packaging: Food waste reduction. *Encycl. Polym. Appl.* **2018**, *3*, 1990–2009.
64. Gabirondo, E.; Melendez-Rodriguez, B.; Arnal, C.; Lagaron, J.M.; de Ilarduya, A.M.; Sardon, H.; Torres-Giner, S. Organocatalyzed closed-loop chemical recycling of thermo-compressed films of poly(ethylene furanoate). *Polym. Chem.* **2021**, *12*, 1571–1580. [CrossRef]
65. Lagarón, J.M. 1-Multifunctional and nanoreinforced polymers for food packaging. In *Multifunctional and Nanoreinforced Polymers for Food Packaging*; Lagarón, J.-M., Ed.; Woodhead Publishing: Sawston, UK, 2011; pp. 1–28.
66. Arrieta, M.P.; Fortunati, E.; Dominici, F.; Rayón, E.; López, J.; Kenny, J.M. PLA-PHB/cellulose based films: Mechanical, barrier and disintegration properties. *Polym. Degrad. Stab.* **2014**, *107*, 139–149. [CrossRef]
67. Sanchez-Garcia, M.D.; Gimenez, E.; Lagaron, J.M. Novel PET Nanocomposites of Interest in Food Packaging Applications and Comparative Barrier Performance with Biopolyester Nanocomposites. *J. Plast. Film Sheeting* **2007**, *23*, 133–148. [CrossRef]
68. Cherpinski, A.; Torres-Giner, S.; Vartiainen, J.; Peresin, M.S.; Lahtinen, P.; Lagaron, J.M. Improving the water resistance of nanocellulose-based films with polyhydroxyalkanoates processed by the electrospinning coating technique. *Cellulose* **2018**, *25*, 1291–1307. [CrossRef]

69. Cao, X.; Wang, Y.; Chen, H.; Hu, J.; Cui, L. Preparation of different morphologies cellulose nanocrystals from waste cotton fibers and its effect on PLLA/PDLA composites films. *Compos. Part B Eng.* **2021**, *217*, 108934. [CrossRef]
70. Jiang, G.; Yu, L.; Zhang, M.; Wang, F.; Zhang, S. Poly(propylene carbonate)/poly(3-hydroxybutyrate)-based bionanocomposites reinforced with cellulose nanocrystal for potential application as a packaging material. *Polym. Adv. Technol.* **2020**, *31*, 853–863. [CrossRef]

applied sciences

MDPI

Article

Multilayer Sheets Based on Double Coatings of Poly(3-hydroxybutyrate-*co*-3-hydroxyvalerate) on Paper Substrate for Sustainable Food Packaging Applications

Eva Hernández-García [1], Pedro A. V. Freitas [1], Pedro Zomeño [2], Chelo González-Martínez [1] and Sergio Torres-Giner [1,*]

1 Research Institute of Food Engineering for Development (IIAD), Universitat Politècnica de València (UPV), 46022 Valencia, Spain
2 Packaging Technologies Department, AINIA, 46980 Paterna, Spain
* Correspondence: storresginer@upv.es

Featured Application: One of the main technological challenges within Circular Economy strategies is to minimize the environmental impact of plastic packaging. In this regard, the use of PHBV films to coat paper sheets represents a highly sustainable strategy to produce food packaging multilayer structures with improved mechanical and barrier properties.

Abstract: This work reports on the development and performance evaluation of newly developed paper sheets coated, on both sides, with thin films of biodegradable poly(3-hydroxybutyrate-*co*-3-hydroxyvalerate) (PHBV) for applications of food packaging. For this, PHBV/paper/PHBV multilayers were first prepared by the thermo-sealing technique, optimizing the process variables of temperature and time. Thereafter, the multilayer sheets were characterized in terms of their morphological, optical, thermal, mechanical, and barrier properties and compared with equivalent paper structures double coated with high-barrier multilayer films of petrochemical polymers. The results indicated that the double coatings of PHBV successfully improved the mechanical resistance and ductility, protected from moisture, and also reduced the aroma and oxygen permeances of paper, having a minimal effect on its optical and thermal properties. Finally, the compostability of the resultant multilayer sheets was analyzed, confirming that the presence of the PHBV coatings slightly delayed the aerobic biodegradation and disintegration of paper.

Keywords: paper; PHBV; multilayers; food packaging; Circular Economy

Citation: Hernández-García, E.; Freitas, P.A.V.; Zomeño, P.; González-Martínez, C.; Torres-Giner, S. Multilayer Sheets Based on Double Coatings of Poly(3-hydroxybutyrate-*co*-3-hydroxyvalerate) on Paper Substrate for Sustainable Food Packaging Applications. *Appl. Sci.* **2023**, *13*, 179. https://doi.org/10.3390/app13010179

Academic Editor: Hyeonseok Yoon

Received: 7 December 2022
Revised: 19 December 2022
Accepted: 20 December 2022
Published: 23 December 2022

1. Introduction

Packaging increases food quality and safety, which is essential to avoid spoilage and reduce food waste [1]. Plastic represents, together with paper, the most widely used material for food packaging applications due to its large-scale availability, low production cost, light weight, transparency, flexibility, good barrier, ease of processing, and versatility [2]. However, despite these advantages, conventional plastics are still based on petrochemical polymers that are synthesized from a non-renewable source and are not biodegradable. Moreover, these plastics are difficult to recycle, particularly when they are found in the form of multilayers. As a result, plastic packaging represents a major source of waste generation and plastics easily accumulate in the environment [3]. In fact, plastic materials can take between 100 and 450 years to disintegrate in the environment, leading to the so-called "white pollution" and to the formation of marine debris and microplastics [4]. In addition, microplastics can be taken up by different species in the food chain, which is a growing concern for human health and natural habitats [5].

Therefore, one of the main current challenges is to minimize the environmental impact of plastic packaging. From this perspective, the use of paper in packaging recently represents a sustainable option due to its biodegradable nature and the fact that it can be more easily recycled than plastic. However, paper lacks moisture resistance and shows poor thermal, mechanical, and barrier performances so that it is habitually coated with plastic materials [6]. Thus, multilayer packaging structures based on paper have expanded tremendously during the past few years, with use of different coating technologies to offer several functionalities, such as hydrophobicity, a gas and moisture barrier, antimicrobial protection, cohesive strength, scratch resistance, etc. [7]. Different petrochemical polymers have been used for coating paper, including polypropylene (PP) and biaxially oriented polypropylene (BOPP), low- and high-density polyethylene (LDPE, HDPE), polyethylene terephthalate (PET), biaxially oriented polyamide (OPA), and more importantly their combinations in the form of multilayers with high-barrier poly(ethylene-*co*-vinyl alcohol) (EVOH) [8]. These plastic multilayers have been demonstrated to increase the barrier against moisture, oxygen, odor, and grease, and offer favorable stiffness, toughness, processability, manufacturing cost, chemical resistance, and thermal stability. Nevertheless, these are not biodegradable and must be properly separated from paper for recycling to avoid impairing the compostability and recyclability of paper [9].

In this context, biopolymers can solve the main environmental constraints and limitations of the food packaging industry [10]. On the one hand, biopolymers can be obtained from natural and renewable sources. On the other hand, the resultant packaging materials can be composted or biodegraded in the environment [11]. Thus, some biopolymers are very promising as paper coating materials, including polysaccharides (e.g., chitosan, starch, lignocellulose derived compounds, and alginates), proteins (e.g., whey, wheat gluten, and zein) and, most relevantly, polyesters such as polylactide (PLA), poly(ε-caprolactone) (PCL), and recently polyhydroxyalkanoates (PHAs) [12]. PHAs are semi-crystalline aliphatic polyesters that offer the advantage of being synthesized by microorganisms using renewable sources, such as sugars and triglycerides, and can also biodegrade under natural conditions [13]. Therefore, PHAs can offer key benefits over petrochemical polymers to reduce packaging's carbon footprint and avoid "white pollution. Poly(3-hydroxybutyrate) (PHB) is the homopolyester and most common type of PHA, which has some physical properties similar to those of PP and PET [14–16]. However, the high crystallinity of PHB results in a highly rigid and brittle material that is also difficult to process due to its narrow processing window and low thermal stability [17,18]. Therefore, the use of microbial copolyesters, for instance poly(3-hydroxybutyrate-*co*-3-hydroxyvalerate) (PHBV), has been extended [19]. As the comonomer ratio of 3-hydroxyvalerate (3HV) increases in PHBV, the flexibility increases and the melting point is reduced [20,21]. For example, PHBV copolyester with 3HV contents above 10 mol% shows lower crystallinity and a larger processing window [22]. From a mechanical point of view, PHBV is more flexible, ductile, and tough [23]. For example, an increase of 3HV from 0 to 28 mol% significantly improves the elongation at break and impact strength [24]. Thus, the use of PHBV to coat paper represents a highly sustainable strategy to be explored in the packaging field.

The overall objective of this work was to develop and ascertain the performance of paper sheets coated, on both sides, with biodegradable PHBV films by heat sealing. To this end, the variables of the thermo-sealing process for obtaining the multilayers were first analyzed in terms of temperature and time. Then, the multilayers obtained from the most optimized conditions were characterized by their morphology, optical, thermal, mechanical, and vapor and gas barrier properties to determine their suitability for different food preservation applications. The performance of the newly developed multilayers was compared with that of the same paper substrate double coated with a commercial high-barrier multilayer film based on PET and also with that of uncoated paper sheets. Finally, the compostability of the paper sheets was analyzed to estimate the effect of PHBV on the biodegradation and disintegration of paper and, thus, its potential organic recycling.

2. Materials and Methods

2.1. Materials

Paper sheets with a grammage of 220 g·m^{-2} and thickness of 290 μm were supplied by Billerudkornäs-CrownBoard Prestige™ (Solna, Sweden). According to the manufacturer, this paper derives from 100% wood fibers and is suitable for use in contact with food. A commercial 10-μm film of PHBV with a 3HV content of 8 mol% was provided by GoodFellow Cambridge Limited (Huntindgon, UK) under the commercial reference BV301010. This grade is also certified by the supplier for its use in contact with food. A 100-μm high-barrier multilayer film based on PET and EVOH was obtained from Cryovac Inc. (Sealed Air Spain, Buñol, Spain) with commercial reference Cryovac® Darfresh® VST300E TOP WEB. This multilayer film is composed of different petrochemical polymers, but it is herein referred to as PET film for simplicity since it is designed by the manufacturer for PET sealant trays in barrier packaging applications.

D-Limonene, with 98% purity, was obtained from Sigma-Aldrich S.A. (Madrid, Spain). Phosphorous pentoxide (P_2O_5) and magnesium nitrate ($Mg(NO_3)_2$) were provided by Panreac Química S.L.U (Castellar del Vallés, Barcelona, Spain). Ripe compost (Abonos Naturales Hnos. Aguado S.L., Toledo, Spain), vermiculite (Leroy Merlin, Valencia, Spain), and microcrystalline cellulose (MCC) (Sigma Aldrich, Madrid, Spain) were used for the biodegradability tests. Refined corn germ oil (Koipe, Córdoba, Spain), sawdust (Productos de limpieza Adrián, Valencia, Spain), saccharose (Azucarera Ebro, Madrid, Spain), rabbit feed (Super Feed S.L., Madrid, Spain), urea (Urea Prill, Antonio Tarazona S.L., Silla, Valencia, Spain), and corn starch (Roquette Laisa, Benifaio, Valencia, Spain) were used for the disintegration tests.

2.2. Thermo-Sealing Process

Before preparing the samples, the paper sheets were dried in a vacuum oven (vacuum TEM-TJP Selecta, SA, Barcelona, Spain) at 60 °C for 2 h and, then, stored in a desiccator with P_2O_5 (0% RH) at 25 °C until processed. Thereafter, the paper sheets were coated on both sides with PHBV films through a heat-sealing process. For this, the commercial paper sheets and PHBV films were cut in sizes of 10 cm × 10 cm. Then, a sheet of paper was placed between two layers of PHBV and subjected to a heat-sealing process in a hydraulic press (Model LP20, Engineering Labtech, Samutprakarn, Thailand). In order to optimize the thermo-sealing process, different temperatures and times were tested. Thus, temperature was increased from 130 to 170 °C, in intervals of 10 °C and for periods of 4, 8, 16, and 20 s, maintaining the pressure constant at 20 bar. Thereafter, the multilayers were cooled at room temperature (25 °C). The same process was carried out using the multilayer films based on PET at 160 °C for 8 s and a pressure of 20 bar since these conditions were the optimal ones provided by the manufacturer for heat sealing. Figure 1 shows the equipment employed (Figure 1a) and the design of the multilayers (Figure 1b). Finally, the resultant PHBV/paper/PHBV and PET/paper/PET multilayers as well as the as-received uncoated paper sheets were stored in a desiccator with P_2O_5 at 25 °C for 15 days prior to characterization. This conditioning was carried out to both adjust the humidity of the samples to dry conditions and to reduce the possible effect of physical aging of PHBV.

2.3. Characterization of Multilayer Sheets

2.3.1. Film Thickness

The sheet thickness was measured using a digital micrometer (Palmer model, Comecta S.A., Barcelona, Spain, accuracy of 0.001 mm). Thickness was measured at ten random points, using eight samples for each formulation.

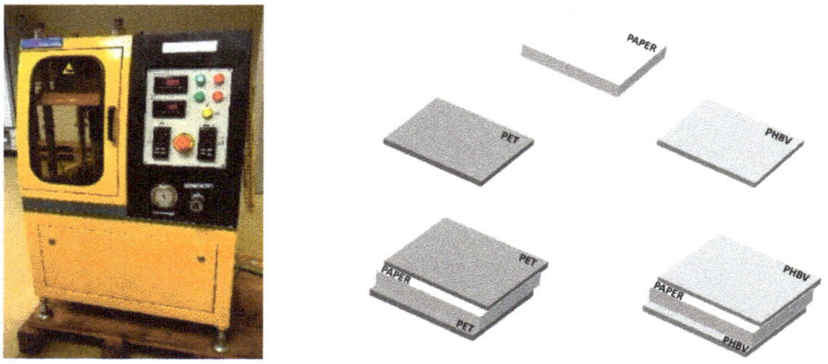

(**a**) (**b**)

Figure 1. (**a**) Hydraulic press used for the heat-sealing process; (**b**) scheme of the developed multilayers based on paper sheets and films of poly(3-hydroxybutyrate-*co*-3-hydroxyvalerate) (PHBV) and polyethylene terephthalate (PET).

2.3.2. Optical Evaluation

The internal transmittance (T_i) was determined for ascertaining the transparency of the monolayer and multilayer samples by the application of the Kubelka-Munk multiple-scattering theory of the reflection spectrum [25]. This theory is based on the fact that light incident on a translucent material can be absorbed or scattered depending on the material's absorption (K) and scattering (S) coefficients. The reflection spectrum (R) of the samples from 360 nm to 700 nm was obtained on white (R_g) and black (R_0) backgrounds using a MINOLTA spectrocolorimeter (CM-3600d, Minolta Co., Tokyo, Japan). Equations (1) and (2) were used to determine the a and b parameters, whereas the reflectance spectra of a sheet with an infinite thickness (R_∞) were calculated according to Equation (3) and T_i was determined using Equation (4). Three measurements were taken on each sample, at different points, from the glossy side of the paper samples.

$$a = \frac{1}{2}\cdot\left(R + \frac{R_0 - R + R_g}{R_0 \cdot R_g}\right) \quad (1)$$

$$b = \left(a^2 - 1\right)^{\frac{1}{2}} \quad (2)$$

$$R_\infty = a - b \quad (3)$$

$$T_i = \sqrt{(a - R_0)^2 - b^2} \quad (4)$$

Film color coordinates L^* (L^* = 0: black, L^* = 100: white), a^* (a^* > 0: red; a^* < 0: green), and b^* (b^* > 0: yellow; b^* < 0: blue) were determined from R_∞ spectra, using 10° observer and D65 illuminant as reference system. The chroma or color saturation (C_{ab}*), expressed from 0 to 100, was determined from Equation (5). The tone or hue angle (h_{ab}*) was calculated from the a^* and b^* coordinates, which are expressed in degrees from 0 to 360°, using Equation (6). Finally, Equation (7) was used to determine the color difference (ΔE_{ab}*), where ΔL^*, Δa^*, and Δb^* correspond to the differences between the color parameters of each multilayer compared to the uncoated paper.

$$C_{ab}* = \sqrt{a*^2 + b*^2} \quad (5)$$

$$h_{ab}* = arctg\left(\frac{b*}{a*}\right) \quad (6)$$

$$\Delta E* = \sqrt{(\Delta L*)^2 + (\Delta a*)^2 + (\Delta b*)^2} \quad (7)$$

The color change was evaluated according to the following criteria: negligible ($\Delta E_{ab}{}^* < 1$), only an experienced observer can tell the difference ($\Delta E_{ab}{}^* \geq 1$ and <2); an inexperienced observer notes the difference ($\Delta E_{ab}{}^* \geq 2$ and < 3.5); there is a clear notable difference ($\Delta E_{ab}{}^* \geq 3.5$ and <5); and the observer notices different colors ($\Delta E_{ab}{}^* \geq 5$) [26].

2.3.3. Microstructural Analysis

The cross-sections of the monolayers and multilayers were obtained by immersion and cryo-fracture in liquid nitrogen. The samples were mounted, using double-sided carbon tape, on the observation holders and covered with a platinum layer (EM MED020 sputter coater, Leica Biosystems, Barcelona, Spain). The cross-sections were then observed by Field Emission Scanning Electron Microscopy (FESEM) in a JEOL model JSM-5410 (Tokyo, Japan), operating at 2.0 kV acceleration voltage. The thicknesses of the internal layers of the samples were determined using the ImageJ v1.53c Program.

2.3.4. Thermal Analysis

The thermal behavior of the different monolayers and multilayers was analyzed by thermogravimetric analysis (TGA) in a TGA 1 Stare System analyzer (Mettler-Toledo GmbH, Greifensee, Switzerland). Approximately 5 mg of the conditioned samples were placed in an alumina pan and heated from 25 to 800 °C at 10 °C·min^{-1} in inert atmosphere under a nitrogen flow-rate of 10 mL·min^{-1}. From the TGA curves and their derivative curves (DTG), the corresponding T_{onset} values (temperature at which thermal degradation begins, corresponding to 5% mass loss), T_{deg} (temperature at maximum degradation rate), percentage of mass lost when the T_{deg} is reached, and percentage of mass remaining at 800 °C were determined. The analysis was performed in duplicate for each formulation.

2.3.5. Tensile Tests

The tensile properties of the samples were determined using a universal testing machine (Stable Micro System TA-XT plus, Haslemere, UK) following the standard method ASTM D882 [27]. To this end, the pre-conditioned samples, previously cut in dimensions of 25 mm × 100 mm, were stretched by two grips, initially separated by 100 mm, at a crosshead speed of 50 mm·min^{-1} until breakage (model A/TG, Stable Micro System, Haslemere, UK). Stress (σ) versus strain (ε) curves were obtained from the force-distance curves by considering sample dimensions and degree of deformation. The Young's modulus (E), stress at yield or elastic limit (σ_y), and percentage of elongation at break (% ε_b) were obtained. Eight samples for each formulation were analyzed.

2.3.6. Permeability Measurements

Permeances to water and limonene vapors of the monolayers and multilayers were determined at 25 °C and 53% RH by the gravimetric methodology according to ASTM E96/E96M [28]. For water vapor, the film and sheet samples (Ø = 3.5 cm) were placed and sealed in Payne permeability cups filled with 5 mL of distilled water (100% RH). Then, the cups were placed into desiccators containing an $Mg(NO_3)_2$ oversaturated solution and, for one week, weighed periodically (ME36S, ±0.00001 g accuracy, Sartorius, Goettingen, Germany). The water vapor permeance was calculated from the water vapor transmission rate (WVTR), determined from the slope of the weight loss vs. time, and corrected for permeant partial pressure. In the case of limonene vapor, the procedure was similar to that described for water vapor but using 5 mL of D-limonene instead of water. Thus, limonene permeation rate (LPR) was obtained from the steady-state permeation slopes of weight loss vs. time and corrected for permeant partial pressure. In both cases, cups with aluminum films were used as control samples to estimate and subtract the vapor loss through the sealing. All the vapor permeance measurements were performed in triplicate.

The oxygen permeance of the monolayers and multilayers was determined using an oxygen permeation analyzer (OxySense® Model 8101e, Systech Illinois, Thame, UK) at 25 °C and 53% RH according to ASTM D3985-05 [29]. The exposed sample area was 50 cm^2

and the permeance values were derived from the oxygen transmission rate (OTR) measurements, which were corrected with the gas partial pressure and recorded in triplicate.

2.4. Controlled Composting Tests

2.4.1. Aerobic Biodegradability

The aerobic biodegradability of the monolayers and multilayers was determined under composting conditions by measuring the amount of CO_2 generated according to ISO 14855 [30]. The ripe compost was mixed with vermiculite (compost/vermiculite ratio: 3:1) to avoid compaction of the compost and to ensure good oxygen access. Glass flasks (2L) containing two PP flasks (60 mL) were used as reactors. One of the flasks contained 3 g of dry compost mixed with 1 g of vermiculite and an amount of sample (previously cut into 2 mm^2 squares) equivalent to 50 mg of carbon, while the other flask contained deionized water to ensure 100% RH. The bioreactors were closed and incubated for 45 days at 58 ± 2 °C. One reactor containing only compost was used as blank and one reactor containing MCC mixed with the compost were used as reference sample. The percentage of CO_2 generated inside the reactors was measured, in triplicate, using a CO_2 analyzer (Gaspace Advance Micro GS3, Systech Illinois, Thame, UK) throughout the biodegradation process. The percentage of biodegradation was calculated using the following Equation (8), assuming that all the carbon in the sample was converted into CO_2:

$$B(\%) = \frac{\sum CO_{2S} - \sum CO_{2B}}{CO_2^{Th}S} \times 100 \tag{8}$$

where ΣCO_{2S} is the accumulative amount of CO_2 produced in the sample bioreactor, ΣCO_{2B} is the accumulative amount of CO_2 produced in the blank bioreactor, and $CO_2^{Th}S$ is the theoretical amount of CO_2 that the test material can produce.

2.4.2. Degree of Disintegration

Disintegration of the monolayers and multilayers in simulated composting conditions was conducted at 58 ± 2 °C and 55% RH to ensure controlled thermophilic conditions, as indicated by ISO 20200 [31]. Film samples sizing 25 mm × 25 mm were placed in mesh bags (1 mm × 1 mm mesh size) and buried in a controlled soil compost made of sawdust (40 wt%), rabbit feed (30 wt%), ripe compost (10 wt%), corn starch (10 wt%), sucrose (5 wt%), corn seed oil (4 wt%), and urea (1 wt%). Samples were periodically unburied from the composting facility. At the initial time and throughout the study, the reactors were weighed and deionized water was added, if necessary, to restore the initial mass, as specified by the standard [31]. Thus, mesh bags, each with one sample square, were extracted from the reactor at different control times in order to carry out the visual morphological analysis and determine mass loss. To this end, the mesh bags containing samples were dried, gently cleaned with a soft brush to eliminate the adhered compost residues, and weighed with an analytical balance. The analysis was conducted for 300 days to ascertain the effect on paper disintegration of the biopolymer coatings for long periods. The weight loss was calculated using Equation (9):

$$Weight\ loss\ (\%) = \frac{Wo - Wt}{Wo} \times 100 \tag{9}$$

where W_0 is the initial dry weight of the sample and W_t is the weight of the sample after a bury time t. All tests were carried out in triplicate to ensure reliability.

2.5. Statistical Analysis

The experimental data were submitted to analysis of variance (ANOVA) using Statgraphics Centurion XVI software (Manugistics Corp., Rockville, MD, USA). Fisher's least significant difference (LSD) procedure was used at the 95% confidence level.

3. Results

3.1. Development of Paper Multilayers

The results of the application of different time and temperature conditions to form the PHBV/paper/PHBV multilayer sheets during the heat-sealing process are shown in Table 1. It gathers the set of values of temperature and time employed to produce the multilayers, ranging from 130 °C to 170 °C and 4 to 20 s, respectively. It also includes an image of the surface of the PHBV/paper/PHBV multilayer attained at each condition, with comments on the quality and/or processing aspects. It can be seen that the application of very low temperatures, that is, 130 °C, was not high enough for heat sealing of the PHBV films on the paper sheet, even if applied for long times, of up to 20 s. Interestingly, increasing the temperature to 140 °C significantly improved the heat-sealability of the PHBV films on the paper substrate. However, for this temperature, processing times of at least 8 s were necessary to obtain uniformity and good adhesion between the substrates and coatings. Moreover, the application of heat-sealing temperatures of 150 °C also resulted in good adhesion between the layers, but the resultant multilayers presented areas full of wrinkles and bubbles. The latter effect can be related to occluded air or evaporated water produced by the effect of high temperature on the paper substrate, which is a very hydrophilic material based on cellulose fibers. Therefore, despite having worked with dried samples, the paper sheets could retain minimal amounts of water and/or adsorbed moisture during handling and processing. As the sealing processing time increased, these effects occurred to a greater extent. Finally, the use of the highest temperatures, and especially at 170 °C, resulted in the partial melting of the PHBV film and a subsequent deterioration of the heat-sealed multilayers, even at short processing times. In this regard, it should be noted that the melting temperature of PHBV copolyesters with 3HV contents of up 20 %mol ranges from approximately 154 °C to 171 °C [32].

Table 1. Optimization of the thermo-sealing process of the poly(3-hydroxybutyrate-*co*-3-hydroxyvalerate) (PHBV) films on both sides of paper.

Temperature (°C)	Time (s)	Multilayer	Comments
130	4	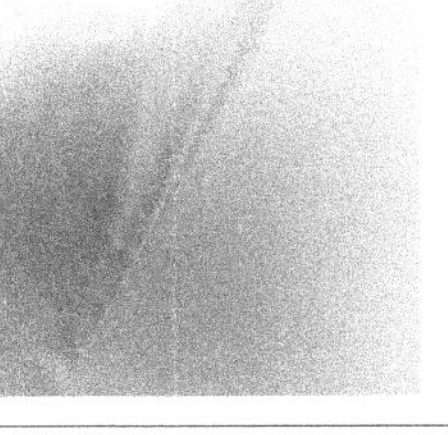	The PHBV film and paper sheet did not stick properly.
	8	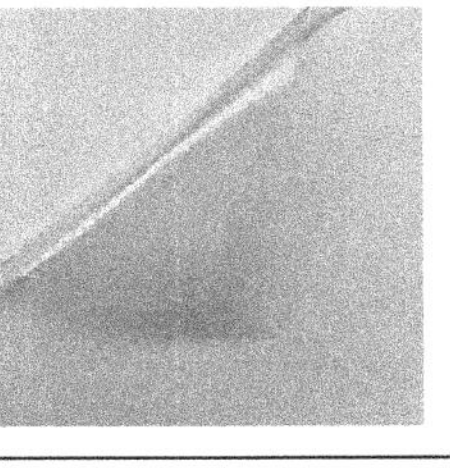	The PHBV film detached easily from the paper sheet.

Table 1. *Cont.*

Temperature (°C)	Time (s)	Multilayer	Comments
	16		The multilayer was homogenous but the PHBV film and paper sheet detached easily.
	20		The multilayer showed some opaque areas where the PHBV film was not properly adhered to paper.
140	4		The PHBV film and the paper sheet sealed, but presented poor adhesion in some areas.
	8		The PHBV film and the paper sheet sealed properly.
	16		The PHBV film and the paper sheet sealed properly.

Table 1. *Cont.*

Temperature (°C)	Time (s)	Multilayer	Comments
	20		The PHBV film and the paper sheet sealed properly.
150	4	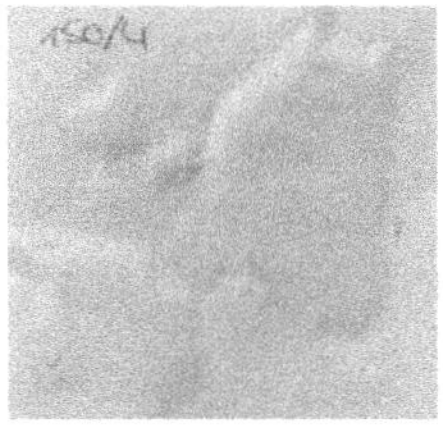	The PHBV film and paper sheet sealed, but the multilayer showed large areas with wrinkles and bubbles due to occluded air or moisture.
	8	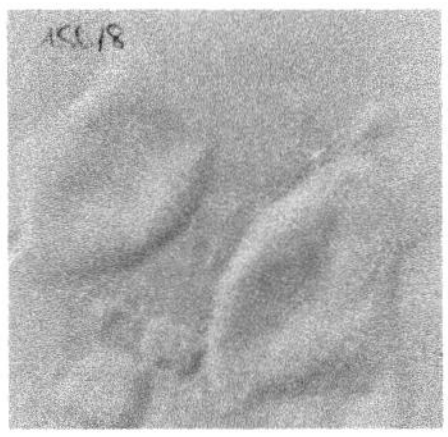	The PHBV film and paper sheet sealed, but the multilayer showed large areas with wrinkles and bubbles due to occluded air or moisture.
	16	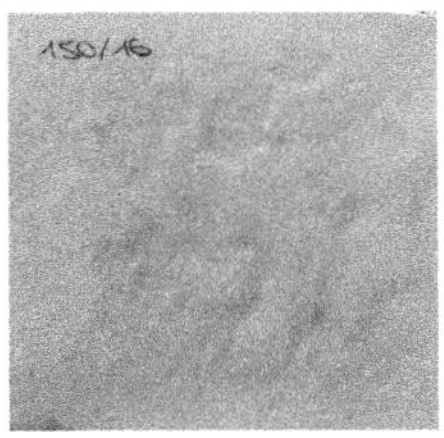	The PHBV film and paper sheet remained sealed, but the multilayer still presented several wrinkles and bubbles.
	20		The PHBV film and paper sheet remained sealed and the surface of the multilayer improved, but it still presented some wrinkles and bubbles.

Table 1. *Cont.*

Temperature (°C)	Time (s)	Multilayer	Comments
160	4		The PHBV film partially melted in several areas and the surface still showed some wrinkles and bubbles. Further times at this temperature were ruled out of the study.
170	4	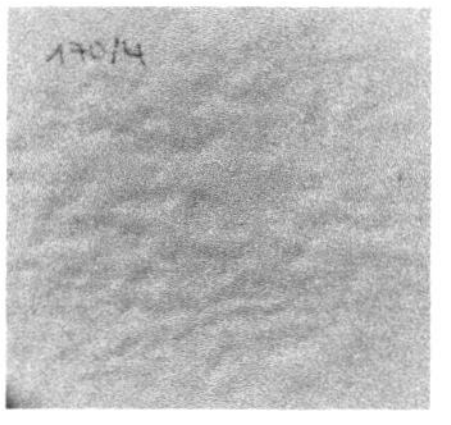	The PHBV film partially melted in several areas and the film still showed some wrinkles and bubbles. Further temperatures and times were ruled out of the study.

Based on the results obtained from the heat-sealing process, it was concluded that the optimal conditions were obtained at a temperature of 140 °C, with times ranging between 8 and 20 s. Since no improvement was observed for times longer than 8 s, the shortest time was selected to minimize costs and reduce energy consumption. Therefore, the heat-sealing process of the PHBV/paper/PHBV multilayers was set at 140 °C and 8 s.

3.2. *Morphology of Paper Multilayers*

Figure 2 shows the cross-sectional morphology observed by FESEM of the three components of the multilayers, that is, the paper sheet and the PHBV and PET films. As can be seen in Figure 2a, the paper sheet presented an average thickness of 291 ± 6 μm and it was composed of micrometer-sized fibers, giving rise to a cross-section with a rough surface. These cellulosic fibers, which showed an average diameter of approximately 20 μm, are responsible for the high level of porosity of the paper [33]. In contrast, one can observe in Figure 2b that the PHBV film showed a continuous section of 12 ± 2 μm, with a brittle fracture, because there were no macroscopic plastic deformations. The observed inorganic microparticles embedded in the biopolymer matrix may correspond to boron nitride, which is habitually employed by the manufacturer as a nucleating agent during the fabrication of PHAs [34]. Finally, Figure 2c shows the section of the commercial PET film, which presented a multilayer structure having an average thickness of 104 ± 1 μm. The external layer, shown at the bottom with a lower thickness, can be ascribed to a sealing layer of polyolefin (e.g., LDPE or PP) or a modified PET (e.g., polyethylene terephthalate glycol, PET-G) with a low sealing temperature. Both the thickest internal layer and the upper external layer would correspond to PET. Moreover, the inner layers with low thickness can be related to EVOH. The layer structure attributed herein is based on that previously observed for a polyolefin-based multilayer, supplied by the same manufacturer for this product range (Cryovac® Darfresh® VST200P TOP WEB), which contained two inner layers of EVOH of nearly 5 μm to provide the film with high oxygen barrier capacity [35].

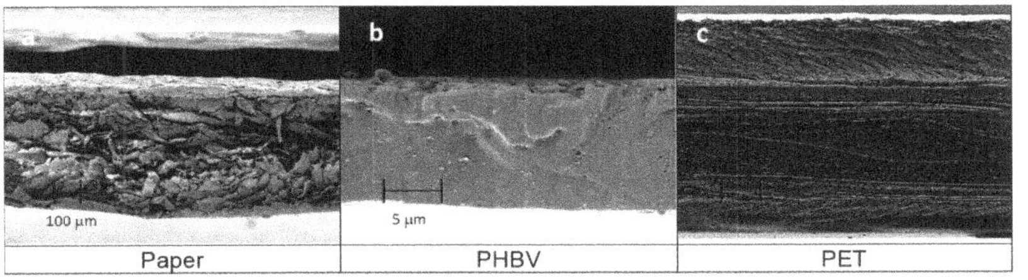

Figure 2. Field emission scanning electron microscopy (FESEM) micrographs taken in the cross-sections of: (**a**) paper sheet, observed at 150×; (**b**) poly(3-hydroxybutyrate-*co*-3-hydroxyvalerate) (PHBV) film, at 3000; (**c**) multilayer film based on polyethylene terephthalate (PET), at 750×.

On the other hand, Figure 3 shows the morphology of the paper-based multilayer structures obtained from the different monolayers, presented in Figure 2, by the heat-sealing process. Thus, Figure 3a shows the resultant three-layer structure based on the paper substrate coated, on both sides, with PHBV films. The average thickness of the multilayer sheet, measured after heat sealing, resulted in a value of 297 ± 8 µm. Likewise, a similar morphology was observed for the paper coated by two PET-based multilayer films, the so-called PET/paper/PET sheet, shown in Figure 3b. This multilayer sheet presented an analogous structure with a higher average thickness, of 473 ± 3 µm, due to the higher thickness of the outer layers of the PET-based films. Both multilayer sheets, that is, PHBV/paper/PHBV and PET/paper/PET, showed good adhesion between the polyester layers and paper substrate, suggesting adequate mechanical resistance for handling and transport in food packaging. Furthermore, it is also worth noting that both multilayers were slightly thinner than the sum of the thicknesses of the individual monolayers. This is due to the radial flow of the polyester monolayers and to the reduction of the free volume of the paper (highly porous) during the heat-sealing step.

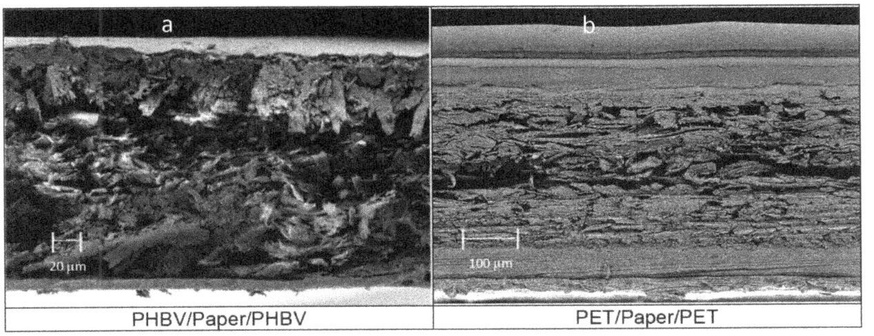

Figure 3. Field emission scanning electron microscopy (FESEM) micrographs taken in the cross-sections of: (**a**) poly(3-hydroxybutyrate-*co*-3-hydroxyvalerate) (PHBV)/Paper/PHBV multilayer sheet, taken at 200×; (**b**) polyethylene terephthalate (PET)/Paper/PET multilayer sheet, at 150×.

3.3. Optical Properties of Paper Multilayers

One of the desired characteristics of packaging materials is that these should protect food from the effects of light, especially ultraviolet (UV) radiation, which is one of the main benefits of paper. Figure 4 shows the spectral distribution curves that represent T_i of the film and sheet samples as a function of the wavelength (λ). Thus, high T_i values are related to high light transmittance of the samples and, therefore, correspond to more transparent films or sheets while, as opposite, low T_i values are related to more opaque samples with lower light passage [36,37]. As can be seen in the graph, both PHBV and PET films presented notably higher T_i values (in agreement with their high transparency) compared to the uncoated paper and multilayer sheets. The slightly lower T_i of the PHBV film compared with the PET multilayer film can be related to the higher crystallinity of the microbial copolyester [38]. As expected, the coating of the paper sheets did not affect the paper's opacity, thus confirming that the good barrier against light was preserved. This may be of interest for the protection of certain foods, such as oil or meat, that must be protected against light-induced oxidative processes [39].

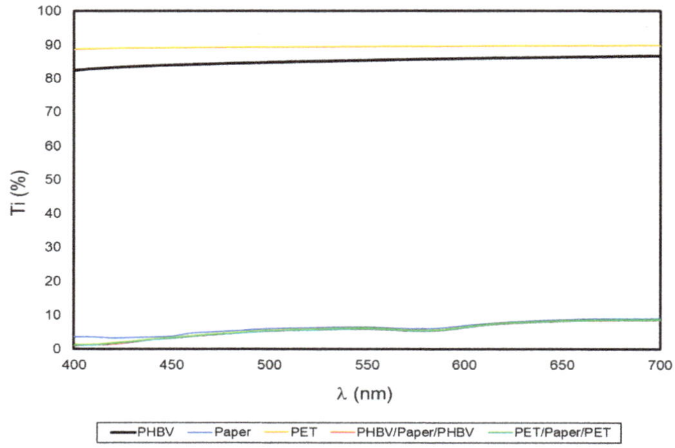

Figure 4. Spectral distribution curves of the percentage internal transmittance (% T_i) of the paper sheet, poly(3-hydroxybutyrate-*co*-3-hydroxyvalerate) (PHBV) and polyethylene terephthalate (PET) films, and PHBV/paper/PHBV and PET/paper/PET multilayer sheets.

Table 2 shows the values of the optical properties of the film and sheet samples. The uncoated paper and paper multilayer sheets presented similar high L^* values, 95–96, while the neat PHBV and PET films showed lower values, 91–92. No remarkable differences were found in the color parameters of the polyester films, with both exhibiting high film lightness in agreement with previous studies [40,41]. Likewise, no significant differences ($p > 0.05$) were found among the L^* values of the paper multilayers and the uncoated paper, so the presence of the polyester films did not alter the lightness of the paper sheets. Nevertheless, multilayers exhibited more bluish (lower h_{ab}^* values) and less saturated (lower C_{ab}^* values) color in comparison with the uncoated paper sample. Nevertheless, these differences, calculated through ΔE_{ab}^* parameter, were not relevant from a practical point of view since color differences below 2 are hardly perceived by the human eye [42].

Table 2. Color parameters in terms of lightness (L^*), color coordinates (a^* and b^*), chroma (C_{ab}^*), hue (h_{ab}^*), and total color difference (ΔE_{ab}^*) of the paper sheet, poly(3-hydroxybutyrate-*co*-3-hydroxyvalerate) (PHBV) and polyethylene terephthalate (PET) films, and PHBV/paper/PHBV and PET/paper/PET multilayer sheets.

Sample	L^*	a^*	b^*	C_{ab}^*	h_{ab}^*	ΔE^*_{ab}
Paper	95.80 ± 0.03^{a}	1.36 ± 0.05^{a}	-5.39 ± 0.20^{a}	5.55 ± 0.21^{a}	284.21 ± 0.09^{a}	-
PHBV	91.40 ± 0.02^{b}	-0.37 ± 0.01^{b}	1.87 ± 0.06^{b}	1.91 ± 0.06^{b}	101.25 ± 0.06^{b}	-
PET	91.59 ± 0.06^{b}	-0.36 ± 0.01^{b}	1.64 ± 0.01^{bc}	1.68 ± 0.01^{c}	102.59 ± 0.01^{c}	-
PHBV/Paper/PHBV	95.54 ± 0.06^{a}	0.75 ± 0.09^{c}	-3.66 ± 0.30^{d}	3.73 ± 0.31^{d}	281.61 ± 0.53^{d}	1.85 ± 0.06^{a}
PET/Paper/PET	95.52 ± 0.05^{a}	0.86 ± 0.02^{cd}	-3.56 ± 0.14^{d}	3.66 ± 0.14^{d}	283.63 ± 0.30^{e}	1.92 ± 0.07^{a}

Mean values and standard deviation. $^{a-e}$ Different superscripts in the same column indicate significant differences between formulations ($p < 0.05$).

3.4. Thermal Properties of Paper Multilayers

Thermal stability was determined by TGA to ascertain the application conditions of the resultant multilayer sheets. The TGA curves of each sample, which show the variation of the mass percentage as a function of temperature, are gathered in Figure 5. Table 3 includes the resultant thermal stability parameters.

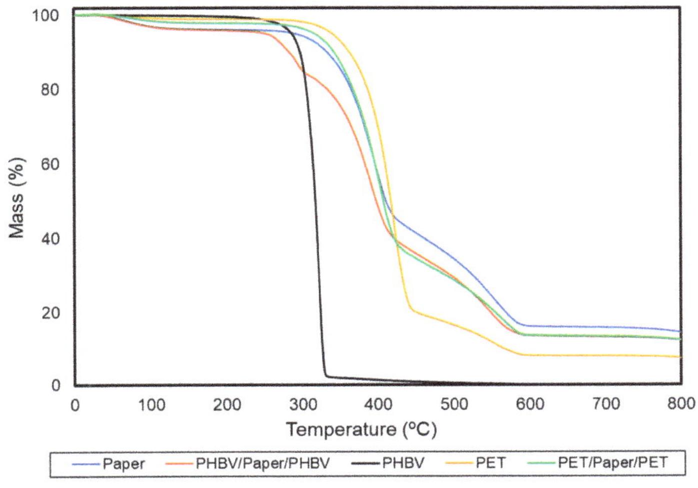

Figure 5. Thermogravimetric analysis (TGA) curves of the paper sheet, poly(3-hydroxybutyrate-*co*-3-hydroxyvalerate) (PHBV) and polyethylene terephthalate (PET) films, and PHBV/paper/PHBV and PET/paper/PET multilayer sheets.

Table 3. Thermal properties in terms of temperature at which thermal degradation begins (T_{onset}), corresponding to 5% mass loss, temperature of maximum degradation rate (T_{deg}), percentage of mass loss at T_{deg}, and percentage of remaining mass at 800 °C of the paper sheet, poly(3-hydroxybutyrate-*co*-3-hydroxyvalerate) (PHBV) and polyethylene terephthalate (PET) films, and PHBV/paper/PHBV and PET/paper/PET multilayer sheets.

Sample	T_{onset} (°C)	T_{deg} (°C)	Mass Loss at T_{deg} (%)	Remaining Mass (%)
Paper	310.8 ± 7.8 ^{a}	353.8 ± 1.6 ^{a}	39.4 ± 1.0 ^{a}	14.0 ± 0.1 ^{a}
PHBV	289.7 ± 5.2 ^{b}	288.2 ± 3.8 ^{b}	60.6 ± 0.5 ^{b}	0.1 ± 0.0 ^{b}
PET	327.3 ± 3.7 ^{c}	367.5 ± 3.2 ^{c}	49.1 ± 8.9 ^{c}	7.2 ± 0.8 ^{c}
PHBV/Paper/PHBV	275.5 ± 4.3 ^{d}	349.6 ± 2.9 ^{d}	54.8 ± 0.2 ^{d}	12.0 ± 1.6 ^{d}
PET/Paper/PET	313.6 ± 4.1 ^{a}	357.9 ± 1.1 ^{e}	55.2 ± 0.4 ^{d}	11.4 ± 0.9 ^{d}

Mean values and standard deviation. $^{a–e}$ Different superscripts in the same column indicate significant differences between formulations ($p < 0.05$).

In the case of paper, three relevant mass losses occurred during heating, which approximately took place at temperatures of 100, 350, and 490 °C. These mass losses have been widely studied in lignocellulosic materials [43]. The first one corresponds to the evaporation of absorbed water in the dried paper sheets, which indicates that the initial moisture content of these samples was 2–3%. The second and main degradation step, between 310 and 410 °C, showed an average mass loss of nearly 55%. This is referred to as the "active pyrolysis zone" since the mass loss rate is high. This mass loss corresponds to the decomposition of hemicellulose and cellulose, which are main components of paper. Both degradation processes are known to involve complex reactions (e.g., dehydration and decarboxylation, among others) as well as the breaking of C–H, C–O, and C–C bonds [44]. The onset of the third stage of degradation overlapped with the previous one and continued progressively until temperatures neared 600 °C. In this step, the mass loss was around 28%, representing the "passive pyrolysis zone" since the mass loss rate was much lower than that observed in the previous one. The latter mass loss can be assigned to the thermal decomposition of lignin, which is known to occur slowly and over a wide temperature range. Furthermore, at 800 °C, the paper showed a remaining mass of 14%. This mass corresponds to both inorganic material, such as silica or titanium dioxide (TiO_2), which is usually applied to give brightness and whiteness to paper, and ashes generated from the decomposition of the organic material in an inert atmosphere [45].

Furthermore, PHBV thermal degradation occurred through a single and sharp degradation step that occurred from 290 to 320 °C. Thermal degradation of PHAs generally follows a random chain scission model of the ester bond, involving a cis-type elimination reaction of −CH and a six-membered ring transition to form crotonic acid and oligomers [46]. The thermal stability of the PET film was significantly higher ($p < 0.05$) than that of PHBV. This thermal degradation occurred in two stages. The first one occurred in the 300−450 °C range that corresponds to the degradation of the polymer chains. The second one was seen between 450 and 600 °C and it is due to the thermo-oxidative degradation of the residual mass produced during the first stage. This thermal degradation of PET is based on a heterolytic cleavage by means of a six-membered intermediate ring. In this process, the hydrogen of a beta (β)-carbon of the ester group is transferred to the carbonyl group of the ester, followed by a breakage of the ester bonds [47]. Whereas the application of the PET films on the paper substrate slightly improved its thermal stability ($p > 0.05$), the thermal resistance of the PHBV/paper/PHBV multilayer slightly but significantly ($p < 0.05$) decreased. In particular, a reduction of around 11% in comparison with the uncoated paper was observed. The latter effect is due to the lower thermal stability of the biopolyester. Nevertheless, the thermal stability of the PHBV/paper/PHBV multilayer can be yet considered adequate for most food packaging applications that do not exceed temperatures above 250 °C, such as microwave heating. In contrast, most heating processes in the oven will be restricted, where the use of neat paper is still very limited [48].

3.5. Mechanical Properties of Paper Multilayers

The mechanical performance of the multilayers was studied by means of tensile tests at room temperature. The resultant σ versus ε curves are shown in Figure 6. The most characteristic values, namely E, σ_y, and ε_b, obtained from these curves, are included in Table 4. It can be observed that the paper sheet showed mechanical parameters that correspond to a rigid and brittle material. This sample broke right after exceeding the elastic deformation zone, with a σ_y value of 31.9 MPa and deformations of nearly 7%. Moreover, the PHBV film also showed a brittle behavior, but it was significantly ($p < 0.05$) more rigid than the paper sample due to its higher value of E, 2928 MPa. The high brittleness of PHBV derives from its high crystallinity, even though the 3HV content in the copolyester was relatively high (10 mol%). This embrittlement takes place by a process of secondary crystallization and/or physical aging that results in the formation of large spherulites acting as stress concentration points [21]. The mechanical behavior of the PET film followed a completely different pattern, being typical of a more stretchable and ductile material. This film sample showed E and σ_y values of approximately 759 and 29.6 MPa, respectively, and broke at a deformation of nearly 53% with a maximum strength above 60 MPa. This mechanical behavior is in agreement with the values reported for PET films in a previous study [9].

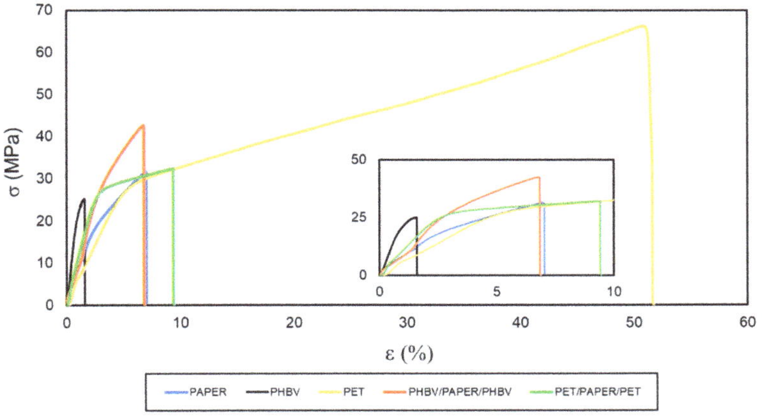

Figure 6. Stress (σ) versus strain (ε) curves of the paper sheet, poly(3-hydroxybutyrate-*co*-3-hydroxyvalerate) (PHBV) and polyethylene terephthalate (PET) films, and PHBV/paper/PHBV and PET/paper/PET multilayer sheets.

Table 4. Mechanical properties in terms of Young's modulus (E), stress at yield (σ_y), and percentage of elongation at break (% ε_b) of the paper sheet, poly(3-hydroxybutyrate-*co*-3-hydroxyvalerate) (PHBV) and polyethylene terephthalate (PET) films, and PHBV/paper/PHBV and PET/paper/PET multilayer sheets.

Sample	E (MPa)	σ_y (MPa)	ε_b (%)
Paper	1787 ± 41 [a]	31.9 ± 1.0 [a]	6.9 ± 0.9 [a]
PHBV	2928 ± 30 [b]	27.0 ± 2.8 [b]	2.3 ± 0.6 [b]
PET	759 ± 18 [c]	29.6 ± 2.1 [c]	52.8 ± 0.6 [c]
PHBV/Paper/PHBV	2591 ± 51 [d]	43.2 ± 2.1 [d]	7.1 ± 1.2 [a]
PET/Paper/PET	1959 ± 20 [e]	29.4 ± 2.3 [c]	9.8 ± 0.9 [d]

Mean values and standard deviation. [a–e] Different superscripts in the same column indicate significant differences between formulations ($p < 0.05$).

Double coating led to a significant enhancement of the tensile properties of the paper substrate, which significantly ($p < 0.05$) depended on the type of polyester used. In the case of the PHBV/paper/PHBV multilayer, it gave rise to a sheet with an E of 2591 MPa and an σ_y of approximately 43 MPa. Thus, the PHBV double coating yielded an improvement in terms of mechanical resistance when compared with the uncoated paper substrate, without affecting significantly ($p > 0.05$) ductility (ε_b of 7.1%). Similarly, for the PET/paper/PET multilayer, values of E of 1959 MPa and σ_y of 29.4 MPa were obtained. However, this multilayer sample broke at deformation values of approximately 10%, indicating a significant improvement ($p < 0.05$) in the flexible properties of paper. In this sense, it is also worth mentioning that, although both films were notably thinner than paper, the PET film used was approximately 10 times thicker than the PHBV film so that the proportion of petrochemical film in the multilayer with respect to paper was higher than in the case of the biopolyester film. Therefore, the PHBV/paper/PHBV multilayer yielded to a greater mechanical resistance, while the PET/paper/PET multilayer was more stretchable than the uncoated paper. According to these results, in terms of food packaging applications, the newly developed PHBV/paper/PHBV multilayers will be mainly restricted to rigid applications. For example, these can be applied to develop trays, lids, or plates, which do not require high deformations but can withstand certain stresses.

3.6. Barrier Properties of Paper Multilayers

The permeance values of the water and limonene vapors and oxygen gas of the uncoated paper sheet, PHBV and PET films, and multilayer paper sheets are shown in Table 5. Permeance is the expression of permeability with the removal of the thickness factor, used to determine the barrier of multilayer structures and the actual performance of a film at given conditions of temperature and %RH. For this reason, the thickness values of each film and sheet samples were also included.

Table 5. Permeance to water and D-limonene vapors and oxygen of the paper sheet, poly(3-hydroxybutyrate-*co*-3-hydroxyvalerate) (PHBV) and polyethylene terephthalate (PET) films, and PHBV/paper/PHBV and PET/paper/PET multilayer sheets.

Sample	Thickness (μm)	Water Vapor		Limonene Vapor		Oxygen	
		Permeance × 10^{10} (kg/m^2·Pa·s)	Permeability × 10^{15} (kg·m/m^2·Pa·s)	Permeance × 10^{10} (kg/m^2·Pa·s)	Permeability × 10^{15} (kg·m/m^2·Pa·s)	Permeance × 10^{15} (m^3/m^2·Pa·s)	Permeability × 10^{19} (m^3·m/m^2·Pa·s)
Paper	291 ± 6 [a]	110 ± 9 [a]	3205 ± 87 [a]	22.3 ± 1.3 [a]	650 ± 9 [a]	> D.L.	> D.L.
PHBV	12 ± 2 [b]	4.8 ± 0.7 [b]	5.7± 0.3 [c]	5.8 ± 0.3 [b]	6.95 ± 0.14 [c]	24.1 ± 5 [a]	2.89 ± 0.17 [a]
PET *	104 ± 1 [a]	0.23 ± 0.04 [d]	2.3 ± 0.4 [d]	0.49 ± 0.01 [e]	4.97 ± 0.14 [d]	0.81 ± 0.01 [b]	0.86 ± 0.17 [b]
PHBV/Paper/PHBV	297 ± 8 [c]	4.5 ± 0.8 [c]	-	4.4 ± 0.8 [c]	-	3.4 ± 0.8 [c]	-
PET/Paper/PET	473 ± 3 [d]	0.19 ± 0.01 [d]	-	0.8 ± 0.2 [d]	-	0.14 ± 0.01 [b]	-

* Assuming a monolayer material. Mean values and standard deviation. [a–e] Different superscripts in the same column indicate significant differences between formulations ($p < 0.05$).

Regarding the water vapor, one can observe that the permeance of the paper sheet was 1.10×10^{-8} kg/m^2·Pa·s, resulting in a permeability value of 3.21×10^{-12} kg·m/m^2·Pa·s. The low water vapor barrier of paper is due to the fact that it is a hydrophilic and highly fibrous material and, therefore, has a high porosity, as observed previously during the FESEM analysis. In relation to the 10-μm PHBV film, it presented a permeance of 4.78×10^{-10} kg/m^2·Pa·s, yielding a permeability of 5.73×10^{-15} kg·m/m^2·Pa·s. This permeability value is slightly higher than that reported for thermo-compressed films of PHB [49] and PHBV with 2–3 mol% 3HV [50], that is, 1.70 and 1.82×10^{-15} kg·m/m^2·Pa·s, respectively. This difference is due to the higher comonomer content in the PHBV tested herein, which induces lower crystallinity and, hence, higher diffusivity to the vapor and gas molecules [18,51]. In the case of the commercial PET multilayer film, the permeance value was 2.32×10^{-11} kg/m^2·Pa·s, which corresponds to a permeability of 2.25×10^{-15} kg·m/m^2·Pa·s, considering it as a monolayer material. The resultant WVTR was approximately 6.3 g/m^2·day, which is in the range but lower than the one reported by the manufacturer at 38 °C and 90% RH, that is, 16 g/m^2·day [52] due to the lower temperature and humidity employed herein. Moreover, this water vapor permeability value is slightly lower than that of PET, that is, 2.3×10^{-15} kg·m/m^2·Pa·s, measured at 38 °C and 90% RH [53]. This can be related to the presence of a layer of polyolefin, such as LDPE, with a lower permeability, that is, 1.2×10^{-15} kg·m/m^2·Pa·s, at 38 °C and 90% RH [54]. This polyolefin layer is generally placed in the external part of the film in order to perform heat sealing at lower temperatures. In a food packaging context, both materials presented water vapor permeability between that of polyamide 6 (PA6), that is, 2.06×10^{-14} kg·m/m^2·Pa·s, and PP, that is, 7.26×10^{-16} kg·m/m^2·Pa·s [54]. One can further observe that the water permeance of the resultant PHBV/paper/PHBV multilayer was in the range of the PHBV film, but still significantly lower ($p < 0.05$), with a value of 4.48×10^{-10} kg/m^2·Pa·s. This enhancement in the permeance can be mainly related to the increase in the thickness sample. However, it can also be ascribed to the fact that the hydrophobic PHBV film successfully protected paper from moisture and, thus, the barrier of paper improved in dry conditions. Similarly, the PET/paper/PET multilayer resulted in a value of 1.91×10^{-11} kg/m^2·Pa·s due to the higher water barrier performance and thickness of the petrochemical film.

As for the permeability of limonene, which is used as a standard to determine the aroma barrier, the paper presented a permeance of 2.23×10^{-9} kg/m^2·Pa·s, corresponding to a permeability value of 6.50 $\times$ 10^{-13} kg·m/m^2·Pa·s. Permeance of the PHBV film was 5.80 $\times$ 10^{-10} kg/m^2·Pa·s, resulting in a permeability to limonene vapor of 6.95 $\times$ 10^{-15} kg·m/m^2·Pa·s. This result indicates that the microbial copolyester was also much less permeable to aroma than paper. This result is relevant since limonene is known to be a good plasticizer for polyesters, such as PHAs and PET, and thus solubility plays a stronger role than diffusion in permeability. For example, it has been previously described that PHBV films of about 100 μm were able to adsorb 12.7 wt.% of limonene [49]. One can also observe that the permeance to limonene vapor of the PET-based multilayer film, the so-called PET film, was 4.91 $\times$ 10^{-11} kg/m^2·Pa·s. This value is equivalent to a permeability of 4.97 $\times$ 10^{-15} kg·m/m^2·Pa·s, assuming a monolayer material. Limonene permeability values for 75-μm PHB and PET films of 6.38 and 6.43 $\times$ 10^{-15} kg·m/m^2·Pa·s have been respectively reported [55], which are relatively similar to the reported herein. In relation to the multilayers, the permeance values of PHBV/paper/PHB and PET/paper/PET multilayers were 4.44 $\times$ 10^{-10} and 7.82 $\times$ 10^{-11} kg/m^2·Pa·s, respectively. Therefore, as it did with water vapor, the use of double coatings improved the aroma barrier of paper significantly ($p < 0.05$). Although the improvement achieved with the PET film was significantly higher ($p < 0.05$), one should consider that both polyesters resulted in a permeance reduction of nearly two orders of magnitude compared with uncoated paper, and the PHBV film thickness was lower.

Regarding the oxygen barrier capacity, it was not possible to determine the oxygen permeance of the uncoated paper since it was above the detection limit (D.L.) of the equipment (OTR of 432.000 cm^3/m^2 · day). The PHBV film presented a permeance to oxygen of 2.41 $\times$ 10^{-14} m^3/m^2·Pa·s. The oxygen permeance was previously determined for a 50-μm film of the same PHBV grade, resulting in a value of 5.78 $\times$ 10^{-15} m^3/m^2 ·Pa·s [56]. This permeance corresponds to an oxygen permeability of 2.9 $\times$ 10^{-19} m^3·m/m^2·Pa·s, which is nearly 50% higher than that of PHBV with 2–3 mol% 3HV; that is, 2.1 $\times$ 10^{-19} m^3·m/m^2·Pa·s [57]. The permeance to oxygen of the PET film was 8.10 $\times$ 10^{-16} m^3/m^2·Pa·s. This corresponds to a permeability of 8.60 $\times$ 10^{-20} m^3·m/m^2·Pa·s and an OTR of approximately 7.2 cm^3/m^2·day at 1 atm, assuming a monolayer material. Similar to WVTR, this value was also lower than the OTR values reported by the manufacturer, namely 14 and 20 cm^3/m^2·day when measured at 0 and 90% RH, respectively, at 1 bar and 23°C (ASTM D3985). This is related to the fact that EVOH and some condensation and relatively hydrophilic polymers, such as PET, present the highest barrier performance at low-to-intermediate moisture conditions (e.g., 20–60 %RH) [50]. At high humidity, the permeability to oxygen gas increases due to an increase in free volume by an effect of water-induced plasticization. At low humidity, the amount of sorbed water is not high enough to reach equilibrium for interchain hydrogen bonding so that the diffusion of the water molecules increases. One can finally observe that both multilayer sheets showed good oxygen-barrier performance, showing values of 3.43 $\times$ 10^{-15} and 1.42 $\times$ 10^{-16} m^3/m^2·Pa·s for the PHBV/paper/PHBV and PET/paper/PET samples, respectively. Interestingly, this represents a respective reduction of approximately 7 and 6 times when compared to the permeance of their respective single monolayer films used for the coatings. This permeance decrease can be ascribed to the use of double layers in the structure as well as the improved barrier performance to oxygen achieved in the paper at dry conditions when it was protected by the hydrophobic external layers. Therefore, the oxygen permeance of the paper sheets was successfully reduced by the application of the double coatings of biopolyester.

3.7. Compostability of Paper Multilayers

According to ISO 17088 [58], composting is the aerobic treatment of the biodegradable plastic parts of packaging waste that consumes oxygen and produces, under controlled conditions and using microorganisms, biomass, inorganic compounds, CO_2, and water, without leaving visible distinguishable or toxic residues. Therefore, the evaluation of compostability includes three main tests, namely disintegration, biodegradation, and ecotoxicity [54,59,60]. Although recycling should be more economically and energetically favorable than composting for paper packaging, it cannot be practical in some situations because of excessive sorting and cleaning requirements [61]. Therefore, compostability of paper multilayers can be relevant from a sustainable point of view of waste management. This is particularly relevant when recycling is not feasible, for instance, in the case of disposable containers contaminated with food residues.

The ripe compost used as inoculum in both the biodegradation and disintegration tests showed an initial organic matter content (VS) of 90% ± 0.9%, expressed as volatile solids with respect to dried solids. The content of total dry solids (DS) was 79% ± 1.3%, whereas the pH value was 8.1, measured according to ISO 14855. Figure 7 shows the results of the aerobic biodegradation test of the films and paper sheets under controlled composting conditions (58 ± 2 °C for 45 days) following the method adapted from the ISO 14855. This test is based on the measurement of the CO_2 generated in the biodegradation process, which is considered proportional to the percentage of biodegradation in organic samples. MCC was also tested and used as positive reference. Moreover, the theoretical maximum quantity of CO_2 that can be produced by the biodegradation of the samples was calculated from their theoretical carbon content.

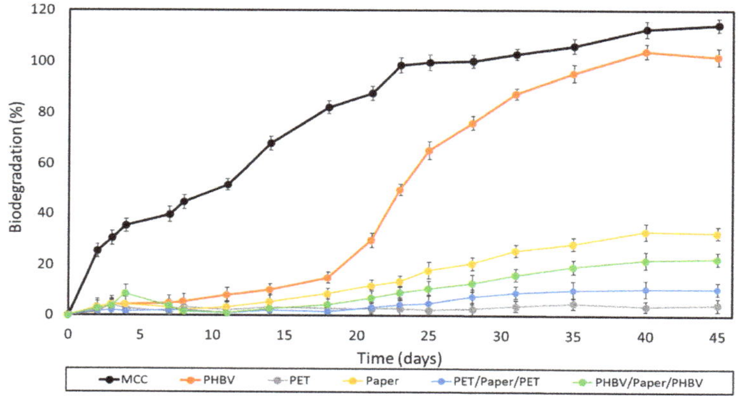

Figure 7. Sample biodegradation as a function of time of microcrystalline cellulose (MCC), the paper sheet, poly(3-hydroxybutyrate-*co*-3-hydroxyvalerate) (PHBV) and polyethylene terephthalate (PET) films, and PHBV/paper/PHBV and PET/paper/PET multilayer sheets.

One can observe that both MCC and the PHBV film completely biodegraded after 28 days and 40 days, respectively. Moreover, these samples exhibited the characteristic sigmoid profiles of respirometric tests, showing three different phases [60,62]. In the case of the PHBV film, it presented an initial log period of 2–15 days, followed by a biodegradation phase that was prolonged up to day 40, and finished with a plateau. PHBV biodegradation is essentially an enzymatic process, where exoenzymes from bacteria and fungi or membrane-bound enzymes, such as proteases, lipases, and esterases, which are present in the composting soil, hydrolyze the biopolyester chains into their corresponding monomers (hydroxy acids) [63]. In this process, the microorganisms first colonize the surface of the biopolymer and then secret depolymerases that hydrolyze the ester bonds. This process yields to low-M_W chains able to pass through the semipermeable external bacterial membranes and are metabolized. MCC, after 45 days, reached a biodegradation percentage of ~115%, which is attributed to the "priming" effect. The latter is an overestimation of the CO_2 released that occurs when the compost inoculum in the test reactor containing the samples generates more CO_2 than the compost inoculum in the blank reactors [64]. This effect has been previously ascribed to the stimulation of organic matter mineralization that takes place after the addition of easily-decomposable organic matter [65]. As expected, the commercial multilayer film showed no biodegradation since this is based on polymers that are not biodegradable. The uncoated paper reached a value of nearly 33%, whereas the PHBV- and PET-double coated papers showed lower biodegradation values, of 23 and 11%, respectively. In all cases, the so-called plateau phase was reached, showing a minor upward slope still noticeable in the curves. This means that biodegradation could continue under the same conditions by degrading the remaining organic carbon. However, since none of the paper samples exceeded the 70% biodegradation limit after 45 days, these cannot be considered as compostable under the conditions established by the standard. This was particularly notable for the PET/paper/PET multilayer due to the non-biodegradability of the petrochemical films. Therefore, it is advisable to avoid its incorporation into compostability plants. These results agree with those of the study performed by López Alvarez et al. [66], who showed that papers do not achieve the same level of biodegradation as MCC after 45 days. This is due to other organic compounds present in commercial paper, for example, lignocellulose and fatty acids, which can retard biodegradation. Furthermore, the large thickness of the paper sheets has to be also considered, 290 μm, which can highly reduce the biodegradation rate of biodegradable materials [67].

The degree of disintegration after 300 days (%D_{300}) of the films and sheets exposed to laboratory-scale composting environmental conditions (58 ± 2 °C) was also analyzed, providing information about the physical breakdown of the samples into smaller fractions with time. The test was validated according to the standard method [31], which establishes a reduction (R) of the volatile-solid content in the sample of the compost of more than 30%, with a standard deviation for %D_{300} values of less than 10 units. In the performance tests, R for the uncoated paper sheet was 57 ± 2, and the standard deviations for %D_{300} values were lower than 10. Figure 8 shows the disintegration values (% mass loss) as a function of time of the different films and sheets. As can be observed, the neat PHBV film fully disintegrated after 44 days, in agreement with previous works [68]. The disintegration rate was low during the first 12 days, then mass loss occurred very fast, and finally the rate slowed down after 40 days. In contrast, the commercial PET-based multilayer film showed mass loss values below 1%, which can be mainly related to losses occurring during the experimental preparation and manipulation of the samples.

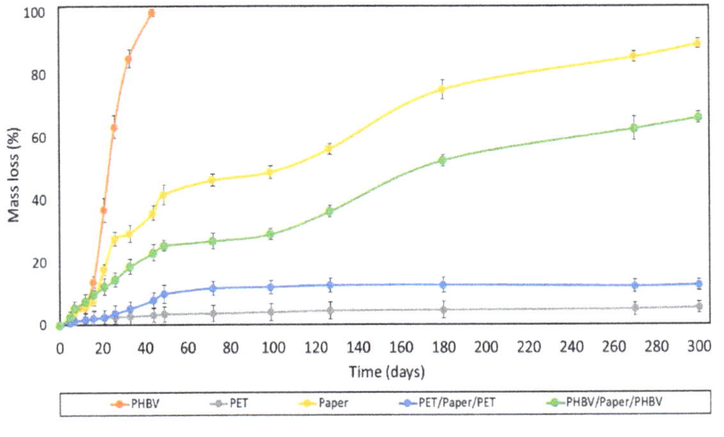

Figure 8. Disintegration rate under composting conditions of the paper sheet, poly(3-hydroxybutyrate-*co*-3-hydroxyvalerate) (PHBV) and polyethylene terephthalate (PET) films, and PHBV/paper/PHBV and PET/paper/PET multilayer sheets.

One can observe that all the paper sheets showed similar disintegration patterns. These presented three stages, clearly distinguished in the tested composting conditions, as similarly reported by Seoane et al. [69]. In the first step, corresponding to the period of 2–15 days, a low degree of disintegration was observed. In the second one, the disintegration rate markedly increased throughout 50 days till reaching a plateau. In this stage, the greatest disintegration rate was observed for the uncoated paper, confirming that the film coatings delayed the composting of paper. This delay can be related to the lower water absorption attained in the coated paper sheets due to the hydrophobic character of the polymers. Thus, the films impaired hydrolysis of paper into smaller molecules and the enzymatic reactions associated with the microbial growth as well as the ingress and colonization of microorganisms. Moreover, the higher compaction and lower porosity of the paper present in the multilayer (as consequence of the thermo-sealing process) could also contribute to the higher resistance to enzymatic degradation of the coated paper. This delay was significantly ($p < 0.05$) higher in the PET/paper/PET sheet due to the lack of biodegradability of the petrochemical polymers. Moreover, the multilayer structures are prone to present a considerably high amount of biodegradation products accumulated on the sheet-film interphases, which could retard degradation due to diffusional limitations [70]. After 300 days, all the paper sheets maintained practically the same rate of degradation, reaching $\%D_{300}$ values of approximately 90%, 70%, and 15% for the uncoated and PHBV- and PET-double-coated paper sheets, respectively. The difference attained between the PHBV/paper/PHBV and PET/paper/PET multilayer sheets can mainly be attributed to the disintegration of the PHBV thin film in the compost. Indeed, PET-double-coated paper exhibited intermediate disintegration behavior between those observed for its counterpart monolayer materials, that is, the paper sheet and PET film, but it was closer to that of the petrochemical film. This can be related to the partial and slow degradation of the inner paper layer, which was only available for microbial attack through the edges of the multilayer sample. In contrast, the biodegradation pattern of the paper sheet double coated with PHBV was similar to that of the uncoated paper sheet, although the disintegration was lower than that expected considering the mass loss observed in the monolayers. In particular, from the linear trend observed during the last disintegration stage of the uncoated and PHBV/paper/PHBV sheet samples, it was estimated that approximately 350 and 475 days would be required, respectively, to reach a 100% reduction of the original mass. In any case, none of the paper-based sheets tested herein were compostable since, after 3 months, their mass did not amount to less than 10% of their original mass.

Finally, Figure 9 illustrates the visual appearance of the recovered films and sheets after the selected times of disintegration, where the different degradable characters of the samples can be observed. At the beginning of the process, all of the film and sheet samples exhibited a continuous structure with no visible holes, but they showed an increase in opacity during the first week. At the third week, signs of erosion were detected in some of the samples. Then, the PHBV film fully disintegrated after 44 days, whereas the PET film developed a brown color without signs of mass loss but with high amounts of compost particles adhered on the surface. All the paper sheets showed certain signs of erosion and small fractures after 100 days of incubation in the compost. One can further observe that after 180 days, the uncoated paper and PHBV/paper/PHBV sheets appeared broken into small parts, particularly the uncoated sample, whereas the PET/paper/PET sheet still maintained its full integrity. Finally, after 300 days, the uncoated paper was hardly distinguishable from compost in the mesh due to the fact that the aggregates were brown. In the case of PHBV/paper/PHBV, some fragments smaller than the mesh size were formed.

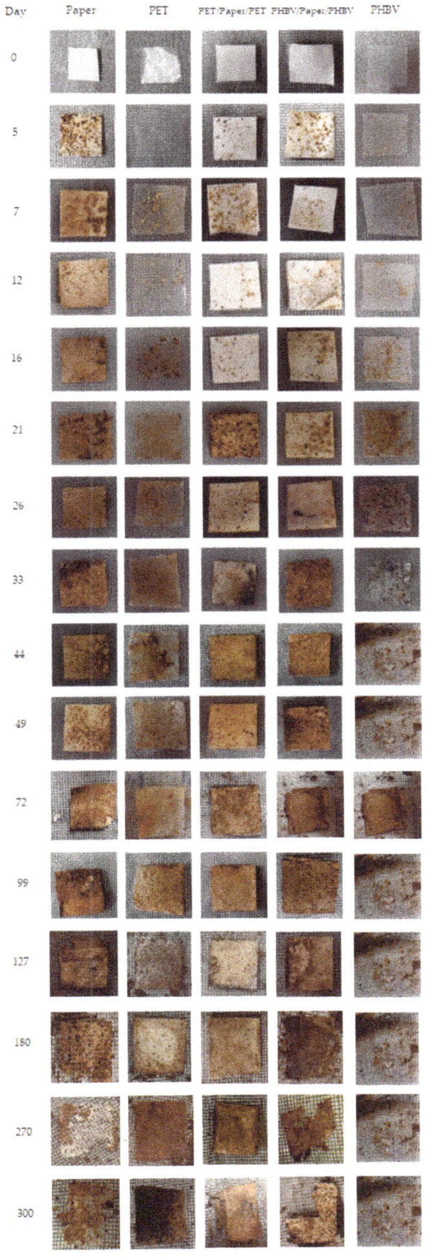

Figure 9. Photographs taken at different composting times of the paper sheet, poly(3-hydroxybutyrate-*co*-3-hydroxyvalerate) (PHBV) and polyethylene terephthalate (PET) films, and PHBV/paper/PHBV and PET/paper/PET multilayer sheets.

4. Conclusions

PHBV has been proven to be an excellent material for paper coating applications. By means of heat-sealing technology, which is currently available in the food packaging industry, for instance in lamination or thermoforming processes, PHBV/paper/PHBV multilayers with improved mechanical and barrier properties were successfully obtained. Moreover, the presence of the double coatings of PHBV did not notably affect the original optical and thermal characteristics of the paper. Although the performance was lower than that of equivalent multilayers developed by the same process using conventional PET-based films with high barrier, the properties presented here by the double-coated biopolymer structures were within the same range. Thus, it can be concluded that PHBV/paper/PHBV multilayer sheets can be excellent candidates to replace currently available paper substrates coated with petrochemical non-biodegradable films, especially in the case of rigid and intermediate barrier packaging materials. However, to meet current requirements to be certified as compostable, paper layers with lower thickness would be required. Future works will focus on developing novel biopolymer/paper trays and their application in packaging, particularly for the preservation of foodstuffs with high water activity, where the use of paper is restricted. Additionally, their food safety assessment will be ascertained by migration tests using food simulants.

Author Contributions: Conceptualization, P.Z. and S.T.-G.; methodology, E.H.-G. and P.A.V.F.; formal analysis, E.H.-G. and P.A.V.F.; investigation, E.H.-G. and P.A.V.F.; data curation, E.H.-G. and P.A.V.F.; writing—original draft preparation, E.H.-G.; writing—review and editing, C.G.-M. and S.T.-G.; supervision, C.G.-M. and S.T.-G.; project administration, S.T.-G.; funding acquisition, P.Z. and S.T.-G. All authors have read and agreed to the published version of the manuscript.

Funding: This research was funded by the Spanish Ministry of Science and Innovation (MICI), grant number PID2021-128749OB-C33 and the Valencian Innovation Agency (AVI) through the TERMOFIB project (INNVA1/2020/46).

Institutional Review Board Statement: Not applicable.

Informed Consent Statement: Not applicable.

Data Availability Statement: Data are contained within the article and also available on request.

Acknowledgments: E.H.-G. and S.T.-G. acknowledge MICI for her predoctoral research grant (BES 2017-082040) and Ramón y Cajal for his contract (RYC2019-027784-I), respectively.

Conflicts of Interest: The authors declare no conflict of interest.

References

1. Torres-Giner, S.; Gil, L.; Pascual-Ramírez, L.; Garde-Belza, J.A. Packaging: Food Waste Reduction. In *Encyclopedia of Polymer Applications*; CRC Press: Boca Raton, FL, USA, 2019. [CrossRef]
2. Lagarón, J.M.; Cabedo, L.; Cava, D.; Feijoo, J.L.; Gavara, R.; Gimenez, E. Improving packaged food quality and safety. Part 2: Nanocomposites. *Food Addit. Contam.* **2005**, *22*, 994–998. [CrossRef] [PubMed]
3. Torres-Giner, S.; Figueroa-Lopez, K.J.; Melendez-Rodriguez, B.; Prieto, C.; Pardo-Figuerez, M.; Lagaron, J.M. Emerging Trends in Biopolymers for Food Packaging. In *Sustainable Food Packaging Technology*; John Wiley & Sons, Ltd.: Hoboken, NJ, USA, 2021; pp. 1–33. [CrossRef]
4. Hale, R.C.; Seeley, M.E.; La Guardia, M.J.; Mai, L.; Zeng, E.Y. A Global Perspective on Microplastics. *J. Geophys. Res. Ocean.* **2020**, *125*, e2018JC014719. [CrossRef]
5. Cverenkárová, K.; Valachovičová, M.; Mackuľak, T.; Žemlička, L.; Bírošová, L. Microplastics in the Food Chain. *Life* **2021**, *11*, 1349. [CrossRef] [PubMed]
6. Nakaya, M.; Uedono, A.; Hotta, A. Recent Progress in Gas Barrier Thin Film Coatings on PET Bottles in Food and Beverage Applications. *Coatings* **2015**, *5*, 987–1001. [CrossRef]
7. Alias, A.R.; Wan, M.K.; Sarbon, N.M. Emerging materials and technologies of multi-layer film for food packaging application: A review. *Food Control* **2022**, *136*, 108875. [CrossRef]
8. Zhong, Y.; Godwin, P.; Jin, Y.; Xiao, H. Biodegradable polymers and green-based antimicrobial packaging materials: A mini-review. *Adv. Ind. Eng. Polym. Res.* **2020**, *3*, 27–35. [CrossRef]

9. Gabirondo, E.; Melendez-Rodriguez, B.; Arnal, C.; Lagaron, J.M.; Martínez De Ilarduya, A.; Sardon, H.; Torres-Giner, S. Organocatalyzed closed-loop chemical recycling of thermo-compressed films of poly(ethylene furanoate). *Polym. Chem.* **2021**, *12*, 1571–1580. [CrossRef]
10. Guillard, V.; Gaucel, S.; Fornaciari, C.; Angellier-Coussy, H.; Buche, P.; Gontard, N. The Next Generation of Sustainable Food Packaging to Preserve Our Environment in a Circular Economy Context. *Front. Nutr.* **2018**, *5*, 121. [CrossRef]
11. Pojanavaraphan, T.; Magaraphan, R.; Chiou, B.-S.; Schiraldi, D.A. Development of Biodegradable Foamlike Materials Based on Casein and Sodium Montmorillonite Clay. *Biomacromolecules* **2010**, *11*, 2640–2646. [CrossRef]
12. Vartiainen, J.; Shen, Y.; Kaljunen, T.; Malm, T.; Vähä-Nissi, M.; Putkonen, M.; Harlin, A. Bio-based multilayer barrier films by extrusion, dispersion coating and atomic layer deposition. *J. Appl. Polym. Sci.* **2016**, *133*. [CrossRef]
13. Arcos-Hernández, M.V.; Laycock, B.; Donose, B.C.; Pratt, S.; Halley, P.; Al-Luaibi, S.; Werker, A.; Lant, P.A. Physicochemical and mechanical properties of mixed culture polyhydroxyalkanoate (PHBV). *Eur. Polym. J.* **2013**, *49*, 904–913. [CrossRef]
14. Bucci, D.Z.; Tavares, L.B.B.; Sell, I. Biodegradation and physical evaluation of PHB packaging. *Polym. Test.* **2007**, *26*, 908–915. [CrossRef]
15. Kamravamanesh, D.; Kovacs, T.; Pflügl, S.; Druzhinina, I.; Kroll, P.; Lackner, M.; Herwig, C. Increased poly-β-hydroxybutyrate production from carbon dioxide in randomly mutated cells of cyanobacterial strain Synechocystis sp. PCC 6714: Mutant generation and characterization. *Bioresour. Technol.* **2018**, *266*, 34–44. [CrossRef]
16. Peelman, N.; Ragaert, P.; De Meulenaer, B.; Adons, D.; Peeters, R.; Cardon, L.; Van Impe, F.; Devlieghere, F. Application of bioplastics for food packaging. *Trends Food Sci. Technol.* **2013**, *32*, 128–141. [CrossRef]
17. Fabra, M.J.; Sánchez, G.; López-Rubio, A.; Lagaron, J.M. Microbiological and ageing performance of polyhydroxyalkanoate-based multilayer structures of interest in food packaging. *LWT-Food Sci. Technol.* **2014**, *59*, 760–767. [CrossRef]
18. Modi, S.; Koelling, K.; Vodovotz, Y. Assessment of PHB with varying hydroxyvalerate content for potential packaging applications. *Eur. Polym. J.* **2011**, *47*, 179–186. [CrossRef]
19. Keshavarz, T.; Roy, I. Polyhydroxyalkanoates: Bioplastics with a green agenda. *Curr. Opin. Microbiol.* **2010**, *13*, 321–326. [CrossRef]
20. Laycock, B.; Arcos-Hernandez, M.V.; Langford, A.; Pratt, S.; Werker, A.; Halley, P.J.; Lant, P.A. Crystallisation and fractionation of selected polyhydroxyalkanoates produced from mixed cultures. *New Biotechnol.* **2014**, *31*, 345–356. [CrossRef]
21. Laycock, B.; Halley, P.; Pratt, S.; Werker, A.; Lant, P. The chemomechanical properties of microbial polyhydroxyalkanoates. *Prog. Polym. Sci.* **2013**, *38*, 536–583. [CrossRef]
22. Anderson, A.J.; Dawes, E.A. Occurrence, metabolism, metabolic role, and industrial uses of bacterial polyhydroxyalkanoates. *Microbiol. Rev.* **1990**, *54*, 450–472. [CrossRef]
23. Ebnesajjad, S. *Handbook of Biopolymers and Biodegradable Plastics: Properties, Processing and Applications*; Elsevier: Amsterdam, The Netherlands, 2013; pp. 1–462. [CrossRef]
24. Savenkova, L.; Gercberga, Z.; Bibers, I.; Kalnin, M. Effect of 3-hydroxy valerate content on some physical and mechanical properties of polyhydroxyalkanoates produced by Azotobacter chroococcum. *Process. Biochem.* **2000**, *36*, 445–450. [CrossRef]
25. Hutchings, J.B. Food and Colour Appearance. In *Chapman and Hall Food Science Book*, 2nd ed.; Aspen Publication: Gaithersburg, MD, USA, 1999.
26. Agüero, A.; Morcillo, M.D.C.; Quiles-Carrillo, L.; Balart, R.; Boronat, T.; Lascano, D.; Torres-Giner, S.; Fenollar, O. Study of the Influence of the Reprocessing Cycles on the Final Properties of Polylactide Pieces Obtained by Injection Molding. *Polymers* **2019**, *11*, 1908. [CrossRef] [PubMed]
27. *ASTM D882*; Standard Test Method for Tensile Properties of Thin Plastic Sheeting. American Society for Testing and Materials: Philadelphia, PA, USA, 2001; pp. 162–170.
28. *ASTM E96/E96M*; Standard Test Methods for Water Vapor Transmission of Materials. American Society for Testing and Materials: Philadelphia, PA, USA, 2005; pp. 406–413.
29. *ASTM 3985-95*; Standard Test Method for Oxygen Gas Transmission Rate through Plastic Film and Sheeting Using a Coulometric Sensor. American Society for Testing and Materials: Philadelphia, PA, USA, 2002; pp. 472–477.
30. *UNE-EN ISO 14855-1*; Asociación Española de Normalización. Determinación de la Biodegradabilidad Aeróbica Final de Materiales Plásticos en Condiciones de Compostaje Controladas. Método Según el Análisis de Dióxido de Carbono Generado. Parte 1; Método General. Asociación Española de Normalización: Madrid, Spain, 2012.
31. ISO 20200. Plastics-Determination of the Degree of Disintegration of Plastic Materials under Simulated Composting in a Laboratory -Scale Test. 2004. Available online: https://www.une.org (accessed on 15 November 2022).
32. Melendez-Rodriguez, B.; Reis, M.A.M.; Carvalheira, M.; Sammon, C.; Cabedo, L.; Torres-Giner, S.; Lagaron, J.M. Development and Characterization of Electrospun Biopapers of Poly(3-hydroxybutyrate-*co*-3-hydroxyvalerate) Derived from Cheese Whey with Varying 3-Hydroxyvalerate Contents. *Biomacromolecules* **2021**, *22*, 2935–2953. [CrossRef] [PubMed]
33. Cherpinski, A.; Torres-Giner, S.; Cabedo, L.; Méndez, J.A.; Lagaron, J.M. Multilayer structures based on annealed electrospun biopolymer coatings of interest in water and aroma barrier fiber-based food packaging applications. *J. Appl. Polym. Sci.* **2018**, *135*, 45501. [CrossRef]
34. Ivorra-Martinez, J.; Quiles-Carrillo, L.; Boronat, T.; Torres-Giner, S.; A Covas, J. Assessment of the Mechanical and Thermal Properties of Injection-Molded Poly(3-hydroxybutyrate-*co*-3-hydroxyhexanoate)/Hydroxyapatite Nanoparticles Parts for Use in Bone Tissue Engineering. *Polymers* **2020**, *12*, 1389. [CrossRef]
35. Hernández-García, E.; Vargas, M.; Torres-Giner, S. Quality and Shelf-Life Stability of Pork Meat Fillets Packaged in Multilayer Polylactide Films. *Foods* **2022**, *11*, 426. [CrossRef]

36. Ortega-Toro, R.; Jiménez, A.; Talens, P.; Chiralt, A. Properties of starch–hydroxypropyl methylcellulose based films obtained by compression molding. *Carbohydr. Polym.* **2014**, *109*, 155–165. [CrossRef]
37. Villalobos, R.; Chanona, J.; Hernández, P.; Gutiérrez, G.; Chiralt, A. Gloss and transparency of hydroxypropyl methylcellulose films containing surfactants as affected by their microstructure. *Food Hydrocoll.* **2005**, *19*, 53–61. [CrossRef]
38. Torres-Giner, S.; Hilliou, L.; Melendez-Rodriguez, B.; Figueroa-Lopez, K.J.; Madalena, D.; Cabedo, L.; Covas, J.A.; Vicente, A.A.; Lagaron, J.M. Melt processability, characterization, and antibacterial activity of compression-molded green composite sheets made of poly(3-hydroxybutyrate-*co*-3-hydroxyvalerate) reinforced with coconut fibers impregnated with oregano essential oil. *Food Packag. Shelf Life* **2018**, *17*, 39–49. [CrossRef]
39. Lennersten, M.; Lingnert, H. Influence of Wavelength and Packaging Material on Lipid Oxidation and Colour Changes in Low-fat Mayonnaise. *LWT Food Sci. Technol.* **2000**, *33*, 253–260. [CrossRef]
40. Costa, M.; Pastrana, L.M.; Teixeira, J.A.; Sillankorva, S.M.; Cerqueira, M. Characterization of PHBV films loaded with FO1 bacteriophage using polyvinyl alcohol-based nanofibers and coatings: A comparative study. *Innov. Food Sci. Emerg. Technol.* **2021**, *69*, 102646. [CrossRef]
41. Velásquez, E.; Garrido, L.; Valenzuela, X.; Galotto, M.J.; Guarda, A.; López de Dicastillo, C. Physical properties and safety of 100% post-consumer PET bottle -organoclay nanocomposites towards a circular economy. *Sustain. Chem. Pharm.* **2020**, *17*, 100285. [CrossRef]
42. Mahy, M.; Eycken, L.; Oosterlinck, A. Evaluation of Uniform Color Spaces Developed after the adoption of CIELAB and CIELUV. *Color Res. Appl.* **1994**, *19*, 105–121.
43. Melendez-Rodriguez, B.; Torres-Giner, S.; Aldureid, A.; Cabedo, L.; Lagaron, J.M. Reactive Melt Mixing of Poly(3-Hydroxybutyrate)/Rice Husk Flour Composites with Purified Biosustainably Produced Poly(3-Hydroxybutyrate-*co*-3-Hydroxyvalerate). *Materials* **2019**, *12*, 2152. [CrossRef]
44. Yang, H.; Yan, R.; Chen, H.; Lee, D.H.; Zheng, C. Characteristics of hemicellulose, cellulose and lignin pyrolysis. *Fuel* **2007**, *86*, 1781–1788. [CrossRef]
45. Souza, B.S.; Moreira, A.P.D.; Teixeira, A.M.R.F. TG-FTIR coupling to monitor the pyrolysis products from agricultural residues. *J. Therm. Anal. Calorim.* **2009**, *97*, 637. [CrossRef]
46. Bugnicourt, E.; Cinelli, P.; Lazzeri, A.; Álvarez, V.A. Polyhydroxyalkanoate (PHA): Review of synthesis, characteristics, processing and potential applications in packaging. *Express Polym. Lett.* **2014**, *8*, 791–808. [CrossRef]
47. Levchik, S.V.; Weil, E.D. A review on thermal decomposition and combustion of thermoplastic polyesters. *Polym. Adv. Technol.* **2004**, *15*, 691–700. [CrossRef]
48. Deshwal, G.K.; Panjagari, N.R.; Alam, T. An overview of paper and paper-based food packaging materials: Health safety and environmental concerns. *J. Food Sci. Technol.* **2019**, *56*, 4391–4403. [CrossRef]
49. Sanchez-Garcia, M.D.; Gimenez, E.; Lagaron, J.M. Morphology and barrier properties of solvent cast composites of thermoplastic biopolymers and purified cellulose fibers. *Carbohydr. Polym.* **2008**, *71*, 235–244. [CrossRef]
50. Melendez-Rodriguez, B.; Torres-Giner, S.; Angulo, I.; Pardo-Figuerez, M.; Hilliou, L.; Escuin, J.M.; Cabedo, L.; Nevo, Y.; Prieto, C.; Lagaron, J.M. High-Oxygen-Barrier Multilayer Films Based on Polyhydroxyalkanoates and Cellulose Nanocrystals. *Nanomaterials* **2021**, *11*, 1443. [CrossRef] [PubMed]
51. Nduko, J.M.; Matsumoto, K.; Taguchi, S. Biological lactate-polymers synthesized by one-pot microbial factory: Enzyme and metabolic engineering. *ACS Symp. Ser.* **2012**, *1105*, 213–235.
52. *ASTM F1249*; Standard Test Method for Water Vapor Transmission Rate through Plastic Film and Shetting Using a Modulated Infrared Sensor. American Society for Testing and Materials: Philadelphia, PA, USA, 2020.
53. Lagarón, J.M. 1-Multifunctional and nanoreinforced polymers for food packaging. In *Multifunctional and Nanoreinforced Polymers for Food Packaging*; Lagarón, J.-M., Ed.; Woodhead Publishing: Sawston, UK, 2011.
54. Balaguer, M.P.; Villanova, J.; Cesar, G.; Gavara, R.; Hernandez-Munoz, P. Compostable properties of antimicrobial bioplastics based on cinnamaldehyde cross-linked gliadins. *Chem. Eng. J.* **2015**, *262*, 447–455. [CrossRef]
55. Fabra, M.J.; Lopez-Rubio, A.; Lagaron, J.M. Nanostructured interlayers of zein to improve the barrier properties of high barrier polyhydroxyalkanoates and other polyesters. *J. Food Eng.* **2014**, *127*, 1–9. [CrossRef]
56. Melendez-Rodriguez, B.; Torres-Giner, S.; Zavagna, L.; Sammon, C.; Cabedo, L.; Prieto, C.; Lagaron, J.M. Development and Characterization of Electrospun Fiber-Based Poly(ethylene-*co*-vinyl Alcohol) Films of Application Interest as High-Gas-Barrier Interlayers in Food Packaging. *Polymers* **2021**, *13*, 2061. [CrossRef]
57. Quiles-Carrillo, L.; Montanes, N.; Lagaron, J.M.; Balart, R.; Torres-Giner, S. In Situ Compatibilization of Biopolymer Ternary Blends by Reactive Extrusion with Low-Functionality Epoxy-Based Styrene–Acrylic Oligomer. *J. Polym. Environ.* **2019**, *27*, 84–96. [CrossRef]
58. ISO 17088. Specifications for compostable plastics. 2012. Available online: https://www.iso.org/obp/ui/#iso:std:iso:17088:ed-2:v1:en (accessed on 4 November 2022).
59. Hernández-García, E.; Vargas, M.; González-Martínez, C.; Chiralt, A. Biodegradable Antimicrobial Films for Food Packaging: Effect of Antimicrobials on Degradation. *Foods* **2021**, *10*, 1256. [CrossRef]
60. Hernández-García, E.; Vargas, M.; Chiralt, A.; González-Martínez, C. Biodegradation of PLA-PHBV Blend Films as Affected by the Incorporation of Different Phenolic Acids. *Foods* **2022**, *11*, 243. [CrossRef]

61. Kale, G.; Kijchavengkul, T.; Auras, R.; Rubino, M.; Selke, S.E.; Singh, S.P. Compostability of Bioplastic Packaging Materials: An Overview. *Macromol. Biosci.* **2007**, *7*, 255–277. [CrossRef]
62. Cano, A.I.; Cháfer, M.; Chiralt, A.; González-Martínez, C. Biodegradation behavior of starch-PVA films as affected by the incorporation of different antimicrobials. *Polym. Degrad. Stab.* **2016**, *132*, 11–20. [CrossRef]
63. Iglesias-Montes, M.L.; Soccio, M.; Luzi, F.; Puglia, D.; Gazzano, M.; Lotti, N.; Manfredi, L.B.; Cyras, V.P. Evaluation of the Factors Affecting the Disintegration under a Composting Process of Poly(lactic acid)/Poly(3-hydroxybutyrate) (PLA/PHB) Blends. *Polymers* **2021**, *13*, 3171. [CrossRef]
64. Shen, J.; Bartha, R. Priming effect of glucose polymers in soil-based biodegradation tests. *Soil Biol. Biochem.* **1997**, *29*, 1195–1198. [CrossRef]
65. Kuzyakov, Y.; Friedel, J.K.; Stahr, K. Review of mechanisms and quantification of priming effects. *Soil Biol. Biochem.* **2000**, *32*, 1485–1498. [CrossRef]
66. López Alvarez, J.V.; Aguilar Larrucea, M.; Arraiza Bermúdez, P.; León Chicote, B. Biodegradation of paper waste under controlled composting conditions. *Waste Manag.* **2009**, *29*, 1514–1519. [CrossRef]
67. Rojas-Lema, S.; Arevalo, J.; Gomez-Caturla, J.; Garcia-Garcia, D.; Torres-Giner, S. Peroxide-Induced Synthesis of Maleic Anhydride-Grafted Poly(butylene succinate) and Its Compatibilizing Effect on Poly(butylene succinate)/Pistachio Shell Flour Composites. *Molecules* **2021**, *26*, 5927. [CrossRef]
68. Lidón Sánchez-Safont, E.; González-Ausejo, J.; Gámez-Pérez, J.; Lagarón, J.M.; Cabedo, L. Poly(3-Hydroxybutyrate-*co*-3-Hydroxyvalerate)/ Purified Cellulose Fiber Composites by Melt Blending: Characterization and Degradation in Composting Conditions. *J. Renew. Mater.* **2016**, *4*, 123. [CrossRef]
69. Seoane, I.T.; Luzi, F.; Puglia, D.; Cyras, V.P.; Manfredi, L.B. Enhancement of paperboard performance as packaging material by layering with plasticized polyhydroxybutyrate/nanocellulose coatings. *J. Appl. Polym. Sci.* **2018**, *135*, 46872. [CrossRef]
70. Bi, S.; Pan, H.; Barinelli, V.; Eriksen, B.; Ruiz, S.; Sobkowicz, M.J. Biodegradable polyester coated mulch paper for controlled release of fertilizer. *J. Clean. Prod.* **2021**, *294*, 126348. [CrossRef]

applied sciences

MDPI

Article

Influence of Temperature on the Composition and Calorific Value of Gases Produced during the Pyrolysis of Waste Pharmaceutical Blisters

Katarzyna Klejnowska [1,*], Krzysztof Pikoń [2], Waldemar Ścierski [2], Krzysztof Skutil [3] and Magdalena Bogacka [2]

1 Łukasiewicz Research Network, Institute of Non-Ferrous Metals, 44-121 Gliwice, Poland
2 Silesian University of Technology, Faculty of Energy and Environmental Enginnering, 44-100 Gliwice, Poland; Krzysztof.pikon@polsl.pl (K.P.); waldemar.scierski@polsl.pl (W.Ś.); magdalena.bogacka@polsl.pl (M.B.)
3 Silesian University of Technology, Faculty of Chemistry, 44-100 Gliwice, Poland; krzysztof.skutil@polsl.pl
* Correspondence: katarzynak@imn.gliwice.pl

Received: 20 November 2019; Accepted: 11 January 2020; Published: 21 January 2020

Featured Application: Such results suggest that when considering a potential technology for WPB processing, the most favorable conditions for the production of gases would be at higher temperatures. After providing the satisfactory composition and calorific value, the gases can be used as a fuel, i.e., to heat the process.

Abstract: Waste pharmaceutical blisters (WPBs) are a type of multimaterial waste that contain layers of plastic and metal connected permanently. The separation of those materials with the application of mechanical methods is impossible. One of the methods that can be used to recover metal from WPBs is pyrolysis—a thermal decomposition process performed in the absence of oxygen. The products are a solid fraction that contain char and metal, liquid fraction, and gases. The gases produced during the process can be used as a fuel, either alone or mixed with another gaseous fuel such as natural gas. The presented research was focused on the determination of the influence of the process temperature on the composition of gases produced during the pyrolysis of two types of pharmaceutical blister waste: pre- and postconsumer. The postconsumer waste contains trace amounts of pharmaceutical products. One of the goals was to determine whether those compounds influence the gas properties. The method of neutralizing these materials can be part of the circular economy, the idea of which is to strive to maximize the use of natural resources and bring them back into circulation. The presented method allows not only to recover char, oil, and metal that can be easily separated from the solid fraction, but also to reuse the process gases as a fuel and, possibly, to form HCl during the decomposition of PVC. The paper includes the analysis of the input material, as well as a detailed chemical analysis of the produced gases.

Keywords: pyrolysis; multimaterial waste; pharmaceutical waste; blisters; WPBs; energy recovery

1. Introduction

Multimaterial waste is a significant group of waste that requires the application of special recycling techniques in order to reduce its negative impact on the environment [1]. In general, it is made of at least two different materials connected permanently [2]. According to the regulations (Polish Act of 13 June 2013 on Packaging and Packaging Waste and Polish Act of 14 December 2012 on Waste), it is treated like the material that is present in the highest amount, i.e., if the material consists of 70% paper and 30% plastic it should be treated like paper and collected in the designated container [3,4].

The European Union's strategy in the plastics field is closely linked to the circular economy. This strategy pays special attention to plastic recycling and the possibilities of reducing it. Currently, 59% of plastic products are plastic packaging, including blisters for drugs, which are very difficult to manage. Packaging recycling is very difficult, but the pyrolysis of waste medicine blisters is one of the answers to the challenges of the circular economy. In the future, the process proposed in this article can be treated as a closed circuit in this group of packaging waste.

Waste pharmaceutical blisters (WPBs) are a type of multimaterial waste. They are the most commonly used type of packaging to store pharmaceutical products [5,6]. They contain approx. 80% thermoformable plastic films (usually polypropylene (PP) and pure polyvinyl chloride (PVC), or polyvinyl with diaphone addition (PVC/PVDC) or chlorotrifluoroethylene PVC/CTFE)) and 20% aluminum alloy (containing approx. 85% Al, 11,5% Mg, and 0,8% Ti) [7,8]. The type of plastic used to produce the blister is usually chosen based on moisture permeability, which has a huge influence on the durability of the products and the price. In order to provide the highest protection from external conditions, the companies sometimes use the Al/Al type blisters, which contain polyamide, aluminum, and PVC. Two layers of aluminum protect the products from moisture, which may decompose the active ingredients and affect the expiration date of the medicine, oxygen, and sunlight [9].

Currently, used blisters are either treated as plastic waste or collected at pharmacies together with the expired medicine and directed to thermal utilization plants, where metal is not recovered [4,10]. As for most multimaterial waste, the mechanical separation of the materials used to produce them is complicated or, sometimes, impossible [11]. Therefore, in order to comply with the circular economy requirements, WPBs require a special treatment to efficiently recycle both metal and plastic fractions [12,13]. Wang et al. [5] analyzed the hydrometallurgical method to recycle waste blisters. The method was based on the leaching of 1 cm x 2 cm pieces of WPBs in hydrochloric acid. Under optimum conditions, this method allowed to separate as much as 100% of plastic from the metal. The obtained aluminum was of excellent quality, but the metal recovery required additional steps, i.e., precipitation, solvent extraction, adsorption, or the application of the ion-exchange method. Also, large amounts of HCl were required to separate the materials. Yin et al. [14] analyzed the thermal method of aluminum recovery, which also generates high-quality gaseous fuel. The material used in the research was a similar multimaterial packaging waste used to store face mask sheets—a postconsumer Neutrogena face mask packaging made of aluminum foil, PET (polyethylene terephthalate), and PP (polypropylene). The produced aluminum was characterized by a high purity and an extremely low surface oxidation level. The process also resulted in the production of 67 wt% of hydrocarbon gases, which can be used as a fuel.

Pyrolysis is a thermal decomposition process that occurs in the absence of oxygen. The products can be divided into a solid, a liquid, and a gaseous fraction. The solid fraction may contain the materials that do not pyrolyze, i.e., metals, which can be easily separated from the char and recovered in a simple metallurgical process. In addition, other products can be further used, i.e., as a fuel or carbon black [15]. This paper is focused on the gases generated during the process. If their quality is high enough, the gases can be used as a fuel, i.e., to generate heat for the pyrolysis. Such a holistic approach can contribute to making even a thermal process sustainable.

2. Materials and Methods

The literature provides no detailed information about the composition of the gases generated during the pyrolysis of WPBs. The presented research was focused on the analysis of the influence of the process temperature on the composition of the gases, with a special focus on hydrocarbons. Two waste materials were analyzed in order to determine the potential influence of the trace amounts of medical products on the quality of the gases. The potential of the process gases to be used as a fuel, individually or mixed with natural gas, was analyzed.

2.1. Materials

Two waste materials were analyzed—pre- and postconsumer waste. The preconsumer WPBs, delivered by the manufacturer of a dietary supplement, were collected after the production line. The preconsumer waste had no contact with the product, and the plastic used as a lidding material was transparent.

Postconsumer WPBs included various mixed pharmaceutical blisters after the medicines had been removed. The material had contact with various products and was not washed before the analysis, in order to provide a reliable source of information that could be used for a real process. The analyzed portion of the mixed types of blisters contained:

- White plastic: 38.5%,
- Transparent plastic: 32.7%,
- Orange plastic: 15.0%,
- Al/Al blisters: 9.9%,
- Red plastic: 3.9%.

The illustrative pictures of the pre- and postconsumer waste used for the tests are presented in Figure 1.

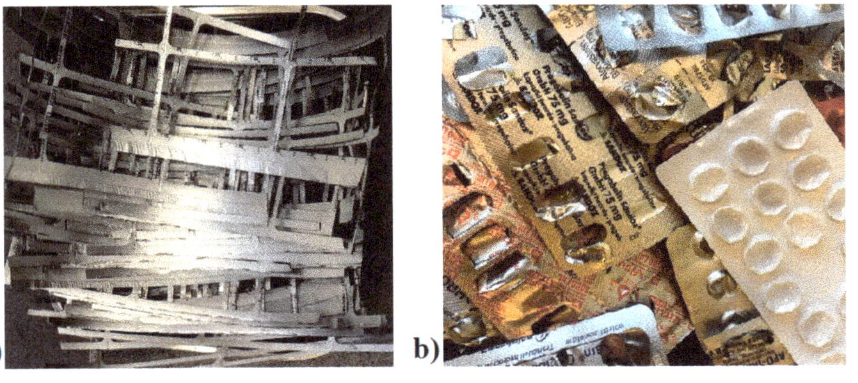

Figure 1. Analyzed materials; (**a**) preconsumer waste, (**b**) postconsumer waste.

Both materials were analyzed for the content of the basic elements. The results of the analysis are presented in Table 1.

Table 1. Chemical analysis of the pre- and postconsumer waste.

Sample	Content (%)				
	C	H	Cl	N	S
Preconsumer	45.46	4.47	43.66	0.60	10.83
Postconsumer	44.27	3.37	35.03	0.35	11.33

The contents of carbon (approx. 45%), hydrogen (approx. 4%), nitrogen (below 1%), and sulfur (approx. 11%) were similar for both materials. The differences were mainly caused by the impurities present in the postconsumer waste as a result of the contact with the pharmaceutical products. The chlorine content in preconsumer waste was much higher (approx. 44%) than that found in postconsumer waste (35%). The reason for such differences was the lidding material. For preconsumer waste, the results suggest that PVC was the main material, while postconsumer waste contained various materials. However, the high chlorine content suggests that PVC was one of the most commonly used materials.

2.2. Experimental Method

The materials were subjected to the pyrolysis process at three temperatures—400 °C, 425 °C, and 450 °C—in a laboratory-scale furnace. The thermocouple was located in the central part of the reaction chamber. The scheme of the reactor is presented in Figure 2. During each test, the sample weight was precisely 150 g. The process temperature was controlled with two thermocouples—one thermocouple, located on the bottom of the reaction chamber, measured the temperature inside the sample, while a second thermocouple, located in the central part of the reaction chamber, measured the process temperature. The thermocouples were automatically calibrated each time the equipment was switched on.

During each test, three gas samples were collected in order to determine their composition. The gases were analyzed for the content of hydrocarbons (CH_4, C_2H_6, C_2H_4, C_3H_8, C_3H_6, C_4H_{10}, C_4H_8, C_4H_4, C5+) and inorganic compounds (H_2, O_2, N_2, CO, and CO_2) in order to determine their potential to be used as a fuel (i.e., natural gas substitute). The gases were collected at the beginning of the process, in the middle of the process, and after the maximum temperature was reached, into 0.5 and 0.2 dm^3 glass pipettes equipped with two cut-off valves. The gases were injected into the analyzers with the application of the hydrostatic pressure method, performed with a liquid (brine) introduced from the bottom of the container. The gases were injected into the analyzer from the second valve.

The gases were analyzed with the application of gas chromatography using two separate sections supplied with a carrier gas (high-purity argon). Hydrocarbons were analyzed on the Alumina packed column (l = 2 m; d = 1 mm) with FID (flame ionization detector). Inorganic ingredients were analyzed on the Carbosphere column (2 m; 3 mm) with a TCD (thermal conductivity detector). The gas composition was determined with the application of the external standard with the relative response rates. The relative uncertainty of the measurement was 2%.

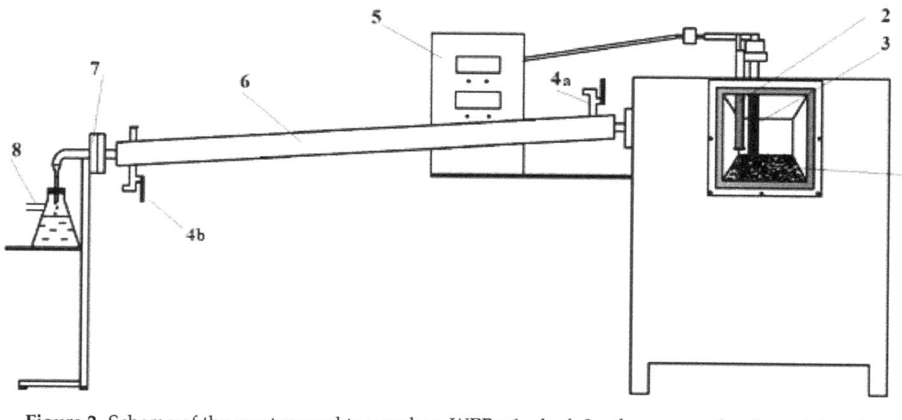

Figure 2. Scheme of the reactor used to pyrolyze WPBs; 1—bed, 2—thermocouples, 3—safety valve, 4—cooling water (a) inlet and (b) outlet, 5—control device, 6—gas cooler, 7—output used to collect process gases, 8—liquid fraction vessel.

3. Results

The tests were performed in two series. Series 1 refers to the preconsumer waste. Series 2 refers to the postconsumer waste. During each series, the tests were performed for three maximum process temperatures: 400 °C, 425 °C, and 450 °C. Also, three gas samples were collected during each test: at the beginning of the process, during the process, when the decomposition was the most rapid, and after the maximum temperature was reached. The moment of most rapid mass loss in the sample was determined based on the thermogravimetric study [9].

The solid and liquid products were weighed after the process, after the reactor was cooled to room temperature. The mass of the generated gases was calculated based on the measurements. The mass diagrams for the products obtained from series 1 and 2 are presented in Figure 3.

The amount of generated gases increased with temperature both during series 1 and 2. At the lowest temperatures, the amount of generated gases was very small. The process performed at lower temperatures is most likely to generate more oil and solid products. This may also suggest that at lower temperatures, not all the material was decomposed. Higher temperatures generate more gases and less oil and solid products.

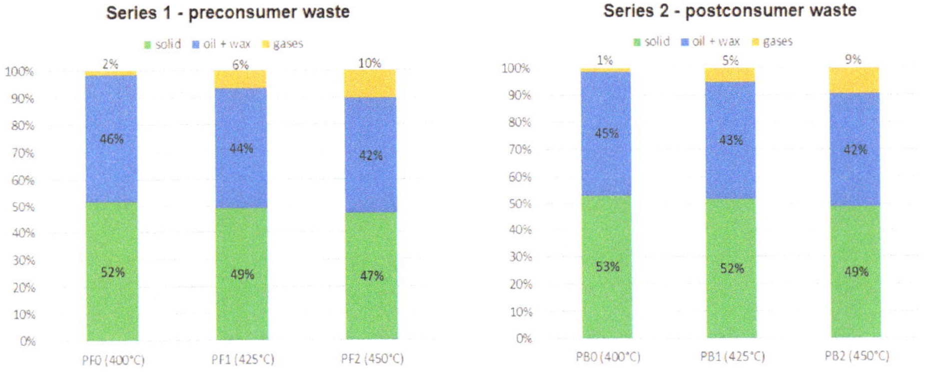

Figure 3. Masses of the products obtained during series 1 and 2.

Slightly more gases and liquid fraction have been generated during the pyrolysis of the preconsumer waste, while the postconsumer waste generated more solid products (aluminum and char). However, those differences were almost negligible. Therefore, there was almost no difference between the yields and the quality of the products generated during the pyrolysis of pre- and postconsumer waste. It seems that the trace amounts of pharmaceutical products left on the used blisters did not affect the process and the quality of the products.

The results of the gas analysis of two materials at 400 °C, 425 °C, and 450 °C are presented in Tables 2–7.

Table 2. Gas composition, preconsumer material, 400 °C.

Temp. (°C)	Hydrocarbons (wt%)									Inorganic Compounds (wt%)						Σ %
	CH_4	C_2H_6	C_2H_4	C_3H_8	C_3H_6	C_4H_{10}	C_4H_8	C_4H_6	C_{5+}	H_2	O_2	N_2	CO	CO_2	Other	
210	0.26	0.11	0.23	0.05	0.07	0.04	0.33	0.06	0.19	19.60	0.00	64.60	1.56	7.72	5.18	100
320	3.94	1.43	0.87	0.57	0.53	0.27	1.17	0.03	0.40	47.50	0.00	20.50	2.90	6.02	13.87	100
400	11.37	2.76	1.18	1.04	0.85	0.45	1.16	0.06	0.68	49.34	0.11	13.30	5.47	4.85	7.38	100

Table 3. Gas composition, preconsumer material, 425 °C.

Temp. (°C)	Hydrocarbons (wt%)									Inorganic Compounds (wt%)						Σ %
	CH_4	C_2H_6	C_2H_4	C_3H_8	C_3H_6	C_4H_{10}	C_4H_8	C_4H_6	C_{5+}	H_2	O_2	N_2	CO	CO_2	Other	
210	0.10	0.04	0.13	0.03	0.04	0.02	0.01	0.00	0.01	1.97	1.48	83.90	1.80	7.45	3.02	100
320	3.38	1.23	1.01	0.44	0.39	0.16	0.40	0.00	0.03	18.00	0.94	57.30	4.90	9.90	1.92	100
425	14.60	4.73	1.58	1.75	1.80	0.42	0.86	0.02	0.18	42.10	0.00	14.50	4.60	6.98	5.88	100

Table 4. Gas composition, preconsumer material, 450 °C.

Temp. (°C)	Hydrocarbons (wt%)									Inorganic Compounds (wt%)						Σ %
	CH_4	C_2H_6	C_2H_4	C_3H_8	C_3H_6	C_4H_{10}	C_4H_8	C_4H_6	C_{5+}	H_2	O_2	N_2	CO	CO_2	Other	
210	0.10	0.04	0.13	0.02	0.04	0.01	0.12	0.00	0.35	7.91	0.88	77.00	1.65	8.77	2.98	100
320	5.20	2.09	1.42	0.83	0.71	0.42	1.48	0.12	1.87	34.80	0.00	29.80	4.21	9.70	7.35	100
450	15.36	3.49	1.38	1.33	1.69	0.39	0.98	0.20	1.34	38.57	0.00	12.76	5.78	9.31	7.42	100

For series 1 (preconsumer waste), the concentration of hydrocarbons increased with the temperature. The total content of hydrocarbons at max. temperature 400 °C was 19.5%, 25.9% at 425°C, and 26.2% at 450°C. The oxygen content at the beginning of the process was neglectable. Nitrogen content in the gases decreases during the process. The amount of CO in the gases was higher at higher temperatures. Based on the results it can be concluded that the quality of the gas as a fuel increases with the process temperature.

Table 5. Gas composition, postconsumer material, 400 °C.

Temp. (°C)	Hydrocarbons (wt%)									Inorganic Compounds (wt%)						Σ %
	CH_4	C_2H_6	C_2H_4	C_3H_8	C_3H_6	C_4H_{10}	C_4H_8	C_4H_6	C_{5+}	H_2	O_2	N_2	CO	CO_2	Other	
210	0.23	0.12	0.14	0.05	0.07	0.02	0.00	0.13	0.00	5.73	0.96	82.80	3.02	3.26	3.47	100
320	2.89	1.05	0.63	0.43	0.48	0.20	0.72	0.21	0.19	32.60	0.00	40.60	3.90	11.20	4.90	100
400	7.61	2.43	1.27	0.99	1.08	0.41	2.18	0.34	0.11	46.60	0.00	10.70	6.66	12.50	7.12	100

Table 6. Gas composition, postconsumer material, 425 °C.

Temp. (°C)	Hydrocarbons (wt%)									Inorganic Compounds (wt%)						Σ %
	CH_4	C_2H_6	C_2H_4	C_3H_8	C_3H_6	C_4H_{10}	C_4H_8	C_4H_6	C_{5+}	H_2	O_2	N_2	CO	CO_2	Other	
210	0.19	0.09	0.13	0.04	0.04	0.03	0.19	0.01	0.05	10.00	0.20	74.10	1.82	9.71	3.40	100
320	1.86	0.66	0.47	0.27	0.27	0.14	0.68	0.08	0.32	29.90	0.40	47.00	3.11	9.98	4.86	100
425	10.91	3.57	1.40	1.33	1.41	0.52	1.96	0.28	1.58	41.85	0.05	13.70	6.51	11.14	3.79	100

Table 7. Gas composition, postconsumer material, 450°C.

Temp. (°C)	Hydrocarbons (wt%)									Inorganic Compounds (wt%)						Σ %
	CH_4	C_2H_6	C_2H_4	C_3H_8	C_3H_6	C_4H_{10}	C_4H_8	C_4H_6	C_{5+}	H_2	O_2	N_2	CO	CO_2	Other	
210	0.28	0.13	0.20	0.06	0.06	0.04	0.26	0.01	0.30	7.30	0.99	75.90	2.99	6.92	4.56	100
320	3.32	1.08	0.69	0.45	0.37	0.25	0.87	0.03	0.87	29.00	0.00	44.70	4.68	9.14	4.55	100
450	12.20	4.10	1.52	1.67	1.74	0.63	1.53	0.05	2.12	39.10	0.00	12.50	6.40	9.78	6.66	100

Similarly to what happened in series 1, in series 2 the content of hydrocarbons increased with the temperature. The concentration at max. temperature 400 °C was 16.4%, 23% at 425 °C, and 25.6% at 450 °C. The nitrogen content decreased during the process. For the mixed waste pharmaceutical blisters, the amount of generated CO at different max. temperatures was almost constant, while the CO_2 content increased. This may have been caused by the fact that the material contained trace amounts of pharmaceutical products that may react more quickly than plastic. Another possible explanation is that some oxygen may have penetrated into the gas cooler and reacted with other gas components.

The diagrams showing the changes in the contents of the selected gas compounds are presented in Figures 4–6. Tests PF0 and PB0 were performed at a maximum temperature of 400 °C. Tests PF1 and PB1 were performed at the max. temperature 425 °C, and tests PF2 and PB2 were performed at the max. temperature 450 °C.

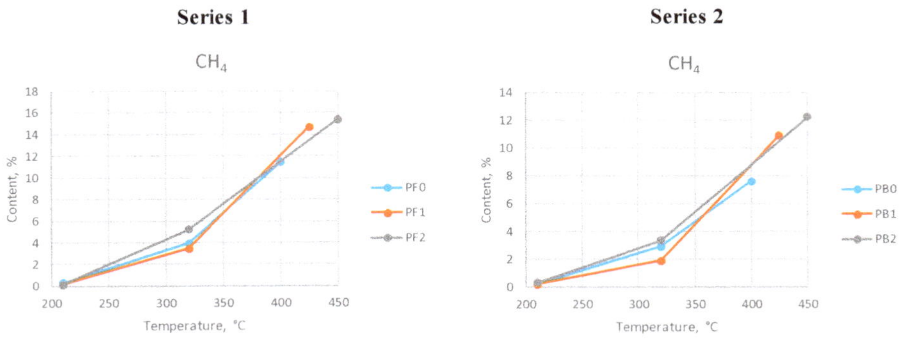

Figure 4. Methane content at different max. temperatures during series 1 and 2.

Methane is the main hydrocarbon generated during the process. Preconsumer waste generated more methane than postconsumer WPBs. The highest amount of this gas was generated at the highest temperature for both series.

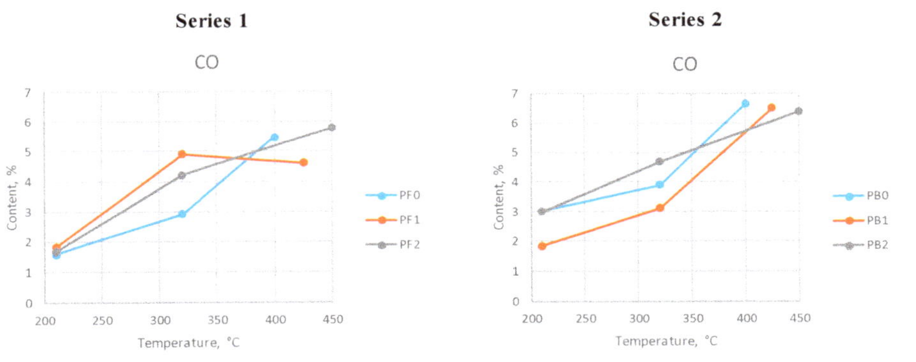

Figure 5. Carbon oxide content at different max. temperatures during series 1 and 2.

More carbon oxide was generated during series 2. The CO content increased at higher temperatures for all tests except in series 1 at 425 °C, where its concentration slightly decreased.

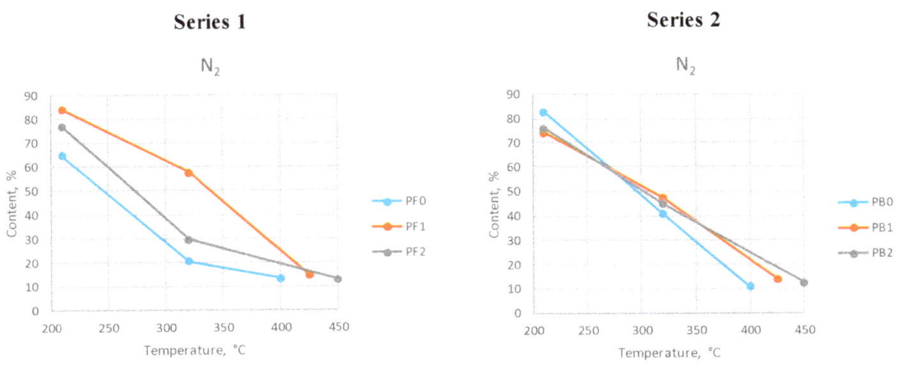

Figure 6. Nitrogen content at different max. temperatures during series 1 and 2.

The nitrogen content decreased with temperature during both series. The minimum level (approx. 10%) has been reached by the end of each process. The level of nitrogen was high at the beginning of the process, which was caused by the presence of this gas in the atmospheric air. As the process proceeded and the gases were generated, nitrogen was removed from the reactor.

In order to confirm that the gas quality as a fuel increases with the max. process temperature, the calorific value was calculated according to the equation:

$$CV = \sum_{i=1}^{n} \mu_i CV_i \tag{1}$$

where:

CV—calorific value of the gas generated during pyrolysis,
M—mass fraction of the gas component,
CV_i—calorific value of the gas component.

The calculated calorific values of the gases for series 1 and 2 at three temperatures—at the beginning of the process, during the process, and after reaching the maximum temperature—are presented in Tables 8 and 9, respectively.

Table 8. Calorific value of the process gases at different temperatures, series 1.

	Series 1		
	Calorific Value (MJ/m^3)		
	T_{max} = 400 °C	T_{max} = 425 °C	T_{max} = 450 °C
t_{start}	3.52	0.69	1.95
t_{mid}	11.70	6.63	15.04
t_{max}	17.33	19.39	20.14

Table 9. Calorific value of the process gases at different temperatures, series 2.

	Series 2		
	Calorific Value (MJ/m^3)		
	T_{max} = 400 °C	T_{max} = 425 °C	T_{max} = 450 °C
t_{start}	1.52	1.93	2.41
t_{mid}	8.55	7.04	9.48
t_{max}	16.42	20.58	21.96

The calorific value of the gases increased with the duration of the process and the maximum process temperature. The highest calorific values for both series could be observed at the maximum process temperatures. The highest value was obtained during the pyrolysis of mixed postconsumer waste. The highest quality of the gaseous products was observed for the tests performed at the highest maximum temperature on the postconsumer waste. Such results suggest that, when considering a potential technology for WPB processing with the production of a gaseous fuel, the most favorable conditions for the production of gases would be at higher temperatures. However, it needs to be stressed that this maximum calorific value cannot be maintained during the whole process. The process needs to be continuous in order to provide the highest calorific value possible. This would require a constant batching of the input material and a simultaneous removal of the solid products.

4. Conclusions

Due to the depleting sources of raw materials and fossil fuels, the recovery of material resources as well as energy is of high importance for the circular economy. Providing a high efficiency of material recovery as well as energy efficiency can help maintain the natural environment. The fractions present in multimaterial waste, due to their composition and production methods, cannot be easily separated only with the application of simple mechanical methods. Therefore, it is highly important to develop a technology for their processing. Pyrolysis is a thermal process that can be used to separate plastic from the metal fraction of waste pharmaceutical blisters in order to recover aluminum and generate gaseous fuel. Within the presented research, two types of waste pharmaceutical blisters were pyrolyzed at three maximum temperatures: 400 °C, 425 °C, and 450 °C. The gases used for the analysis were collected at the beginning and during the process and after the maximum process temperature was reached. The analysis of the chemical composition of the gases was performed with the application of chromatography.

A higher temperature favors the production of larger amounts of process gas rather than oil and wax. In addition, the calorific value of the gas at the maximum temperature 450 °C is higher (21.96 MJ/m^3) for the mixed postconsumer waste than for the clean preconsumer material (20.14 MJ/m^3).

One of the goals of the study was to determine whether the presence of trace amounts of medical products as well as the more complex character of the postconsumer material influence the process products. No significant differences in the process performed on the clean preconsumer waste and postconsumer waste contaminated with the trace amounts of pharmaceutical products have been observed. The results suggest that there is no necessity to wash the blisters before thermal process, since the impurities do not affect the quality of the gas. In order to provide the optimal conditions for the production of gaseous fuel, the process should be carried out at 450 °C.

Most EU-level documents show that there is a need for multisectoral research in value chains, going beyond waste management, the environment, and active building of consumer awareness toward the acceptance of a circular economy. Each element, which is part of the concept of qualitative recycling, directed at waste that is difficult to manage, will be an indispensable element in building a full circular economy [16–20]. The presented research results have the potential to be implemented and are part of the circular economy.

Author Contributions: Conceptualization, K.P., K.K., and W.Ś.; methodology, W.Ś.; validation, K.S.; investigation, K.K. and K.S.; resources, K.K. and K.P.; data curation, K.K., K.S., and M.B.; writing and original draft preparation, K.K.; writing, review, and editing, W.Ś. and M.B.; supervision, K.P. and W.Ś. All authors have read and agreed to the published version of the manuscript.

Funding: The study was funded by the Ministry of Science and Higher Education under the project "Charakterystyka produktów termicznego rozkładu wielomateriałowych odpadów opakowań po produktach farmaceutycznych (Characterization of the products of thermal decomposition of multimaterial pharmaceutical packaging waste)" and research-funded by subsidies granted for the year 2019 to the Department of Technology and Installations for Waste Management, Silesian University of Technology.

Conflicts of Interest: The authors declare no conflicts of interest.

References

1. Klejnowska, K.; Lewandowski, D. Aluminum Recovery From Multimaterial Tetra-Pak Waste Pyrolysis. *Civ. Env. Eng. Rep.* **2019**, *29*, 39–52. [CrossRef]
2. Wzorek, M.; Koziol, M.; Ścierski, W. Emission characteristics of granulated fuel produced from sewage sludge and coal slime. *J. Air. Waste Manag. Assoc.* **2010**. [CrossRef] [PubMed]
3. Rosik-Dulewska, C. *Podstawy Gospodarki Odpadami*; Wydawnictwo Naukowe PWN: Warszawa, Poland, 2007.
4. Landrat, M. Termiczne właściwości odpadowych tworzyw sztucznych. *Przem. Chem.* **2018**, *97*, 1496–1498. [CrossRef]
5. Wang, C.; Wang, H.; Liu, Y. Separation of aluminum and plastic by metallurgy method for recycling waste pharmaceutical blisters. *J. Clean. Prod.* **2015**, *102*, 378–383. [CrossRef]
6. Czop, M. Select polymer wastes as possible sources of solid recovered fuels. *Pol. J. Environ. Stud.* **2014**, *23*, 1369–1372.
7. Pilchik, R. Pharmaceutical Blister Packaging, Part I and II. *Pharm. Technol.* **2000**, *24*, 56–60.
8. Dobrucka, R. Recent trends in packaging systems for pharmaceutical products. *Sci. J. Logist.* **2014**, *10*, 393–398.

9. Klejnowska, K.; Pikoń, K.; Ścierski, W.; Zajączkowski, A. Termiczny rozkład wielomateriałowych odpadowych opakowań po produktach farmaceutycznych. *Przem. Chem.* **2019**, *98*, 1414–1416. [CrossRef]
10. Czop, M.; Pikoń, K.; Bogacka, M. Optimization of polyethylene waste utilization methods from disposable packaging. *Przem. Chem.* **2015**, *94*, 1503–1505.
11. Landrat, M. Possibilities of polyphenol floral foam waste utilization. *Przem. Chem.* **2017**, *96*, 1704–1706.
12. Piontek, W.; Jarzębowski, S. *Analiza Warunków Przetwarzania Odpadów Opakowaniowych Po Środkach Niebezpiecznych Oraz Odpadów Po Opakowaniach Wielomateriałowych w Polsce*; National Fund for Environmental Protection and Water Management: Warszawa, Poland, 2015.
13. Nawrocki, T. Blistry—Najpopularniejsze opakowania leków. *Aptekarz Pol.* **2014**, *2*, 30–32.
14. Yin, S.; Rajarao, R.; Gong, B.; Wang, Y.; Kong, C.; Sahajwalla, V. Thermo-delamination of metallised composite plastic: An innovative approach to generate Aluminium from packaging plastic waste. *J. Clean Prod.* **2019**, *211*, 321–329. [CrossRef]
15. Plis, A.; Lasek, J.; Skawińska, A.; Zuwała, J. Thermochemical and kinetic analysis of the pyrolysis process in Cladophora glomerata algae. *J. Anal. Appl. Pyrolysis.* **2015**. [CrossRef]
16. WRAP. Economic Growth Potential of More Circular Economies. 2015. Available online: https://www.researchgate.net/profile/Peter_Mitchell21/publication/284187253_ECONOMIC_GROWTH_POTENTIAL_OF_MORE_CIRCULAR_ECONOMIES/links/564f40fb08ae1ef9296e83f3.pdf (accessed on 1 December 2019).
17. IAI. *Global Aluminium Recycling: A Cornerstone of Sustainable Development*; International Aluminium Institute: London, UK, 2009.
18. EEA. *Intensified Global Competition for Resources (GMT 7)*; The European Environment Agency: Luxembourg, 2017.
19. EEA. Report No 6/2017. In *Circular by Design. Products in the Circular Economy*; European Environment Agnecy: Luxembourg, 2017.
20. EEA. Report No 2/2016, Directive 2008/98/EC of the European, Parliament and of the Council of 19 November 2008, on waste and repealing certain Directives. In *Circular Economy in Europe Developing the Knowledge Base EU*; European Environment Agnecy: Luxembourg, 2008; pp. 3–30.

applied sciences

MDPI

Article

The Investigation of Key Factors in Polypropylene Extrusion Molding Production Quality

Dyi-Cheng Chen, Der-Fa Chen *, Shih-Ming Huang and Wen-Jye Shyr *

Department of Industrial Education and Technology, National Changhua University of Education, No. 1, Jin-De Road, Changhua 500, Taiwan; dcchen@cc.ncue.edu.tw (D.-C.C.); hsm2014y@gmail.com (S.-M.H.)
* Correspondence: dfchen@cc.ncue.edu.tw (D.-F.C.); shyrwj@cc.ncue.edu.tw (W.-J.S.)

Abstract: This study took food-grade polypropylene packaging products as the research project and discussed how to control the polypropylene extrusion sheet thickness and vacuum thermoforming quality and weight. The research objective was to find the key factors for reducing costs and energy consumption. The key aspects that may influence the polypropylene extrusion molding quality control were analyzed using literature and in-depth interviews with scholars and experts. These four main aspects are (1) key factors of polypropylene extrusion sheet production, (2) key factors of the extrusion line design, (3) key factors of polypropylene forming and mold manufacturing, and (4) key factors of mold and thermoforming line equipment design. These were revised and complemented by the scholar and expert group. There are 49 subitems for discussion. Thirteen scholars and experts were invited to use qualitative and quantitative research methods. A Delphi questionnaire survey team was organized to perform three Delphi questionnaire interviews. The statistical analyses of encoded data such as the mean (M), mode (Mo), and standard deviation (SD) of various survey options were calculated. Seeking a more cautious research theory and result, the K-S simple sample test was used to review the fitness and consistency of the scholars' and experts' opinions on key subitem factors. There are ten key factors in the production quality, including "A. Main screw pressure", "B. Polymer temperature", "C. T-die lips adjustment thickness", "D. Cooling rolls pressing stability", "E. Cooling rolls temperature stability", "F. Extruder main screw geometric design", "G. Heating controller is stable", "H. Thermostatic control", "I. Vacuum pressure", and "J. Mold forming area design". The key factors are not just applicable to classical polypropylene extrusion sheet and thermoforming production but also to related process of extrusion and thermoforming techniques in expanded polypropylene (EPP) sheets and polylactic acid (PLA). This study aims to provide a key technical reference for enterprises to improve quality to enhance the competitiveness of products, reduce production costs, and achieve sustainable development, energy savings, and carbon reductions.

Keywords: Delphi technique; extrusion molding; key factors; polypropylene; production quality

Citation: Chen, D.-C.; Chen, D.-F.; Huang, S.-M.; Shyr, W.-J. The Investigation of Key Factors in Polypropylene Extrusion Molding Production Quality. *Appl. Sci.* **2022**, *12*, 5122. https://doi.org/10.3390/app12105122

Academic Editor: Sergio Torres-Giner

Received: 23 April 2022
Accepted: 18 May 2022
Published: 19 May 2022

1. Introduction

The prevalence of COVID-19 has severely impacted economic activities and common people's lives worldwide [1]. The anti-epidemic measures have made many restaurants use portable food packages. When the epidemic situation became severe in 2021, many restaurants or shops stopped their dine-in services for epidemic control. Some stores went further to provide takeaway in disposable food packaging containers only. They refused to put food in the environmentally friendly containers of the customers. The epidemic prevention and environmental protection issues fall into a dilemma [2].

Most food packaging containers in the world are plastic and composite paper packaging materials. The annual consumption of different kinds of beverage paper cups in Taiwan is estimated at more than 2 billion pieces. According to the investigation of Greenpeace International in 2021 [3], it is hard to estimate the quantity increase during the COVID-19 pandemic. The issue of plastic reduction has been the trend and consensus of environmental protection globally [4]. In 2019, the European Union appealed to reduce the use of single-use plastics (SUP). They proposed the single-use plastics-related policy (EU) 2019/904 and SUP instruction. The EU required the member states to prohibit using some single-use plastics products since July 2021, which were generally regarded as the most familiar marine garbage [5], mostly dinnerware and beverage packing containers. To reduce pollution in the transition period, scientific methods are required to control the source, increase manufacturing efficiency, improve quality, and reduce the weight of products. There are many environmentally friendly materials for substituting plastic raw materials, especially the alternative products of biomass material [6] or degradable plastics [7]. However, there are problems in sanitary preservation and the pollutant plastic recycling system to be overcome, for example, using the biomass material polylactic acid (PLA) [8] to replace some plastic packaging materials for food. The replacement of material is not a radical solution. It can be confused with the existing plastics recycling system in Europe, e.g., polyethylene terephthalate (PET), resulting in more pollution and losses.

This study suggested that environmentally friendly disposable food packaging containers should be a simple material, recyclable, or biodegradable. In compliance with the global trend of plastic reduction [9], the UK government collects a plastic packaging tax. It implements the deposit return scheme (DRS), which directly collects petty environmental protection deposits from beverage manufacturers while encouraging them to be involved in recycling. As the name suggests, extended producer responsibility (EPR) extends the producers' responsibility [10], meaning that relevant manufacturing enterprises bear all expenses related to recycling and processing packages. To ensure the recycling of more plastic products is a scientific and feasible method, Lavoisier, a French chemist, proposed the "law of conservation of matter" in 1787 [11]. He stated that industrial products consume energy, and the environment will be inevitably polluted. The merits and demerits should be discussed using scientific and objective methods. How to reduce the resource consumption and pollution resulting from developing and manufacturing plastic packaging containers is the actual aspect conforming to the "plastic reduction" reality. This study discusses assisting enterprises to develop more conveniently recyclable and environmentally friendly plastic products, enhancing key technologies and manufacturing machines and molds for high-efficiency bulk production, meeting the demand for market competition cost and finding quality improvement methods.

The current recyclable materials development focus in Europe is mainly on polypropylene (PP) [12]. It is characterized by a higher impact resistance, robust mechanical properties, resistance to multiple organic solvents, and resistance to acid/alkaline corrosion. It is extensively used in food containers (routine food containers), such as meal boxes and beverage cups. The food-grade PP container is free of oral toxicity, and it will not increase the probability of getting cancer. The productivity of PP is expected to be increased by almost six times in 2024. Therefore, PP is a relatively ideal and environmentally friendly and recyclable food packing material among the present plastics [13].

This study used food-grade PP as a raw material in the production of extrusion sheets and food packaging containers of vacuum thermoforming and discussed how to control quality factors. The key factors of production quality can be obtained from the research. The results are analyzed by sophisticated PP production techniques and manufacturing machines and molds. This provides a reference for academic discussion or the development of automatic intelligent machines and technical enhancement. The intelligent automated machine production and quality control can obtain some suggestions or conclusions. The quality of recyclable plastic products can be further upgraded, and the enterprise cost can be reduced. This can further help environmental protection, energy savings, carbon reductions, and sustainable operations.

The purposes of this study are concluded as follows: (1) to probe into the key technology of producing PP extrusion sheets and molded products, to find the important influencing factors in stable production quality, to provide some suggestions or opinions, and to provide references for the researchers and managers of related production techniques; (2) to provide some key advantageous factors in designing and manufacturing special machinery equipment and molds, to provide innovative directions for business circles to improve and develop intelligent machinery and mold equipment; and (3) to discuss the reference indicators of upgrading quality and reducing the production cost for enterprise managers to improve technology and business performance.

2. Literature Review

2.1. Polypropylene

Polypropylene [14] is derived from the reaction of petrochemical propylene gas with a density of 0.89–0.94 g/cm^3. It is a lightweight plastic material with a semitransparent white color and has a lower cost than other plastic materials. It is also one of the recyclable plastic materials. In 1954, Italian chemist Giulio Natta successfully synthesized low-molecular-weight PP [15]. PP is a thermoplastic that is polymerized of propylene monomers, which are resistant to −20–120 °C. Its melting point is 145–162 °C, which is higher than HDPE [16]. Food packaging containers made of PP can be sterilized by steam. It has higher impact resistance and is resistant to multiple kinds of organic solvents and acid/alkaline corrosion. PP is the second largest polymer plastic in Europe. It accounted for 19.1% of the plastic output of the European Union in 2015 [17] and accounts for 21% of global production output [18]. PP is resistant to acid, alkali, and high temperatures (100–120 °C). Therefore, it is extensively used for food packaging containers. Food manufacturers pay attention to packages' thermal performance or temperature requirements [19]. They generally use PP plastic raw material containers to protect the consumers' food safety and health. Due to the COVID-19 pandemic, wearing a mask has become a basic consensus in epidemic prevention worldwide [20]. The primary material of non-woven fabrics for masks is PP, and one mask uses about 4 g of PP raw material [21]. Medical masks and protective clothing are closely related to PP.

The sample of the purpose for this study is producing coffee cup lids of polypropylene material, as shown in Figure 1. The features of the chosen material are different, as shown in Table 1. Although it is general purpose, the object uses polypropylene homopolymer as a material with several features [22]: (1) high transparency, (2) good moldability, (3) high stiffness, and (4) high heat deflection temperature. Recently, newly developed expanded polypropylene (EPP) sheets, in the place of expanded polystyrene (EPS), are used generally in the automobile industry [23]. Furthermore, although the additives or agents in the extrusion process for expanded or non-expanded polypropylene sheets have different uses [24], the key factor of product quality control stays the same. Similarly, the processes of vacuum forming and mold design for both are different as well. Expanded polypropylene sheets for food packaging can achieve less material usage, better isolation purpose, and higher heat resistance under the slight changing in the parameter setting on the basis of their different shrinkage ratio. As long as changing the extruded die and cooling mandrel, etc. is a key factor, it can also achieve the good result in production quality control to use isobutane or CO_2 as a blowing agent injected in the extruder to increase the foaming ratio under the physical foaming process, as shown in Figure 2 [25]. The study can correspond to the key factor in the quality of thermoforming processing. Many manufacturers stopped producing foamed (Styrofoam) food containers due to the prohibited policy in Taiwan in July 2022, so no further explanation in expanded polypropylene was needed.

Figure 1. Polypropylene coffee lid with molds.

Table 1. Properties of raw materials for producing polypropylene coffee cup lids.

Selection of Reference Values for Raw Material Properties			
Typical Property	**Test Method**	**Unit**	**Reference Value**
Melt flow rate (230 °C, 2.16 kg)	ASTM D1238	g/10 min	1.6 ± 0.1
Density	ASTM D792	g/cm^3	0.901 ± 0.001
Tensile strength at yield	ASTM D638	kg/cm^2	370 ± 10
Flexural modulus	ASTM D790	kg/cm^2	15,500 ± 1000
Rockwell hardness	ASTM D785	R scale	100 ± 5
Heat deflection temperature (4.6 kg/cm^2)	ASTM D648	°C	105 ± 2
Izod impact strength, notched 23 °C	ASTM D256	kg-cm/cm	5.0 ± 0.5
Mold shrinkage	ASTM D955	%	1.5 ± 0.1

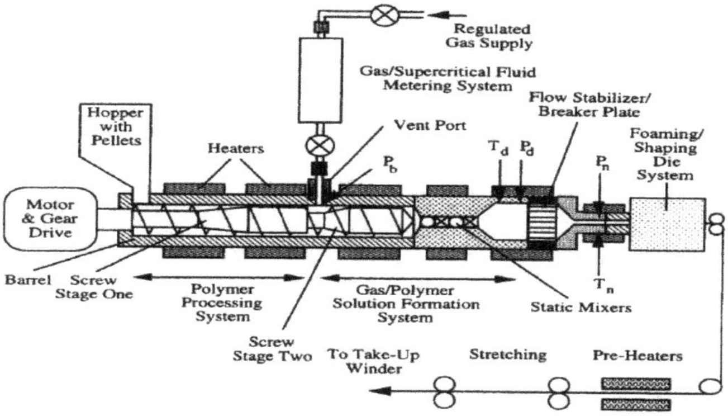

Figure 2. The schematic of polypropylene physical foaming process (Reprinted with permission from Ref. [25]. Copyright 2022, Chen and Lin).

2.2. Extrusion Molding

Extrusion vacuum forming [26,27] is a standard plastic heating method. The granular or sheet polymer material of a plastic fiber is heated and melted by an extruder [28]. After the extrusion sheet is obtained, the plate [29] or sheet skin material roll of the melt is continuously heated by a mold [30] or molding module or vacuum pressed [31]. After cooling modeling, the plate or sheet skin is then cut into plastic products.

In recent years, constant innovative techniques to develop special-purpose machines with higher demand in the market for food containers continuously keeps going. The machine, as shown in Figure 3 [32], for mass production in extrusion in-line with a thermoforming system reduces the heating time in the heat tunnel before sheets in the forming process to raise the cycle time, increase efficiency, and decrease energy consumption in manufacturing, leading to cost savings. In the meantime, intelligent control systems and robotic arms conducted in automatic machinery lessen the risk of operation error and manual movement. This achieves manpower savings, increases operator safety and reduces the risk of accident. Currently the product process of a polypropylene in-line system is extrusion connected with a thermoforming machine, such as a cut-in-mold thermoforming machine, and further in connection with an automatic regrind recycling system, as shown in Figure 4 [33].

Figure 3. Full-automatic polypropylene extrusion line (Reprinted with permission from Ref. [32]. Copyright 2022, Sunwell Global).

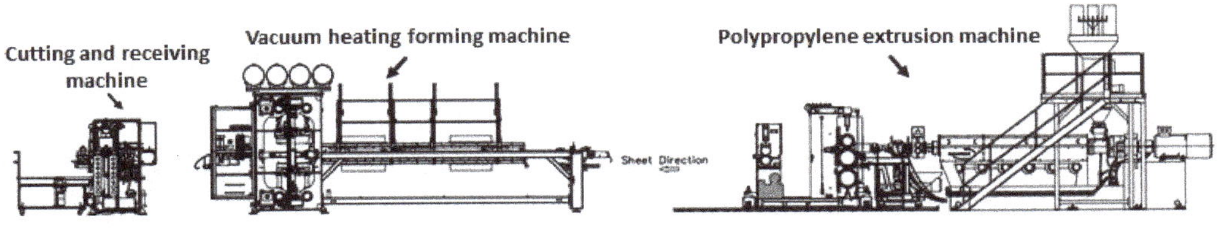

Figure 4. The schematic of polypropylene extrusion and thermoforming in-line system (Reprinted with permission from Ref. [33]. Copyright 2022, Sunwell Global).

2.3. Key Success Factors

Professional quality controllers and production technologists were interviewed using a half-structured method to discuss the key factors in extrusion molding quality [34,35]. The key factor data were judged and collected according to the experience in real production and inspection. The data were scientifically analyzed to determine specific key success factors and obstacles [36]. Four main aspects were induced. The questionnaire content contained 49 subitem key factors, and the key success factors in extrusion molding quality were discussed.

2.4. Delphi Technology

The Delphi technology [37] is used in different areas of academic research [38], including curriculum planning and capability indexing. It can probe into or understand the assumptions and related decisions of many disciplines. If the background knowledge of Delphi technology is used, 10 to 15 experts and scholars for concluding the consensus of Delphi technology can meet the quantity standard of experts and scholars. Through repeated written discussion by anonymous and structured group interviews [39], the scholars and experts establish a consistent consensus with their expertise or technical experience and opinions to solve complex issues [40]. Anonymity is required in the Delphi interview process to avoid the complex factors of interpersonal communication disturbing objective respondence and avoid the psychological factor of authoritarian submission or bandwagon effect induced by leaders or senior executives of the group [41]. The anonymity enables expert group members to express their real opinions further and leads to different levels of considerations. The experts' opinions are fairer. Murry and Hammons [42] indicated that the result of experts' collective discussion and shared decision making should be more comprehensive than the conclusion of individual thinking. The expert group majorly consists of professors and scholars in plastic technique, including two university professors in chemical engineering, 11 scholars in mechanical design in enterprises, and polypropylene product engineers in manufacturing. The Delphi technique group is composed of method analysis scholars including four factory managers and eleven senior engineers with 21 years of experience. The Delphi technique members mainly work in mechanical manufacturing-related or research and development-related plastic extrusion thermoforming processing. These experts were chosen from homogeneous fields of experience and technique with less error and higher reliability.

3. Methodology

The opinions of experts and scholars of Delphi are evaluated according to the five-point Likert scale [43]. The scale represents 5 as very important, 4 as important, 3 as normal, 2 as unimportant, and 1 as very unimportant.

3.1. Research Structure and Method

This study was divided into in-depth interviews with scholars and experts and structured expert questionnaire interviews of the Delphi technique [44]. The selected experts were interviewed with the Delphi questionnaire three times [45]. The key success factors in PP extrusion molding quality were induced by statistical analysis.

3.1.1. Research Methods

The research methods included qualitative and quantitative research [46]. The qualitative research integrated the scholars' and experts' opinions from in-depth interviews through semi-structured interviews. The first Delphi research questionnaire was discussed. The quantitative research selected the structured questionnaire of Delphi for three interviews. A total of four key success factors were developed as the research structure and discussed. After interviews, the experts' opinions were collected to induce the result shown in Figure 5.

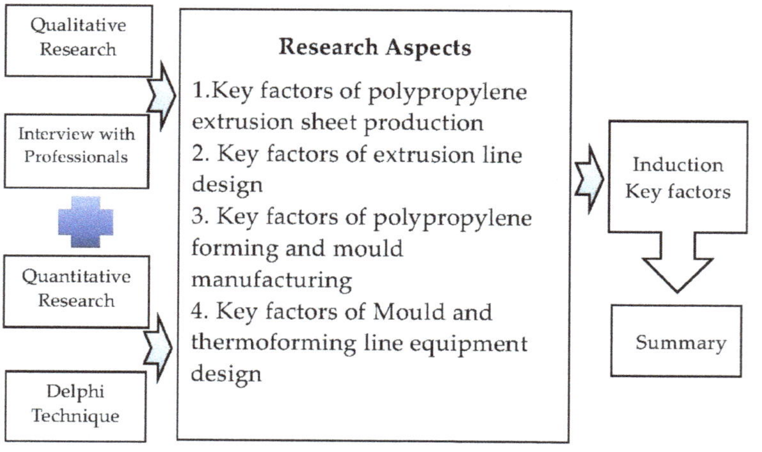

Figure 5. Research structure.

3.1.2. Process of Delphi Technique

The process of three questionnaires of the Delphi technique is shown in Figure 6.

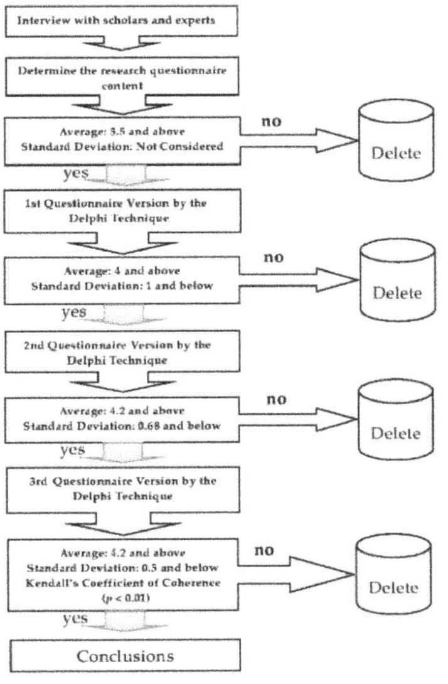

Figure 6. Steps to prepare the questionnaire by the Delphi technique.

3.2. Delphi Research Method and Design

This study established the Delphi questionnaire aspect content. The interview subject team comprised three categories of professionals: (1) scholars of chemical plastic production, (2) senior engineers developing special machines, and (3) technical executives of production manufacturing. Discussions were made according to their expertise, and the Delphi research questionnaire with the four aspects of specific key quality factors and 49 subitems was formulated.

According to the expert interview team's opinion, the Delphi research questionnaire had 13 subitems in dimension 1, "key factors in PP extrusion sheet production technique" [47], as shown in Table 2.

Table 2. Key factors of polypropylene extrusion sheet production (13 subitems).

1-1 The ratio of PP virgin pellets and reclaimed pellets
1-2 PP extruder main screw output pressure control
1-3 PP extruder polymer temperature control
1-4 T-die lips adjustment depends on sheet thickness
1-5 Cooling rolls pressing stability
1-6 Cooling rolls flatness
1-7 Cooling rolls temperature stability
1-8 T-die output and cooling rolls pressing position
1-9 Processing according to extrusion and raw material characteristics
1-10 PP extruder output (experts complement)
1-11 PP sheet thickness (experts complement)
1-12 PP sheet width (experts complement)
1-13 PP extruder main screw geometric design (experts complement)

According to the expert interview team's opinion, there were 12 subitems in dimension 2, "extrusion sheet manufacturing machinery equipment design", of the Delphi research questionnaire [48], as shown in Table 3.

Table 3. Key factors of extrusion line design (12 subitems).

2-1 Extruder main motor coordinate with main screw output
2-2 Gearbox coordinate with main screw output
2-3 Extruder main screw geometric design
2-4 The relationship between the metering pump and main screw output
2-5 The relationship between the metering pump and motor reducer
2-6 Screen changer and main screw diameter output
2-7 The max./min. width of the sheet and die design
2-8 The maximum sheet width and cooling rolls design
2-9 The maximum output of the screw and the cooling design of the cooling rolls
2-10 Cooling rolls and motor reducer design
2-11 Cooling rolls and cooling mechanism strength design
2-12 Maximum sheet width/maximum winding roll diameter and winding mechanism

According to the expert interview team's opinion, there were 12 subitems in dimension 3, "key factors in PP molding [49] and mold manufacturing [50]", of the Delphi research questionnaire, as shown in Table 4.

Table 4. Key factors of polypropylene forming and mold manufacturing (12 subitems).

3-1 Material of vacuum aluminum alloy forming mold
3-2 Requirements for mold vacuum holes and pressure holes
3-3 Feeding stability of forming heater
3-4 Core/cavity and product modeling design
3-5 Forming heating controller element stability
3-6 The deviation of forming heating constant temperature control area
3-7 Near scenic sport or night markets
3-8 Collective residential community, enterprise office building, or factory concentration area
3-9 Mold cooling mode
3-10 Mold clamping to prevent leakage
3-11 Product geometric design
3-12 Product and mold cutter shrinkage

According to the expert interview team's opinion, there were 12 subitems in dimension 4, "mold and molding trimmer equipment design", of the Delphi research questionnaire, as shown in Table 5.

Table 5. Key factors of mold and thermoforming line equipment design (12 subitems).

4-1 Depends on the clamping force of the thermoformer
4-2 The product and mold maximum layout design arrangement
4-3 The maximum forming height of the thermoformer and product mold design
4-4 The maximum trimming width of the trimmer and product mold design
4-5 The maximum clamping force of the thermoformer and mold forming area design
4-6 Thermoformer motor education ratio
4-7 Trimmer motor reducer and trimming mold weight cutting force
4-8 Trimmer platen opening and trimming mold
4-9 Maximum product depth and trimmer opening
4-10 Thermoforming mold and thermoformer platen center positioning clamping
4-11 Trimming mold and trimmer platen center positioning clamping
4-12 Thermoformer cooling pump flowrate

3.3. The Statistical Analysis

For the statistical analysis of the interviews with scholars and experts [51], besides the in-depth interviews with two professor scholars, there were 13 effective questionnaires of interviews with experts of the circle. The statistical results of the first Delphi questionnaire are as follows: (1) the mode was above 4, (2) the high fitness selection criteria were adopted with the mean $\geq$4, and (3) the consistency selection standard deviation was $\leq$1. Forty subitems met the above three requirements, which were kept to establish the second Delphi research questionnaire. According to the statistical result of the second Delphi research questionnaire, the high fitness selection criteria were employed with a mean of $\geq$4.2, and the consistency selection standard deviation was $\leq$0.68. Sixteen subitems meeting the requirements were kept to establish the third Delphi research questionnaire. A statistical analysis of the key subitems of the third questionnaire was conducted via a Kolmogorov–Smirnov one-sample test. The statistical analysis result shows that the expert opinions on each subitem were significant (* $p < 0.05$). This means that the third questionnaire corrected by Delphi was more reliable and valid than the first one. The mean, mode, and standard deviation of the subitems, the Kolmogorov–Smirnov one-sample test, and progressive significance results are shown in Table 6.

Table 6. Statistical analysis of the third Delphi questionnaire.

No.	Item	Mo	M	SD	K–S Z-Test	Choice
1. Key Factors of Polypropylene Extrusion Sheet Production						
1–2	PP extruder main screw output pressure control	5	4.54	0.499	1.941 **	Keep
1–3	PP extruder polymer temperature control	4	4.46	0.499	1.941 **	Keep
1–4	T-die lips adjustment depends on sheet thickness	4	4.38	0.487	2.219 **	Keep
1–5	Cooling rolls pressing stability	4	4.38	0.487	2.219 **	Keep
1–6	Cooling rolls flatness	4	4.23	0.576	1.525 *	Delete
1–7	Cooling rolls temperature stability	4	4.31	0.462	2.496 **	Keep
1–12	PP sheet width (experts complement)	4	4.07	0.615	1.248	Delete

Table 6. *Cont.*

No.	Item	Mo	M	SD	K–S Z-Test	Choice
2. Key Factors of the Extrusion Line Design						
2–3	Extruder main screw geometric design	4	4.23	0.421	2.774 **	Keep
2–4	The relationship between the metering pump and main screw output	4	4.23	0.576	1.525 *	Delete
2–9	The maximum output of the screw and the cooling design of the cooling rolls	4	4.08	0.474	1.525 *	Delete
3. Key Factors of Polypropylene Forming and Mold Manufacturing						
3-4	Core/cavity and product modeling design	4	4.23	0.576	1.525 *	Delete
3-5	Forming heating controller element stability	4	4.46	0.499	1.941 **	Keep
3-6	The deviation of forming heating constant temperature control area	4	4.23	0.421	2.774 **	Keep
3-7	Near to scenic sport or night markets	4	4.46	0.499	1.941 **	Keep
4. Key Factors of Mold and Thermoforming Line Equipment Design						
4-5	The maximum clamping force of the thermoformer and mold forming area design	4	4.23	0.421	2.774 **	Keep
4-9	Maximum product depth and trimmer opening	4	4.15	0.533	1.525 *	Delete

* $p < 0.05$, ** $p < 0.01$.

4. Results

From September 2021 to January 2022, three Delphi structured questionnaire interviews were performed. The statistical results of the third Delphi questionnaire revealed that (1) the high fitness selection criteria was employed with the mean of ≥4.2, (2) the consistency selection standard deviation was ≤0.5, (3) the consensus of the K-S test has reached consistency, and (4) the subitems of progressive significance have reached the significance level of ** $p < 0.01$. The subitems would be kept only when the above four requirements were met. Based on the final statistical analysis result of the third Delphi research questionnaire, the subitems of ten key control factors in production quality were obtained, as shown in Table 7.

Table 7. Statistical analysis of the final Delphi technique.

No.	Item	Mo	M	SD	K–S Z-Test	Choice
1. Key Factors of Polypropylene Extrusion Sheet Production						
1–2	PP extruder main screw output pressure control	5	4.54	0.499	1.941 **	Keep
1–3	PP extruder polymer temperature control	4	4.46	0.499	1.941 **	Keep
1–4	T-die lips adjustment depends on sheet thickness	4	4.38	0.487	2.219 **	Keep
1–5	Cooling rolls pressing stability	4	4.38	0.487	2.219 **	Keep
1–7	Cooling rolls temperature stability	4	4.31	0.462	2.496 **	Keep
2. Key Factors of the Extrusion Line Design						
2–3	Extruder main screw geometric design	4	4.23	0.421	2.774 **	Keep

Table 7. *Cont.*

No.	Item	Mo	M	SD	K–S Z-Test	Choice
3. Key Factors of Polypropylene Forming and Mold Manufacturing						
3-5	Forming heating controller element stability	4	4.46	0.499	1.941 **	Keep
3-6	The deviation of forming heating constant temperature control area	4	4.23	0.421	2.774 **	Keep
3-7	Near to scenic sport or night markets	4	4.46	0.499	1.941 **	Keep
4. Key Factors of Mold and Thermoforming Line Equipment Design						
4-5	The maximum clamping force of the thermoformer and mold forming area design	4	4.23	0.421	2.774 **	Keep

** $p < 0.01$.

4.1. Test Key Factors in Production Quality

The 16 key factors in the third Delphi research questionnaire were analyzed and verified by the Kolmogorov–Smirnov Z test. When the extreme differences and progressive significance (p) were analyzed, it was found that the test result matched the progressive significance of the two-tailed $p < 0.01$ test. The analysis result indicated ten key subitems of quality. The key factors for their being retained in the questionnaire were the consensus of scholars and experts who considered these subitems the most critical factors in extrusion molding quality.

4.2. The Delphi Research Analysis Result

According to the Delphi research analysis result, ten key success factor items were induced from the four main aspects of key factors in PP extrusion molding production quality. A fish-bone diagram was used [52] to illustrate these factors in PP extrusion molding production quality, as shown in Figure 7.

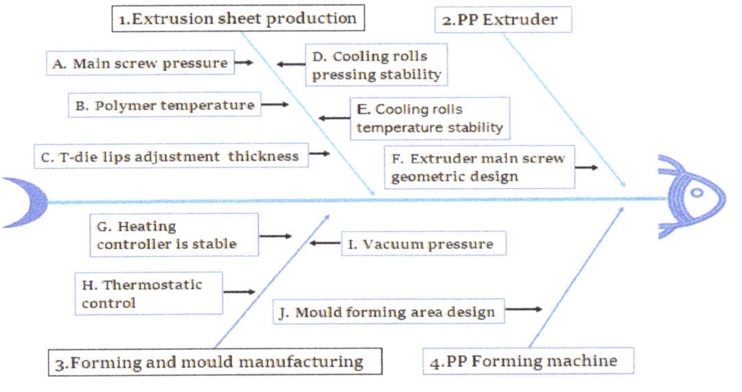

Figure 7. Important key factors affecting the quality of polypropylene extrusion molding.

5. Conclusions

According to the statistical results, the key factors in PP extrusion molding quality control had the following ten key subitems: (1-2) PP extrusion driving screw feed pressure control (correction: PP extrusion driving screw discharge pressure control), abbreviated as A. Main screw pressure; (1-3) PP extrusion resin temperature change control, abbreviated as B. Polymer temperature; (1-4) T-die lips adjustment sheet stability (correction: die lips adjustment depends on sheet thickness), abbreviated as C. T-die lips adjustment thickness; (1-5) mirror wheel pressing stability, abbreviated as D. Cooling rolls pressing stability;

(1-7) mirror wheel temperature stability, abbreviated as E. Cooling rolls temperature stability; (2-3) extrusion driving screw geometric design, abbreviated as F. Extruder main screw geometric design; (3-5) molding heating controller element stability, abbreviated as G. Heating controller is stable; (3-6) molding heating thermostatic control area error value, abbreviated as H. Thermostatic control; (3-7) molding vacuum and compressed air system stability, abbreviated as I. Vacuum pressure; and (4-5) molding machine maximum clamping force and molding area relationship design, abbreviated as J. Mold forming area design. Therefore, the key factors in the PP extrusion molding production quality are abbreviated to "A. Main screw pressure", "B. Polymer temperature", "C. T-die lips adjustment thickness", "D. Cooling rolls pressing stability", "E. Cooling rolls temperature stability", "F. Extruder main screw geometric design", "G. Heating controller is stable", "H. Thermostatic control, "I. Vacuum pressure", and "J. Mold forming area design", totaling ten key control factors in quality. The said key quality factors had a few differences in setting the technical parameters of real production due to different specifications of manufacturing machines and molds. Through the manufacturing and production of S company's intelligent modern extrusion molding machine, with the chain restaurant McDonald's drinks the cup lid as a production case, its production efficiency is more than 1.5 times that of injection molding products, and the production cost is only injection molding. The market price of molded products is about 60%. At present, the innovative and optimized automatic machinery and production technology are not comparable to the general traditional semi-automatic extrusion and cutting molding machines. According to the product weight, the plate thickness is about 0.45 mm~0.65 mm, corresponding to the key control factors of fishbone diagram quality: "A. Main screw pressure control: Speed set at 480~500 rpm", "B. Polymer temperature control at 200~240 °C", "C. T-die lips adjustment thickness is controlled at 0.45~0.69 mm ", "D. Cooling rolls pressing stability pressing control at ±0.05 mm ", "E. Cooling rolls temperature stability is controlled at 20~45 °C", "F. Extruder main screw geometric design adopts American style. Standard", "G. Heating controller is stable controller at 230~250 °C", "H. Thermostatic control error of molding heating ±1 °C", "I. Stability of molding vacuum pressure ±0.01 MPa", "J. Mold forming area design between the mold and the area is designed from 60% to 90% of the size". In the production of food packaging, to reach a good result in production quality control by using isobutane or CO_2 as a blowing agent injected in the extruder to increase the foaming ratio under the physical foaming process can result in less material usage, better isolation purpose, and higher heat resistance, as long as the extruded die and cooling mandrel, etc., are changed. The study can correspond to the key factor in the quality of thermoforming processing. It is confirmed in practical manufacture processing in the experimental factory for an obvious quality and energy-saving improvement. The production processing of the regrind recycling system is assisted in conveying scrapped material from the edge trimmer or crusher back to the extruder loading system in collocating with the automatic volumetric dosing device. The returned recycled material mixes with new material. At the present, the extruder and thermoforming system with precise vacuum forming molds is intelligent automatic machinery. The division of management and procurement are most concerned about having reliable quality and less error. As there are few related literature reviews of the Delphi technique, this study applied science to actual operators engaging in plastic production or clients with plain and specific key factors or data. Worldwide polypropylene sheet production is over 30 million metric tons. The key to the market competition is the condition that the percentage raises up in 1–2% for certified products and regrind recycling material to produce polypropylene coffee cups.

Nonetheless, the findings indicated that the manufacturing quality and energy savings could be improved effectively in the experimental plant. The study findings are expected to improve intelligent manufacturing and quality for production engineers and researchers, enhance the market competitiveness of products, and reduce enterprise management costs. As a result, the findings may contribute to achieving sustainable development, environmental protection, energy savings, and carbon reductions.

Author Contributions: All authors contributed meaningfully to this study. D.-F.C. and S.-M.H., research topic; D.-C.C., S.-M.H. and D.-F.C., data acquisition and analysis; D.-F.C., S.-M.H. and W.-J.S., methodology support; S.-M.H., D.-C.C. and W.-J.S., original draft preparation; D.-F.C. and W.-J.S., writing—review and editing. All authors have read and agreed to the published version of the manuscript.

Funding: This research received no external funding.

Institutional Review Board Statement: Not applicable.

Informed Consent Statement: Not applicable.

Data Availability Statement: Not applicable.

Conflicts of Interest: The authors declare no conflict of interest.

References

1. Guan, W.J.; Ni, Z.Y.; Hu, Y.; Liang, W.H.; Ou, C.Q.; He, J.X.; Liu, L.; Shan, H.; Lei, C.L.; Hui, D.S. Clinical characteristics of coronavirus disease 2019 in China. *N. Engl. J. Med.* **2020**, *382*, 1708–1720. [CrossRef] [PubMed]
2. da Costa, J.P. The 2019 global pandemic and plastic pollution prevention measures: Playing catch-up. *Sci. Total Environ.* **2021**, *774*, 145806. [CrossRef]
3. Walther, B.A.; Yen, N.; Hu, C.S. Strategies, actions, and policies by Taiwan's ENGOs, media, and government to reduce plastic use and marine plastic pollution. *Mar. Policy* **2021**, *126*, 104391. [CrossRef]
4. Clayton, C.A.B. Building Collective Ownership of Single-Use Plastic Waste in Youth Communities: A Jamaican Case Study. *Soc. Sci.* **2021**, *10*, 412. [CrossRef]
5. Elliott, T.; Gillie, H.; Thomson, A. European Union's plastic strategy and an impact assessment of the proposed directive on tackling single-use plastics items. *Plast. Waste Recycl.* **2020**, 601–633. [CrossRef]
6. Velvizhi, G.; Balakumar, K.; Shetti, N.P.; Ahmad, E.; Pant, K.K.; Aminabhavi, T.M. Integrated biorefinery processes for conversion of lignocellulosic biomass to value added materials: Paving a path towards circular economy. *Bioresour. Technol.* **2022**, *343*, 126151. [CrossRef]
7. Pizarro-Ortega, C.I.; Dioses-Salinas, D.C.; Severini, F.; López, A.F.; Rimondino, G.N.; Benson, N.U.; De-la-Torre, G.E. Degradation of plastics associated with the COVID-19 pandemic. *Mar. Pollut. Bull.* **2022**, *176*, 113474. [CrossRef]
8. Wei, H. Optimisation on Thermoforming of Biodegradable Poly (Lactic Acid), (PLA). *Polymers* **2021**, *13*, 654. [CrossRef]
9. European Commission. *Communication from the Commission, The European Green Deal*; European Commission: Brussels, Belgium, 2019. Available online: https://eur-lex.europa.eu/legal-content/EN/TXT/?qid=1588580774040&uri=CELEX%3A52019DC0640 (accessed on 16 May 2022).
10. Hilton, M.; Sherrington, C.; McCarthy, A.; Börkey, P. *Extended Producer Responsibility (EPR) and the Impact of Online Sales*; OECD Environment Working Papers, No. 142; OECD Publishing: Paris, France, 2019. [CrossRef]
11. Johnson, H.A. Revolutionary Instruments, Lavoisier's Tools, as Objets d' Art. *Chem. Herit.* **2008**, *26*, 30–35.
12. Wu, M.H. Preparation of Maleated and Ionically Modified Polypropylene and Its Application as an Additive for Functional Polypropylene. Ph.D. Thesis, National Cheng Kung University, Taiwan, China, 2020.
13. Maddah, H.A. Polypropylene as a promising plastic a review. *Am. J. Polym. Sci* **2016**, *6*, 1–11.
14. Lau, H.C.; Bhattacharya, S.N.; Field, G.J. Melt strength of polypropylene: Its relevance to thermoforming. *Polym. Eng. Sci.* **1998**, *38*, 1915–1923. [CrossRef]
15. Tripathi, D. *Practical Guide to Polypropylene*; Rapra Technology Limited: Shrewsbury, UK, 2002; pp. 2–82.
16. Ahmedzade, P.; Demirelli, K.; Günay, T.; Biryan, F.; Alqudah, O. Effects of Waste Polypropylene Additive on the Properties of Bituminous Binder. *Procedia Manuf.* **2015**, *2*, 165–170. [CrossRef]
17. Plastics Europe. *Plastics the Facts-2016, an Analysis of European Plastics Production, Demand, and Waste Data*; Plastic Europe: Brussels, Belgium, 2016.
18. Geyer, R.; Jambeck, J.R.; Law, K.L. Production, use, and fate of all plastics ever made. *Sci. Adv.* **2017**, *3*, e1700782. [CrossRef] [PubMed]
19. Brasileiro, L.; Moreno-Navarro, F.; Tauste-Martínez, R.; Matos, J.; del Rubio-Gámez, M.C. Reclaimed Polymers as Asphalt Binder Modifiers, A review. *Sustainability* **2019**, *11*, 646. [CrossRef]
20. Rahman, M.Z.; Hoque, M.E.; Alam, M.R.; Rouf, M.A.; Khan, S.I.; Xu, H.; Ramakrishna, S. Face Masks to Combat Coronavirus (COVID-19)—Processing, Roles, Requirements, Efficacy, Risk and Sustainability. *Polymers* **2022**, *14*, 1296. [CrossRef]
21. China Futures Daily. More than 3000 New Mask and Protective Clothing Companies Are Added, Has the PP Market Felt It? *China Futures Daily*, 11 February 2022.
22. Tammaro, D.; Ballesteros, A.; Walker, C.; Reichelt, N.; Trommsdorff, U. Expanded Beads of High Melt Strength Polypro-pylene Moldable at Low Steam Pressure by Foam Extrusion. *Polymers* **2022**, *14*, 205. [CrossRef]
23. Taiwan Plastics Industry Technology Development Center. *High Melt Strength Polypropylene Materials for Physical Foaming, Plastics Column*; Taiwan Plastics Industry Technology Development Center: Taichung, Taiwan, 2022.

24. Kuhnigk, J.; Standau, T.; Dörr, D.; Brütting, C.; Altstädt, V. Progress in the development of bead foams—A review show less. *J. Cell. Plast.* **2022**, *in press*. [CrossRef]
25. Chen, C.H.; Lin, K.C. The application of new developed expanded polypropylene, Department of Innovative Design and Entrepreneurship Management, Far East University, Tainan, Taiwan, 2013. Available online: https://cheers-longterm-cdn.cwg.tw/upload/1449060140.pdf (accessed on 16 May 2022).
26. George, G. *Thermoforming: A Plastics Processing Guide*; Routledge: London, UK, 2018.
27. Maier, C.; Calafut, T. *Polypropylene: The Definitive User's Guide and Databook*, 1st ed.; Plastics Design Library; Elsevier: Amsterdam, The Netherlands, 1998; pp. 205–221.
28. Eldridge, M. *Mount III Extrusion Processes Applied Plastics Engineering Handbook*; William Andrew Publishing: Norwich, NY, USA, 2017; pp. 217–264.
29. Cantor, K. *Extrusion Overview Blown Film Extrusion*, 2nd ed.; Hanser Publications: Munich, Germany, 2011.
30. Lamont, P.R. Equipment and Processing Considerations for Thin Gauge PP Sheet. *J. Plast. Film. Sheeting* **1998**, *14*, 256–267. [CrossRef]
31. Giles, H.F., Jr.; Mount, E.M., III; Wagner, J.R., Jr. *Extrusion: The Definitive Processing Guide and Handbook*, 1st ed.; Elsevier: Amsterdam, The Netherlands, 2004.
32. Global, S. HIPS/PP/APET Rigid Sheet Extrusion Lines. 2022. Available online: https://www.sunwellglobal.com.tw/tw/product/plastic-sheet-extrusion-lines.html (accessed on 16 May 2022).
33. Global, S. Marketing and Sales. 2002. Available online: https://www.sunwellglobal.com.tw/en/about/marketing-and-sales.html (accessed on 16 May 2022).
34. Gan, Y.; Shen, L.; Chen, J.; Tam, V.W.Y.; Tan, Y.; Illankoon, I.M.C.S. Critical Factors Affecting the Quality of Industrialized Building System Projects in China. *Sustainability* **2017**, *9*, 216. [CrossRef]
35. Long, T.B.; Looijen, A.; Blok, V. Critical success factors for the transition to business models for sustainability in the food and beverage industry in the Netherlands. *J. Clean. Prod.* **2018**, *175*, 82–95. [CrossRef]
36. Abeykoon, C.; McMillan, A.; Nguyen, B.K. Energy efficiency in extrusion-related polymer processing: A review of state of the art and potential efficiency improvements. *Renew. Sustain. Energy Rev.* **2021**, *147*, 111219. [CrossRef]
37. Rowe, G.; Wright, G. The Delphi technique as a forecasting tool: Issues and analysis. *Int. J. Forecast.* **1999**, *15*, 353–375. [CrossRef]
38. Linstone, H.A.; Turoff, M. The Delphi method: Techniques and applications. *Read. Addison-Wesley* **1975**, *29*, 3–12.
39. Jonassen, D.H.; Tessmer, M.; Hannum, W.H. Structured Group Interviews: Delphi Technique. In *Task Analysis Methods for Instructional Design*; Routledge: London, UK, 1998; pp. 275–278.
40. Lima Filho, L.R.; Bastos, J.V.F.; Macêdo-Júnior, R.O.; Silva, L.S.; Santos, B.L.P.; Ruzene, D.S.; Vasconcelos, C.R.; Silva, D.P. Evaluation of the Bibliometric Scenario of the Delphi Method with Brazilian Affiliations. *Int. J. Innov. Educ. Res.* **2020**, *8*, 225–236. [CrossRef]
41. Nworie, J. Using the Delphi technique in educational technology research, Techtrends Link. *Res. Tech. Trends* **2011**, *55*, 24–30. [CrossRef]
42. Murry, J.W., Jr.; Hammons, J.O. Delphi: A Versatile Methodology for Conducting Qualitative Research. *Rev. High. Educ.* **1995**, *18*, 423–436. [CrossRef]
43. Lahiri, K.; Pulungan, Z. Racial/Ethnic Health Disparity in the U.S.: A Decomposition Analysis. *Econometrics* **2021**, *9*, 22. [CrossRef]
44. Ying, L.; Yupin, A.; Liyan, S.; Joko, G.; Dejian, Z. Construction of evaluation indexes of nursing students' quality and safety competencies: A Delphi study in China. *J. Prof. Nurs.* **2021**, *3*, 501–509.
45. Nor, M.Z. Developing a preliminary questionnaire for the faculty development programme needs of medical teachers using Delphi technique. *J. Taibah Univ. Med. Sci.* **2019**, *6*, 495–501. [CrossRef]
46. Chen, D.C.; Chen, D.F.; Huang, S.M.; Huang, M.J.; Shyr, W.J.; Chiou, C.F. Critical Success Factors to Improve the Business Performance of Tea Drink Chains. *Sustainability* **2021**, *13*, 8953. [CrossRef]
47. Tingrui, F.; Barry, H.; Leno, M. Analysis of process parameters related to the single-screw extrusion of recycled polypropylene blends by using design of experiments. *J. Plast. Film. Sheeting* **2017**, *33*, 168–190.
48. Wagner, J.R., Jr.; Spalding, M.A.; Crabtree, S.L. Data Analysis of an Extrusion Experiment. *J. Plast. Film. Sheeting* **2008**, *24*, 137–157. [CrossRef]
49. Kumar, P.S.; Kumar, G.K.; Kommoji, S.; Banerjee, R.; Ghosh, A.K. The effect of material characteristics and mould parameters on the thermoforming of thick polypropylene sheets. *J. Plast. Film. Sheeting* **2014**, *30*, 162–180. [CrossRef]
50. Johansson, K.T.; Reber, D.H. In-Line Extrusion/Thermoforming Techniques for Deep Draw Polyolefin Containers. *J. Plast. Film. Sheeting* **1989**, *5*, 140–153. [CrossRef]
51. Shyr, W.-J.; Shih, F.-Y.; Liau, H.-M.; Liu, P.-W. Constructing and Validating Competence Indicators for Professional Technicians in Fire Safety in Taiwan. *Sustainability* **2021**, *13*, 7058. [CrossRef]
52. Luo, T.; Wu, C.; Duan, L. Fishbone diagram and risk matrix analysis method and its application in safety assessment of natural gas spherical tank. *J. Clean. Prod.* **2018**, *174*, 296–304. [CrossRef]

MDPI
St. Alban-Anlage 66
4052 Basel
Switzerland
Tel. +41 61 683 77 34
Fax +41 61 302 89 18
www.mdpi.com

Applied Sciences Editorial Office
E-mail: applsci@mdpi.com
www.mdpi.com/journal/applsci

www.ingramcontent.com/pod-product-compliance
Lightning Source LLC
Chambersburg PA
CBHW061502210926
51850CB00008B/1793
9783036582719